Representation in Scientific Practice Revisited

CW00486424

Inside Technology
edited by Wiebe E. Bijker, W. Bernard Carlson, and Trevor Pinch

A list of books in the series appears at the back of the book.

Representation in Scientific Practice Revisited

edited by Catelijne Coopmans, Janet Vertesi, Michael Lynch, and Steve Woolgar

The MIT Press
Cambridge, Massachusetts
London, England

MIT Press books may be purchased at special quantity discounts for business or sales promotional use. For information, please email special_sales@mitpress.mit.edu.

This book was set in Stone Sans and Stone Serif by Toppan Best-set Premedia Limited, Hong Kong. Printed and bound in the United States of America.

Library of Congress Cataloging-in-Publication Data

Representation in scientific practice revisited / edited by Catelijne Coopmans, Janet Vertesi, Michael Lynch, and Steve Woolgar.
    pages cm. — (Inside technology)
    Includes bibliographical references and index.
    ISBN 978-0-262-52538-1 (pbk. : alk. paper) 1. Research—Methodology. 2. Science—Methodology. 3. Technology—Methodology. I. Coopmans, Catelijne, 1976– editor of compilation.
    Q180.55.M4R455 2014
    502.2′2—dc23
                        2013014968

10 9 8 7 6 5 4 3 2 1

Contents

Preface

Michael Lynch and Steve Woolgar

Thirty years ago, a workshop on "Visualization and Cognition" was held in Paris.[1] It was attended by an eclectic group of scholars including art historians, historians of science and engineering, semioticians, cognitive scientists, and ethnographers of scientific laboratories. The workshop marked a recent and rapidly growing scholarly interest in the production, use, and dissemination of maps, engravings, photographs, micrographs, and other pictorial and graphic displays in science and technology. In the decade prior to the meeting, historians such as Martin Rudwick, Samuel Edgerton, Martin Kemp, and Svetlana Alpers had already begun to establish that artistic and graphic techniques and technologies did not simply produce *images* that were secondary to logical reasoning and mathematical reckoning in the sciences. Instead, their research demonstrated that visual and graphic materials were crucial for enabling discovery and establishing the properties of natural phenomena. By then, sociologists and anthropologists had also been conducting ethnographies of laboratory practices that highlighted the pragmatic shaping of raw materials into polished and publishable exhibits of "facts." One of these ethnographers, Bruno Latour, organized the Paris workshop and presented the keynote address. His address integrated a down-to-earth focus on the work of constructing and inscribing the results of scientific investigations with a more sweeping overview of the importance of "immutable mobiles"—the fixed and transportable literary products of scientific work—in the history of science. Latour argued that, to a large extent, the scientific imagination was a matter of "thinking with eyes and hands." He and others who participated in the workshop preferred the term "visualization" over that of "perception" or "observation," because of the way it connoted practices of *making visible*—fashioning and exhibiting witnessable and accountable material and virtual displays. This emphasis on making visible also downplayed the supposed importance of cognitive and perceptual attributes in doing representation.

Several years later, we were invited to guest-edit a special issue of *Human Studies: A Journal for Philosophy and the Social Sciences*. The editor of the journal at the time, George Psathas, was particularly interested in recent work on science that exemplified an ethnographic and ethnomethodological treatment of scientific practices. After some discussion, we decided to focus the issue on *representation* in scientific practice. Inspired in part by the growing interest in visualization, we also wanted to bring into play close studies of verbal interaction at the lab bench (or field site), as well as analyses of the literary and pragmatic relations among texts, depictions, and activities. The special double issue (Lynch and Woolgar 1988) included one article that had been presented at the Paris workshop (Lynch 1988) and several other studies of laboratory practices and expository discourse. Authors of the different chapters deployed semiotic, ethnomethodological, conversation-analytic, and discourse-analytic approaches to the practical, interactional, and textual organization of representation in science, and they also drew inspiration from and critically reexamined historical and philosophical conceptions of representation. The MIT Press agreed to publish a volume (Lynch and Woolgar 1990) that included papers from the special issue, supplemented by English translations of Latour's keynote from the 1983 Paris workshop and a paper by the late Françoise Bastide, which had originally been presented there.[2]

Representation in Scientific Practice was not the first book, and certainly not the last, to address representation in the sciences, but it established a distinctive approach to that topic which examined and elucidated the temporal and practical working and reworking of materials that (sometimes) culminate in the presentation and re-presentation of scientific facts, models, and ordered regularities. This approach became a familiar reference point in and beyond science and technology studies. At the same time, scientific visualization and representation became an increasingly established topic in other fields in the social sciences and humanities, including art history and visual studies, literary criticism, feminist and gender studies, anthropology, cognitive science, and the history and sociology of science.

Two decades after its publication, *Representation in Scientific Practice* continued to be read and cited, but by then a new edition seemed long overdue. After discussing plans for the new edition with each other, and with our former students Catelijne Coopmans and Janet Vertesi, we decided to publish an entirely new set of chapters rather than reprinting and revising those that had been published in the earlier volume. A key reason for this decision was that there had been a resurgence of interest in representation in the sciences among younger scholars. Many of them were interested in the uses of novel technologies: fMRI, probe microscopes, and digital visualization and image-processing technologies of all kinds. At the same time, by the second decade of the twenty-first century, the very question of representation in scientific practice had become situated in a different theoretical and conceptual landscape than it had been in the 1980s—a landscape colored by discussions of mediation, ontology, enactment, materiality, and the discursive "performance" of images, among other things. In addition to full-length chapters written mainly by younger scholars, *Representation in Scientific Practice Revisited* also includes commentaries by more established scholars who were invited to reflect upon changes in the field during the more than twenty years since the publication of the original volume, and thirty years since the Paris workshop.

Notes

1. The workshop on "Visualization and Cognition" was held at the Centre de Sociologie de l'Innovation at the Ecole Nationale Supérieure des Mines de Paris on 12–15 December 1983.

2. See Latour (1990) and Bastide (1990). These and other articles from the 1983 workshop had been published in French, in a special issue of *Culture Technique* (Latour and de Noblet 1985).

References

Bastide, Françoise. 1990. The iconography of scientific texts: Principles of analysis. In Lynch and Woolgar 1990, 187–230.

Latour, Bruno. 1990. Drawing things together. In Lynch and Woolgar 1990, 19–68.

Latour, Bruno, and Jocelyn de Noblet, eds. 1985. Les "vues" de l'esprit. Special issue of *Culture Technique*, no. 14 (June).

Lynch, Michael. 1988. The externalized retina: Selection and mathematization in the visual documentation of objects in the life sciences. *Human Studies* 11 (2/3):201–234. Reprinted in Lynch and Woolgar 1990, 153–186.

Lynch, Michael, and Steve Woolgar, eds. 1988. Representation in scientific practice. Special issue of *Human Studies* 11(2/3).

Lynch, Michael, and Steve Woolgar, eds. 1990. *Representation in Scientific Practice*. Cambridge, MA: MIT Press.

1 Introduction: Representation in Scientific Practice Revisited

Catelijne Coopmans, Janet Vertesi, Michael Lynch, and Steve Woolgar

1 Introduction

Over the past three decades, representation in scientific practice has become an established topic in science and technology studies (STS). From anatomical to astronomical
illustrations, from protein gels to atlases, from remote-sensing imagery to brain scans,
a rich field of inquiry spanning historical, sociological, and philosophical approaches
has produced analyses of scientific efforts to "capture," "render," and otherwise make
available aspects of the world. To examine the full richness of these efforts, STS scholars situate historical and contemporary notions of a representation's "truth to nature"
within the contingent activity of locally grounded and discipline-specific, yet also
mobile and powerful, practices. As the first volume to bear the name *Representation in
Scientific Practice* (Lynch and Woolgar 1990; hereafter, *RiSP*) demonstrated, representation involves lengthy struggles with research materials to reconstruct them in a way
that facilitates analysis, for example through coding and highlighting key features of
interest and aligning them with particular concepts and theories. This treatment of representation in and as *practice* has since spurred a rich body of ethnographic, historical,
and discourse-analytic inquiries that demonstrate how the circumstances of knowledge
production are folded into epistemological claims and ontological orderings.

Enter a scientific workplace today and representations of all kinds continue to play a
dominant role. Now, however, they are not only exchanged on printed pages or visible
as protein gels in scientists' hands. Computer screens have pride of place in laboratories
and scientific offices, where researchers' attentions are as likely—or more likely—to
be focused on colorful digital images, simulations, software suites, databases, or lines
of code as on unruly specimens or instruments. Biomedical imaging enrolls fMRI and
PET scans alongside X-rays, which themselves are frequently digitally manipulated to
produce new modes of vision. Planetary image processing and financial analysis rely
on massive datasets (or streams) with their own concomitant visualization tools and
skills. As "the laboratory" extends to other spaces and places via collaborative ventures,
shared data centers, and information and communication technologies, this expansion

challenges the very distinction between laboratory and field. Still, alongside new computational practices continue to sit older representational forms in scientific work: chalk, marker, and pen scribbles decorating blackboards, whiteboards, and napkins; models of complex phenomena perched atop bookshelves; and glossy, retouched photographs in journal pages.

In part due to the proliferation and, perhaps, intensification of representational technologies and representational forms, STS research on the topic shows no signs of abating. The present volume is a response to a resurgence of such research in recent years, research that grapples with change and continuity in representational practices, and which also bears testimony to the way STS itself has changed. Contributors to the original *RiSP* volume made use of historical, sociological, ethnographic, literary, ethnomethodological, and conversation-analytic investigations, and sought to respecify "representation" as practical action in social and material contexts. They stressed the roles of instruments and textual formats, and the interactional and interpretive work surrounding them. Their emphasis on such public, practical, communicative, and textual work was set off against an established philosophical picture of representation as mental, verbal, or pictorial reference to features of an independent world.[1]

The interest in practice and social interaction remains strong today, but there have been many changes of theoretical emphasis and disciplinary location in the field. STS has become a robust and diverse field, with constituencies in anthropology and cultural studies, communication and information studies, geography, political science, economic sociology, and management and organization studies, in addition to history, philosophy, and sociology of science. Concomitantly, actor-network theory (Callon 1986; Latour 1987), which was just beginning to coalesce when the chapters in *RiSP* were drafted, is now a pervasive approach in STS. Inspired by actor-network theory as well as feminist and cultural studies of science (Haraway 1991), a "turn to ontology" emphasizing material enactments as well as embodied action and social interaction (Mol 2002; Woolgar and Lezaun 2013) has supplemented the "practice turn" (Schatzki et al. 2001) that many of the studies in *RiSP* exemplified.

These and other shifts in analytical interests, expository themes, and research sites are exemplified and elaborated in this new volume on representation in scientific practice. In light of new approaches and thematic interests, the chapters in the volume revisit the question of how we should study and understand (visual) representation, while also building upon prior scholarship on scientific, technical, and clinical practice.

2   The Concept Formerly Known as Representation

When the first volume of *RiSP* was composed more than twenty years ago, most contributors took up one or both of two main analytical objectives. The first, much less prominent today, was to create distance from idealized descriptions of scientific procedure. At the time, and related to the development of STS as a field, there was a strong emphasis on showing how scientific practice differs from established versions of scientific method in mid-twentieth-century history, philosophy, and social studies of science. Those versions depicted science as a historically unique, logically governed, and socially exceptional method for attaining truth (or eliminating error), which differed from "commonsense" knowledge and everyday practices. The "practice turn" depicted science in its everyday modes as immanently practical, locally organized, and infused with interpersonal trust and tacit knowledge. This reconception of science through its everyday practice has since become so successful that it is now largely taken for granted in STS, if not in philosophy.[2]

The second analytical objective was to extend the analysis and critique of representation from language and logic to nonlinguistic, often visual practices and formats and to instrumental interventions (Hacking 1983). A key move was to reframe representation from an expectation that visual traces and numerical measurements were references to independent objects and properties, to a series of open-ended inquiries into the many different kinds of relations, reference among them, that are accomplished (or dismantled) in the work people do with representational forms. The production and presentation of *scientific* representations served as a particularly revealing source of the dynamics of demonstration and disputation, because of the cultural weight assigned to such representations. In the realm of science, understandings of representations as referential forms were considered particularly tenacious (albeit more among commentators than among practitioners) and thus in need of empirical reframing.

Given the philosophical baggage associated with the term "representation," it may be fair to ask whether it would not be better to abandon rather than to revisit representation in scientific practice. Has not the investment in representation or reference as a key philosophical problem been criticized to death? Even critical antipositivist treatments may have run out of steam by now (Latour 2004). Indeed, we have noted a tendency in STS scholarship to move away from the use of the term "representation." Instead, some authors prefer "mediation" (Pasveer 2006), while others adopt notions associated with the turn to ontology, such as "enactment" (Woolgar and Lezaun 2013). Perhaps most pervasive has been the substitution of "visualization" for "representation" (see, for example, Burri and Dumit 2008; also Wise 2006). It is clear that there is now an abundance of STS research on the effortful accomplishments through which images, graphs, and models are produced, and on how these come to speak for a phenomenon (or a set of relations) and are discursively deployed (Burri and Dumit 2008); but do we need an alternative word to designate *the concept formerly known as representation*?

While we are sympathetic to the argument that "representation" is a problematic term, we have chosen to stick with "representation in scientific practice" as an organizing theme for the present volume.[3] Despite the concern about philosophical baggage, there is in our view no unproblematic way of designating the practices described and analyzed in this volume. Like "representation," "visualization" is a loaded term, as are closely related concepts such as "observation" and "perception."[4] Perhaps the best way to use such terms, then, is not as purportedly neutral summary descriptions of the scientific, technical, or medical work that is made the subject of analysis, but as unsettled concepts: *Is* it representation we are dealing with? Does what we are dealing with prompt us to extend, or expand, or rethink what we mean by this term? Can we, for instance, following Rheinberger's (1995, 51) provocative suggestion, conceive of "the activity of scientific representation . . . as a process without 'referent' and without 'origins'"?

The new contributions brought together around this organizing theme span a variety of empirical settings, place their emphasis in various ways, and ground themselves in anthropology, sociology, philosophy, history, and permutations of those fields. Some chapters discuss the articulation of particular phenomena, such as adolescence, soil on Mars, or human anatomy. Others concentrate on representational conventions and the work and negotiations associated with them in fields such as nanotechnology and molecular biology. Some chapters are concerned with the tropes that animate the production and presentation of visual representations, while others identify features in the practices they study that call for new analytical repertoires. Several contributions pay attention to the embodied interactions that constitute representational work in science, technology, and medicine. Some contributors to the volume treat representation as a noun, focusing on the "outputs" of scientific endeavor, while others talk of representing as a verb. In some settings, and not exclusively those characterized by interactive digital technologies, this very distinction itself dissolves as screen displays or physical inscriptions are manipulated to expose or enact the reality they make tractable.

3 "New" Studies of Representation in Scientific Practice

The chapters in this volume allow us to ponder the question: How do we understand representational practice today? Rather than provide either an encyclopedic view or a snapshot of a moment when particular visualization technologies (such as various digital systems) are "new," the volume's chief aim is to articulate conceptual issues that promise to outlast any particular example. Consequently, as editors we have not attempted to encompass all forms of scientific representation or all possible analytical approaches to that topic (an impossible task, in any case). Instead, each of the contributions we selected for the volume aims to identify certain key concerns, constituents, mechanisms, or animating features of representation in scientific practice that illuminate and allow reflection upon recent developments in STS. We outline some broad themes here.

The first volume of *RiSP* emphasized interactional elements of practice: many of the chapters examined interactions among researchers, often while working with visual data displays. Such interactions between persons and with things continue to play a key role in our understanding of representation today. However, the notion of practice employed in the present volume must also be understood in the context of current enthusiasms for the notion of *materiality*. Some versions of this notion stress the embodied nature of scientific work, as well as the tools, objects, technologies, and environments in and through which science is practiced. Such concerns have long been a central focus of STS; so, then, what is the importance today of emphasizing "the material"? One way in which some of the chapters in this volume answer this question is by characterizing practices that involve, for example, digital or mathematical phenomena and fields as no less material than, say, a tabletop experiment with vials and burners. In line with current scholarship in STS and media studies, authors move away from commonplace notions of digitality that treat the virtual as ephemeral, and instead attend to the material conditions of digital work expressed in specific entanglements of language, visual evidence, embodied actions, and worldly phenomena. More broadly, the chapters attend to a variety of different material domains: gestures that make sense of visual materials, multimodal interactional environments, and suites of technologies (including extensive databases and software scripts but also blackboards and scrap paper) that constitute an infrastructure for scientific engagement with worldly phenomena.

Many of the representational techniques or technics featured in this volume involve *new* computational systems using digital technology. But rather than establish a discontinuous or a simple skeumorphic relationship between contemporary digital practices and analog forms, the authors focus on continuities, and complicate distinctions between the old and new. Innovation is treated not as a revolutionary break, but as a question of working with, across, and through established representational conventions, technologies, and communities. This applies from the way data are rendered in nanoscience all the way to the low-tech environments in which theoretical mathematicians and economists work toward new approaches and theorems. Another way in which distinctions between the old and new are complicated is entailed in the suggestion in some chapters that insights gained from observing the specificities of digital manipulation also apply to other historical times and places.

Recent STS research has called attention not only to the practices and technological infrastructures of representational production, but also to their entanglement with the *dynamics of reception and circulation* (though see Shapin and Schaffer 1985 for a now-classic treatment of this theme). There is now, for example, an increased sensitivity to shifting notions of objectivity (Daston and Galison 2007) and to the situated production and reception of expert witness accounts (Jasanoff 1998), where issues of trust, expertise, and accountability are very much in play. These issues sometimes remain internal to a scientific field, but are more often engendered by the circulation of scientific representations beyond the settings in which they were produced, settings in which the audience may be comprised of scientists in different fields, clinicians, or various publics engaged through popular culture or the politico-legal sphere. Several chapters in the volume explore how expertise is produced and contested in and through representational practices that entangle both scientific practitioners and their audiences. In various contexts and with different analytical approaches, they examine how the visual outputs of such practices generate trust (and sometimes mistrust) for particular witnesses in particular contexts.

What is "new" about this volume of *RiSP*, then, is not simply a consequence of focusing on novel technologies, sciences, or institutional arrangements. Instead, it has to do with revisiting and respecifying recurrent themes in light of developments over the past two decades in the sciences studied as well as in our own fields of research. In the process of studying practices and practical entanglements, chapters in this volume trouble many presumed clear-cut boundaries, for example between visual and nonvisual representations, and between epistemic and ontological work. Similarly, the boundary between "science" and "nonscience" is blurred, with discussions related to the domains of surgery and business analytics that raise key questions regarding the nature of technology-mediated seeing (and intervening), and with contributions that trace the circulation of representations across scientific and nonscientific domains.

To be sure, chapters in the present volume invoke some of the classic staples of sociological inquiry, such as trust, value, community norms, and status, as well as some of the now-established concepts of the original volume of *RiSP*, such as inscription devices and the public and discursive production of "perceptual" activity. However, they also extend these conceptual repertoires. Classic questions of visual epistemology are reimagined by reference to contemporary material configurations. Orienting concepts such as "seeing as," modeling, mediation, objectivity, phenomenology, or conceptual hybridity are worked through by reference to particular practices, instruments, and communities. This is the sense in which this volume *revisits* the conceptual themes and analytical perspectives associated with the 1990 volume, presenting a fresh analytical perspective on themes of continuing importance to the contemporary study of scientific representation.

4   The Arc of the Work

The chapters that follow begin with a focus on the detailed practices with screens, data, and visualization algorithms that craft viewing experiences in the digital era. The chapters by Janet Vertesi and Catelijne Coopmans deal with the work of *revealing* that draws digital data into valuable and sensible configurations. Revealing, here, evokes the notion of "making visible" in order to be readily witnessed in a communal perceptual space. These chapters discuss empirical settings in which participants "make visible" by manipulating large quantities of digital data. **Vertesi's** chapter discusses practices of image construal in NASA's Mars Exploration Rover mission, focusing on how Martian soil is made seeable as a phenomenon of interest. Vertesi stresses that revealing is intense and effortful work: "seeing as" experiences are not limited to an observer's perceptual field but also are crafted with visual materials, and are hence better captured in the notion of *drawing as*. **Coopmans's** chapter interrogates the claim that new data visualization software can help users "see" hitherto hidden insights in datasets. Tracing how this claim is bolstered in and through online software demonstrations that portray visual analysis as a complex interplay between "artfulness" and "revelation," Coopmans argues for analytical attention to the ways in which long-standing epistemological tropes animate and are animated through new practices.

Zooming out from the practices at the screen, the next two chapters by Morana Alač and Rachel Prentice draw our attention to the embodied nature of work with digital visual technologies. These authors insist that the cognitivist notion of "looking at" bodies that are "visually represented" on a screen is wholly inadequate to understand the nature of working with brain scans or doing remotely mediated surgery. Only by fully inhabiting the setting, using gesture and touch alongside visual information, are practitioners able to make present what is salient to their work. **Alač** describes the multimodal coordination work—the screen work, gestures, and talk—through which brain-related objects and features of note are enacted by scientists in a cognitive neuroscience laboratory. Rather than understanding fMRI data patterns as visual "representations" that are being "interpreted" by practitioners, she sees them as *materials for enactment* through dynamic, interactive, and embodied engagement on the part of practitioners. **Prentice** explores "how surgeons and trainees at various levels come to acquire surgical means of perceiving and acting, especially perceiving and acting with technological mediation." She shows how sight and touch merge in the technical and social actions that constitute surgical skill. Prentice argues that the now-widespread use of remotely mediated surgery has brought about an intriguing change in how surgeons inhabit the operating space: they safeguard the coherence of that space by locating their own bodies *inside* the body parts they are operating on.

These discussions of embodied engagement also highlight the technological interfaces and material infrastructures that enable the work of representing (and, for Prentice, intervening): a theme that is taken further in the next two chapters. These focus closely on the constitutive role of materials and technologies in the production of new scientific knowledge. Michael **Barany** and Donald **MacKenzie** discuss the role of chalk, blackboards, and scrap paper in the development of theoretical concepts and approaches in research mathematics. The mundane and modest nature of these materials, according to the authors, is precisely what makes them so important to the "performative unfolding" of mathematical argument. They further contend that, contrary to the notion that inscription practices in the natural sciences are designed to discipline or tame unruly phenomena, the symbolic objects of mathematics are substantially freed and allowed to morph and change through their rendering in material form. This resonates with Sarah **de Rijcke** and Anne **Beaulieu's** discussion of brain atlases comprised of collections of brain scans that are powered by, and remain linked to, dynamic databases. Brain scans, the authors argue, do not represent a vision of the brain at a static moment in time. Each image viewable on a computer screen stands for a statistical dataset that derives its meaning from its relation to a database that is continually changing. Practitioners thus handle and manipulate these images as *interfaces* to a digital infrastructure, and it is through this configuration that new knowledge and understandings of the brain can be achieved.

Such material infrastructures bring with them conceptual tools and analytical practices that animate ways of thinking and working, and these are discussed in the next three chapters. Natasha **Myers** in her chapter draws on the work of Donna Haraway to show how molecular models *render* protein structures as machines. In her account, machine metaphors are rendered into material form through the development of models that serve as tools for thinking and acting with biological phenomena. This focus on rendering marks a point of continuity between the original *RiSP* and the present volume (see Lynch's [1990] treatment of "renderings" of electron micrographs [also see Lynch 1985, 64n] and Vertesi's focus on "drawing as" in the present volume). Myers argues that machine metaphors are highly productive: they support the enactment of objects of research, bring people together, and even drive entire research programs. Martin **Ruivenkamp** and Arie **Rip** see a similar mobilizing function in the images associated with nanotechnology. They characterize nanoimages as "hybrid monsters" that mix representational conventions. Rip and Ruivenkamp argue that the hybridity of nanoimages is productive for organizing and creating a space for nanotechnology by spurring different imaginations of what the nanoscale might look like, as well as what we might do with it in future (see also de Ridder-Vignone and Lynch 2012). Annamaria **Carusi** and Aud Sissel **Hoel** also discuss hybridity in their chapter on computational biology, here in relation to what they identify as an intertwining of qualitative and quantitative methods in visual practice. Drawing on the later work of Maurice Merleau-Ponty, Carusi and Hoel argue that the new configurations of vision, computational technologies, and objects evident in computational biology necessitate an "ontological reframing" that also has repercussions for how scientific vision is conceived in other domains.

The emphasis then shifts to how the status and significance of scientific imagery are negotiated within communities of practice. Cyrus **Mody** looks historically at the development of a scientific community around a novel instrument that converts haptic apprehension of surface electronics into visual topography. The development of visual styles with the rise of the scanning tunneling microscope and atomic force microscope in nanotechnology shows tensions in the ways the results of representational work are understood within the relevant communities: as conventional or iconoclastic. Emma **Frow** discusses recent concerns voiced by editors of leading biology journals about the exploitation of programs such as Photoshop to manipulate digital data when preparing images for publication. Frow points out that the editors' efforts to develop rules against illegitimate data manipulation tend to ignore the extent to which, as STS research has shown, data manipulation is a normal feature of expository science. Both Frow and Mody discuss the tensions entailed in stipulating what scientific images are supposed to look like; and both suggest that scientists' perceived trustworthiness or innovativeness is bound up with emerging visual conventions.

The final two full-length chapters in the volume consider the status of particular representations as they become widely disseminated beyond the circumstances of their initial development. Both authors suggest that science is subordinated to popular culture in the deployment of such representations. Yann **Giraud** traces how the Laffer curve—allegedly drawn initially on a restaurant napkin to suggest how government revenues vary with tax rates—became a celebrated (and much criticized) icon for supply-side economics in the 1980s. Giraud shows how a representation that started life as a propaganda tool was subsequently translated into an object of economics research. Curiously, despite extensive modification and criticism, it is the original version that continues to surface in economics textbooks to this day. Joseph **Dumit** provides an account of the role of brain scans in recent disputes about the legal status of adolescence. Dumit shows how brain scan images were configured and juxtaposed to address a legal distinction between degrees of criminal culpability assigned to adults and adolescents. He cautions that "neuroscience has come to have explanatory power far in excess of its confirmatory ability" and that the flexible use of brain images as "scientific" backings for established moral categories should be resisted, despite the temptation they present for lawyers, journalists, and neuroscientists.

The thirteen chapters in the book offer detailed case studies and their elucidation in terms of thematic, theoretical, or methodological implications for studies of representation in scientific practice. These chapters are complemented by short reflections from Lorraine **Daston**, Michael **Lynch**, Steve **Woolgar**, Lucy **Suchman**, John **Law**, Martin **Kemp**, and Bruno **Latour**. Many of these authors were included in the original *RiSP*; others are equally well known for their contributions to research on visual representation in science and other domains. Each of their reflections provides broader commentaries on past, present, and future scholarship on representation in science studies. With topics as diverse—and sometimes as provocative—as the authors themselves, these reflective and reflexive pieces inspire our continued attention to the changing hows and whys of studying representation in scientific practice.

Through these diverse contributions, both chapters and reflections, the volume attempts to raise new questions and revisit old ones, to open up investigative possibilities, and to reinspire engagement with representational practice—or with the concept formerly known as representation—in and beyond science and technology studies. We do not claim the present volume to be exhaustive or conclusive in its contributions.[5] We simply invite our readers to explore the collection, place its pieces in conversation, and bring new questions, new answers, and new challenges to the fore.

Notes

1. Prevalent at the time, what is sometimes called the "correspondence theory of knowledge" presumed a fundamental distinction between a natural order "out there" and efforts to approximate that order (more or less accurately) with representations in the form of measurements, equations and graphs, verbal descriptions, and visual images. Pervading the diverse contributions to *RiSP* was the insistence that, instead of investing in the correspondence theory, with its established problems and ongoing efforts to overcome them, STS researchers should attend to the "contextually organized and contextually sensitive way" (Lynch and Woolgar 1990, vii) in which particular representational forms are composed and used.

2. See, for example, the recent special issue of *Studies in History and Philosophy of Science* titled "Model-Based Representation in Scientific Practice: New Perspectives," which aims to "explore ways in which close attention to scientific practice . . . can shed light on the philosophical issues raised by scientific representation" (Gelfert 2011).

3. It should be noted that Latour has made a particular effort to reinvigorate the study of "representation" by insisting that its epistemological connotations should be considered in tandem with artistic and religious representational practices, as well as with political meanings of the term. This effort has borne fruit in two exhibitions called "Iconoclash" (Latour and Weibel 2002) and "Making Things Public" (Latour and Weibel 2005), in which a wide range of contributions were brought together "to foster a new respect for mediators" (Latour and Weibel 2005, 29). In contrast to Latour's explicit situating of the question of representation at the intersection of distinct domains of public life, the present volume—in continuity with the earlier one—maintains "representation in scientific practice" as a classic STS concern to be revisited.

4. Lynch (1994) has argued that "representation is overrated," which was a play on Hacking's (1983, 137) earlier statement that "observation is overrated." At the workshop on "Visualization and Cognition" in Paris in 1983, "perception" and "observation" were criticized for being too cognitivist, while "visualization" was considered a less troubled term (see the preface to this volume). Visualization, however, has been associated with its own set of problems, ranging from an uncritical privileging of sight (Garforth 2012) to the "mimetic . . . obsession for an image as a copy" (Latour, this volume) that draws our attention to particular, singular images, graphs, models, and so on, rather than tracing the dynamic way reference is constituted through multiple conversions of form. None of these critiques has been a significant deterrent; recent years have seen the publication of edited collections on *Histories of Observation* (Daston and Lunbeck 2011), *Visual Cultures of Science* (Pauwels 2006), and *Skilled Visions* (Grasseni 2007).

5. Notably absent, for example, are questions of colonialism, non-Western approaches to representation, or cultural modes of representing differences in gender or race (see, for example, Verran 2001; Raj 2007; Anderson 2008).

References

Anderson, Warwick. 2008. *The Collectors of Lost Souls: Turning Kuru Scientists into Whitemen*. Baltimore: Johns Hopkins University Press.

Burri, Regula Valérie, and Joseph Dumit. 2008. Social studies of scientific imaging and visualization. In *The Handbook of Science and Technology Studies*, ed. Edward J. Hackett, Olga Amsterdamska, Michael Lynch, and Judy Wajcman, 297–318. Cambridge, MA: MIT Press.

Callon, Michel. 1986. The sociology of an actor-network: The case of the electric vehicle. In *Mapping the Dynamics of Science and Technology: Sociology of Science in the Real World*, ed. M. Callon and A. Rip, 19–54. Basingstroke, UK: Macmillan.

Daston, Lorraine, and Peter Galison. 2007. *Objectivity*. New York: Zone Books.

Daston, Lorraine, and Elizabeth Lunbeck, eds. 2011. *Histories of Observation*. Chicago: University of Chicago Press.

De Ridder-Vignone, Kathryn, and Michael Lynch. 2012. Images and imaginations: An exploration of nanotechnology image galleries. *Leonardo* 45 (5):447–454.

Garforth, Lisa. 2012. In/visibilities of research: Seeing and knowing in STS. *Science, Technology and Human Values* 37 (2):264–285.

Gelfert, Axel. 2011. Model-based representation in scientific practice: New perspectives. *Studies in History and Philosophy of Science Part A* 42 (2):251–252.

Grasseni, Christina, ed. 2007. *Skilled Visions: Between Apprenticeship and Standards*. Oxford: Berghahn Books.

Hacking, Ian. 1983. *Representing and Intervening*. Cambridge: Cambridge University Press.

Haraway, Donna J. 1991. *Simians, Cyborgs, and Women: The Reinvention of Nature*. New York: Routledge.

Jasanoff, Sheila. 1998. The eye of everyman: Witnessing DNA in the Simpson trial. *Social Studies of Science* 28:713–740.

Latour, Bruno. 1987. *Science in Action: Following Scientists and Engineers through Society*. Cambridge, MA: Harvard University Press.

Latour, Bruno. 2004. Why has critique run out of steam? From matters of fact to matters of concern. *Critical Inquiry* 30 (2):225–248.

Latour, Bruno, and Peter Weibel, eds. 2002. *Iconoclash: Beyond the Image Wars in Science, Religion, and Art*. Cambridge, MA: MIT Press.

Latour, Bruno, and Peter Weibel, eds. 2005. *Making Things Public: Atmospheres of Democracy*. Cambridge, MA: MIT Press.

Lynch, M. 1985. Discipline and the material form of images: An analysis of scientific visibility. *Social Studies of Science* 15 (1): 37–66.

Lynch, M. 1990. The externalized retina: Selection and mathematization in the visual documentation of objects in the life sciences. In Lynch and Woolgar 1990, 153–186.

Lynch, M. 1994. Representation is overrated: Some critical remarks about the use of the concept of representation in science studies. *Configurations* 2 (1):137–149.

Lynch, M., and S. Woolgar, eds. 1990. *Representation in Scientific Practice*. Cambridge, MA: MIT Press.

Mol, Annemarie. 2002. *The Body Multiple: Ontology in Medical Practice*. Durham: Duke University Press.

Pasveer, Bernike. 2006. Representing or mediating: A history and philosophy of X-ray images in medicine. In *Visual Cultures of Science*, ed. Luc Pauwels, 41–62. Lebanon, NH: Dartmouth College Press.

Pauwels, Luc, ed. 2006. *Visual Cultures of Science*. Lebanon, NH: Dartmouth College Press.

Raj, Kapil. 2007. *Relocating Modern Science*. Basingstoke, UK: Palgrave Macmillan.

Rheinberger, Hans-Jörg. 1995. From microsomes to ribosomes: "Strategies" of "representation." *Journal of the History of Biology* 28 (1):49–89.

Schatzki, Theodore, Karin Knorr Cetina, and Eike von Savigny, eds. 2001. *The Practice Turn in Contemporary Theory*. New York: Routledge.

Shapin, Steven, and Simon Schaffer. 1985. *Leviathan and the Air-Pump: Hobbes, Boyle, and the Experimental Life*. Princeton: Princeton University Press.

Verran, Helen Watson. 2001. *Science and an African Logic*. Chicago: University of Chicago Press.

Wise, M. Norton. 2006. Making visible. *Isis* 97 (1):75–82.

Woolgar, Steve, and Javier Lezaun. 2013. The wrong bin bag: A turn to ontology in science and technology studies? *Social Studies of Science* 43 (3):321–340.

Chapters

2 *Drawing as*: Distinctions and Disambiguation in Digital Images of Mars

Janet Vertesi

The image never changes, but you can manipulate the image, and everyone sees something different.

—Mars Exploration Rover team member

1 Introduction: A Martian Discovery

In the summer of 2006, Susan Lee, a geochemist at a midwestern US university, saw something unusual in an image of Mars. A participating scientist on NASA's Mars Exploration Rover mission, Susan was working with the Panoramic Cameras, an instrument on board the Rovers that takes digital photographs in color and in stereo. The Rover *Spirit* had recently been stuck in a sandy patch, and had left behind some unusually deep tracks during its extraction process. As Susan examined the Rover's images of these tracks using digital image-processing software, she started to recommend to her colleagues on the mission that they return to the area—which they called Tyrone—to investigate further.

At a teleconferenced science team meeting in October of that year, Susan explained why. She made a presentation with thirty-two PowerPoint slides, using the same few images of Tyrone to produce a variety of visual transformations (figure 2.1). She used these images to argue that while the Rover was driving away from the Tyrone area, it had exposed two different kinds of light soil that were compositionally different from the rest of the reddish-brown soil in the area. Transforming the same digital images into graphs and then into false color, Susan showed that the soils were some kind of salt deposit, that one was layered deeper than the other, and that the soil's spectral characteristics were changing gradually over time. At this point, this was still Susan's observation, made possible through her work of image processing. But her colleagues were convinced that her preliminary results were intriguing, and commanded the Rover to return to Tyrone to take more images for analysis.

At a team meeting a few months later, Susan presented results based on these newly acquired images. Her fifty colorful slides transformed Tyrone into a rich variety of

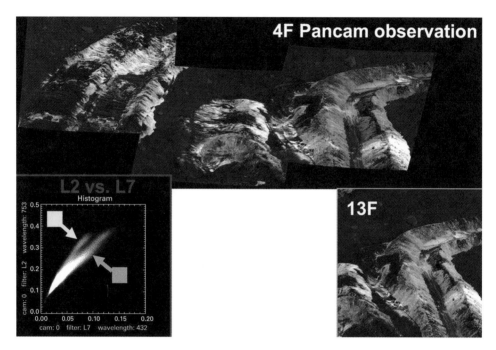

Figure 2.1
Tyrone. Slides from Susan Lee's presentation, October 2006. Used with permission.

visual forms and applied these same visual transformations to eight different pictures of Rover tracks from across the region. Now the team not only *saw* the two-toned light soil at Tyrone: *they saw it everywhere.* An hour-long discussion ensued with scientists rapidly trading hypotheses about what the salts were, whether water had transported them to distribute the deposits around the region, and what other observations they would need to resolve their questions. Susan's techniques of digital image processing had not only revealed an otherwise invisible phenomenon; they had also transformed an observation into a collective vision and ultimately a mission discovery, garnering a NASA press release and an eventual publication in *Science* magazine (Squyres et al. 2008). How were images enrolled in such a discovery?

There are many ways to analyze this brief story to illuminate the role of representation in scientific practice in the digital era, several of which are the subject of other chapters in this volume. We might focus on the model of objectivity Susan adheres to in her image manipulation practices. Or we might examine how her screen work also enrolls talk and gesture to make sense of the digital image of Tyrone. We could also look into how she and her colleagues manage the interplay between the visual and statistical modes of representation. In this chapter, however, I will approach this story in a way that maintains our focus on Susan's representational practices with images of

Tyrone, but also draws our attention to exactly how she and her colleagues came to see the two-toned salty soil at Tyrone—with its implications for the presence of past water on Mars—and to see it everywhere. To do so, I turn to classic work in the philosophy of scientific observation to develop a novel conceptual tool for analyzing scientific representation.

Many influential selections in the original *Representation in Scientific Practice* volume took their cue from empirical cases, working primarily in an analytical vein to establish theoretical contributions, from the externalized retina (Lynch 1990) to the immutable mobile (Latour 1990). This chapter draws on two years of ethnographic fieldwork with the Mars Exploration Rover mission team to examine the use of images in the daily work of the mission team. I bring together strands from the philosophy of science, the history of science, and ethnomethodological studies of work to elaborate how visualization practices with digital tools both construct knowledge of an alien planet and inscribe this same knowledge into an image, with consequences for future visions, representations, and interactions. Throughout, I will articulate how this theory of representation in scientific practice provides an analytical framework that we may usefully apply to imaging in a variety of periods of scientific work. I call this framework, *drawing as*.

2   *Drawing as* in Theoretical Context

Although a novel analytical frame, *drawing as* synthesizes—or perhaps draws together—a range of existing theoretical tools within the history, philosophy, and sociology of representation in science. Its specific contribution may best be examined through the choice of this somewhat awkward turn of phrase.

Drawing as purposefully recalls the phrase *seeing as* to invoke prior work on theory-laden observation by philosopher of science Norwood Russell Hanson (1958). While Hanson's formulation focuses on individual perception and cognition and does not encompass the individual and collective activity with materials, graphic instruments, and language involved in observation, *drawing as* attends instead to *theory-laden representation* in a form that maintains our focus on practice. The epistemological question of theory-ladenness is recast as an interest in those representational practices through which actors' epistemic commitments are enacted and produced.[1] That is, if the theory in "theory-laden observation" is anywhere, it is produced through practices of representation. The phrase *drawing as* returns the interrelated issues of salience, expectation, and visual expertise to the forefront of the conversation, made available to the analyst through observation of the activities of purposeful visual construal.

Practice-based analyses of scientific observation have produced a variety of case studies examining visual salience, expectation, and expertise. For example, Law and Lynch's study of ornithological field guides suggests that the observational question is one of establishing a "language game" of recognition and proper use of the drawings in a field context (Law and Lynch 1990). Other studies draw attention to how speech and gesture are used to make sense of an object: Amann and Knorr Cetina call this "optical induction," or *"visual operations carried out through talk"* (1990, 100; emphasis in original), while Morana Alač (2008 and this volume) demonstrates how gestures produce visual legibility. Recently, Catelijne Coopmans (2011) has developed the term "face value" to demonstrate the role of artful revelation in commercial software demonstrations. And Charles Goodwin's classic formulation of "professional vision" (1994) also articulates the role of interactions among people with different forms of expertise in producing visual claims.

In these examples and others, the observed image is not innocent. Instead, analyzability" is *built into* the record from the beginning" (Amann and Knorr Cetina 1990, 107; emphasis in original) in both the design of the observation and in the resulting image's digital processing. Like the traces of analytical production still visible in a document (Latour 1995) or the conflation of several degrees of externality in an observational report (Pinch 1985), skilled visual distinctions are inscribed into the image itself (Lynch 1990, 1985). The scientific image, I argue, gains analyzability only when it can present the relevant features to analysis. It can only be recognizable if it has been *drawn as* something recognizable: a presentation of a particular kind of object, or an object with particular features. Representation in scientific practice, I claim, is always a question of *drawing* a natural object *as* an analytical object; of conflating epistemological and ontological work in the world through purposeful visual construal. Such a stance brings the practices of drawing and of seeing ever closer together.

I use the term "drawing" to emphasize that this phenomenon is not unique to digital image processing: it applies across a variety of domains and time periods. A suggestive example is that of Galileo's images of the moon in *Sidereus Nuncius* (1610) as drawn through a telescope (figure 2.2A). Historians of science would call it historiographically unsound to guess at Galileo's perceptual experience, and this is where simply appealing to "theory-laden observation" becomes problematic. But his visual production presents an interesting point of comparison. In Galileo's images, there can be no ambiguity about what the patches of shadow on the moon's surface are. A simple and widely recognized shading technique (*chiaroscuro*) is employed to represent craters and pockmarks, to *draw* the moon *as* a topographical body. Due to the longstanding Ptolemaic assumption that superlunary physics was fundamentally different from terrestrial physics, identifying the features of a heavenly body as craters and other Earthly imperfections reveals profound Copernican commitments. That is, by depicting a planet with imperfections and topography, Galileo *drew* the moon *as* a Copernican object. The images in *Sidereus Nuncius* are thus an excellent example of *drawing as*. They demonstrate how visual and theoretical insight is produced in and through representational technique. The drawing is not just a projection of what Galileo saw.

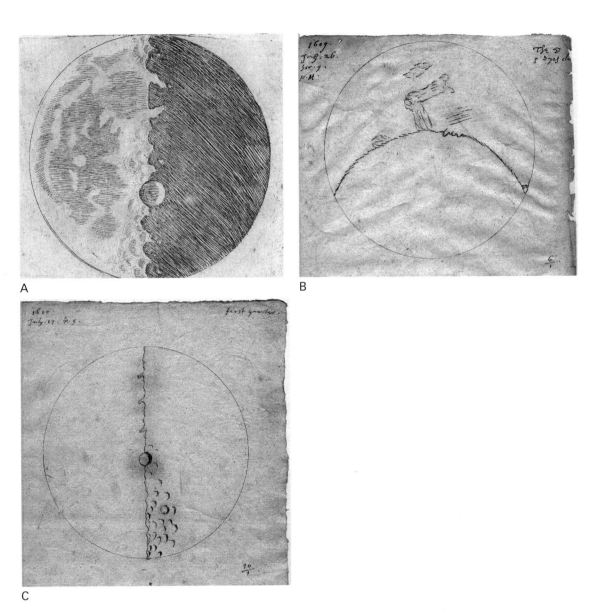

Figure 2.2

(A) Galileo draws the moon as a topographical body. *Sidereus Nuncius* (1610), p. 10 recto. Courtesy of the Institute for Advanced Study, Princeton, NJ, USA. (B) Harriot draws the moon in 1609 and (C) Harriot draws the moon in 1610. With permission of Lord Egremont.

In an important sense—like Susan Lee's digital image work—it is where the discovery emerges. Instead of talking about the great idea that occurs to Galileo's *prepared mind*, we can speak of a novel inscription produced by his *prepared hand*.[2]

Following from this sense of the prepared hand, "drawing" brings our attention to the craft[3] and the intentional work of purposeful image construal. As Amann and Knorr Cetina say, "seeing is *work*" (1990, 90; emphasis in original), and producing images that make such seeing possible, that set up a narrative that makes sense of objects, and that "fixate" visual evidence comprises much of this work on the Rover mission. Producing such images is not a question of finding an ambiguous image in/of the world and interpreting it. It is a question of skilled eyes and hands working in concert to disambiguate an imaged object. As I will show, when Susan constructs her image of Tyrone, each of her mouse clicks reveals or conceals an aspect of an object at hand. Thus the skills of visual interpretation arise from and are enmeshed in skills of image manipulation: in this case a kind of drawing with digital tools.

The term "drawing" should not limit the analyst to pencil and paper. We draw meaning out of objects, draw readers into texts, draw curtains to a close, and draw water from wells. In these cases "drawing" can mean to pull or guide, to reveal and to conceal, to work with and around material objects, to produce new configurations of space or movement. We may also explore "drawing things together," in terms of both inscribing and associating elements of a network (Latour 1990, 1995), or drawing distinctions (Kaiser 2005). Such definitions bring the material-semiotic configurations of representation in scientific practice more clearly into view. Yet objects cannot be *drawn as* anything: they exert their own push and pull. Following a feminist or a posthumanist line of thought in science studies (e.g., Haraway 1991; Barad 2007), we may inquire into the mutual relationships between who is drawing and what is being drawn, and how objects and subjects may be constituted through *drawing as* work. The mid-twentieth-century limitations of *seeing as* need not impede the more reflexive position required of the twenty-first.

Drawing as therefore presents analytical opportunities in the philosophy, history, and sociology of scientific imaging. Situated firmly in the study of representational practices, its novel turn is in articulating the reciprocal relationship between scientific representation and observation, between *drawing as* and *seeing as*. As an important part of that relationship, we must note that when Susan *drew* Tyrone *as* composed of two-tone salts, her colleagues could see Tyrone as she saw it too. Similarly, Galileo's image of the moon in 1610 was remarkable not only as a singular drawing, but because it clearly showed others a new way of *seeing* the moon *as* a topographical object, and *drawing* it that way ever after.

A powerful example of this phenomenon is in the drawings of Thomas Harriot. Following a tour to the New World where he had mapped the territory of Virginia, Queen Elizabeth I's geometer Harriot turned his telescope to the moon in 1609 and, presumably, drew what he saw: a crescent, some shading, and a dark patch near the center (figure 2.2B). But following the release of Galileo's images, Harriot produced a dramatically different set of drawings of the moon the next year, this time clearly emulating the Galilean view: a pockmarked moon, divided perpendicularly into light and shade, with a giant crater in the center (figure 2.2C). The story here is not that knowledge of the technique of chiaroscuro shading helped Galileo to uncover the moon's "true" nature (Edgerton 1984), but rather that it was a tool that enabled Galileo's knowledge of the moon to travel and be reproduced. The result is that Harriot didn't *see* the moon *as* a topographical body, but "just saw" and "just drew" the moon. Galileo's drawing therefore not only "founded visual astronomy" (Winkler and Van Helden 1992): it also influenced future viewings, depictions, and theoretical understandings of the moon, blinding viewers to other aspects such that the European scientific community might "just see" and draw the moon according to his vision. Contemporary visual work produces the same kind of *drawing as* and *seeing as* experiences using digital tools, as we will see in the discussion of image-processing work on the Mars Exploration Rover mission.

3 "Making It Pop Out": Image Work and the Dawn of Aspect

Mars Rover images are never singular views. In front of the two Panoramic Cameras (or, "Pancams"; Bell et al. 2003), often referred to as the Rover's "eyes," thirteen carefully chosen color filters rotate on a wheel to present different filtered possible visions of Mars taken from the same camera angle. Each camera pointing can return up to thirteen filtered black-and-white photographs of the same object, a measurement of the object's ability to reflect light in that limited range of wavelengths. As photons hit the photographic detector, they are tallied into pixel values. These values can be displayed along a gradient from white (many photons) to black (none). When a scientist combines three images of the same object in an image processor, that gradient changes to shades of red, green, and blue, producing a colorful image that may or may not align with what can be seen with the human eye. As mission scientists explain, the minerals common to Mars appear red to the human eye, but they reflect and absorb other wavelengths of light differently, which is diagnostic of their mineralogy. Selecting and combining specific filtered frames is the first step in image-processing techniques used across the mission.

On the Rover mission, standardized views of Mars have emerged involving the combination of particular sets of filters or processing choices, distinct from the human eye's sensitivity. These filter sets are combined and recombined depending on which features individual scientists are most interested in seeing. For example, a soil scientist interested in the composition of the terrain of Cape Verde, a crater promontory, assembled the Pancam's second, fifth, and seventh filters on its left-mounted

A

B

Figure 2.3
(A) False-color (L257) image of Cape Verde; the use of color demonstrates compositional differences in the soil. Opportunity Sol 952. Courtesy NASA/JPL/Cornell. (B) Approximate True Color image of Cape Verde, with adjusted saturation to reveal striations and stratigraphy. Opportunity Sol 952. Courtesy NASA/JPL/Cornell.

camera (abbreviated "L257") into a false-color image. This combination was judged helpful for revealing a wide range of textural and compositional differences. The resulting picture was well received by soil scientists and doubled as a "good image" for planning a drive into the crater, because it highlighted different types of soil that might be hazardous or safe for Rover wheels (figure 2.3A). But a geomorphologist pointed me to the same transformed image, saying: "we think we're getting all this [great data] but look, what do we get? [points to shadowed region] Artifact Soup." This scientist was most interested in the crater's stratigraphy: for him, "lighting and geometry" were more important than compositional difference, as they would allow him to measure the exact shapes, sizes, and depths of the crevices on the cliff face. He therefore combined the filtered frames that showed the least variation in pixel values and adjusted the lighting saturation to better reveal these distinctions (figure 2.3B).

In these two renderings of the same photographic frame we witness a switch between the artifact and the object of analysis: the rich colors in figure 2.3A show composition and texture at the expense of lighting, while the lightness in 2.3B reveals stratigraphy at the expense of compositional information. The images also demonstrate how the selection and combination of multiple raw data products vary based on the image processor's intent: whether to show stratigraphy or soil properties. But the flexibility to see it both ways is crucial to the mission's science and operations. The geomorphologist would not be satisfied with the soil scientist's picture, and a Rover driver could not identify slippery soil in the geologist's image. Like Goodwin's (1995) example of interdisciplinary scientists on an oceanographic vessel, the two observers are attuned to different elements in their experience. But in this case, both representations are produced from exactly the same data set, the same set of pixels. It is only because of the image processor's choices that a different set of features is revealed or subdued each time. The result of this plethora of possibilities is that one is often confronted with an image of Mars repeated through different filters or processing algorithms. With so many possible viewings (and recalling that True Color and false color refer to results of filter combinations, and not to any claim to representational primacy), it is clear that there is no "one best way" of picturing Mars. Rather, such images represent different ways of seeing and knowing the Martian surface.

In fact, it is often *necessary* to see different things in the same image. For example, when calibrating images that return from the Panoramic Cameras, a human operator must go through several computational steps to locate and eliminate dust in the atmosphere that scatters the light in the Pancam frame. The equation that results from this activity is applied across the board to an entire suite of images to systematically subtract a value from all pixels, such that the images are standardized, "corrected for" dust and atmospheric opacity on any given day. But one person's artifact is another's data. The mission's atmospheric scientists rely on these dust values to understand the atmosphere and Martian weather patterns, and soil scientists try to understand the optical qualities of the dust itself. They therefore use the output from the calibration procedure to get the dust information, and would rather see the dust than the image it obscures. The multiple views that result are therefore not an attempt to hone in on a better representation of Mars in an absolute sense; nor are they necessarily views that invoke a different moral or epistemic status (Lynch 1991). They are the result of multiple, purposeful construals of a single visual field, for different purposes.

Multiplicity is by now a familiar topic in science studies. We know that interactions with objects enact multiple ontologies (Mol 2002); and that viewings from multiple standpoints or "views from somewhere" can produce alternative objectivities (Longino 1990; Haraway 1991). Here I want to draw our attention to another aspect of multiplicity: the opportunities of ambiguity. Indeed, seeing the same data in different ways recalls the ambiguous gestalt images that were central to mid-twentieth-century accounts of observation. The most famous example is the duck-rabbit, in which the gestalt switch from *seeing it as a rabbit* to *seeing it as a duck* is an example of "the expression of a new perception and at the same time of the perception's being unchanged" (Wittgenstein 1953, 167). In his *Philosophical Investigations,* Ludwig Wittgenstein explores the conditions under which it makes sense to say "I see it *as* x" as opposed to "I see x." He notes that usually people do not say "I see it as" about their visual experiences—they just *see*—but the ability to say "I see it as . . ." arises in situations where there is some ambiguity or doubt as to which features are salient: which elements form the background and which the foreground. Wittgenstein then describes the "dawning of aspect," the change in the organization of visual experience by means of which the foreground and the background, or the artifact and the object, shift. Although the object does not change, this change of aspect produces a different observation, "quite as if the object had altered before my eyes" (Wittgenstein 1953, 195).[4]

Norwood Russell Hanson draws upon Wittgenstein's discussion of *seeing as* in his formulation of theory-laden observation in science. In *Patterns of Discovery* (1958), Hanson claims that scientific seeing is not a question of freeing observations from bias, but rather of acquiring a theoretical and practical orientation that enables the scientist to *see as*. Thus the physicist must learn to *see* the glass object *as* a cathode ray tube, and Kepler's Copernican commitments enable him to *see* a sunrise *as* the sun standing still in the sky while the Earth rotates toward it. After acquiring an aspect, scientists can thenceforth distinguish foreground from background and signal from noise, giving the visual field coherence, recognizability, and meaning. Observational reports in science thus involve interpretation at their most basic level, as "theories and interpretations are 'there' in the seeing from the outset" (Hanson 1958, 10). This learning to see, as it is deployed in the practice of skilled observation, is also described with respect to bird watching (Law and Lynch 1990), and has powerful resonances with "professional vision" (Goodwin 1994), among other concepts.

The skill of *seeing as* that Hanson identifies as essential to scientific practice is also evidenced in the above images of Mars. As with the duck-rabbit, we might *see* Cape Verde *as* a stratified cliff face (figure 2.3B), or *see* it *as* composed of different soils (figure 2.3A). Unlike in the duck-rabbit example, however, these *seeing as* experiences are not "found" experiences but purposefully crafted ones. They are the result of directed image-processing activities that compose the image into something meaningful, distinguishing foreground from background or object from artifact and highlighting key features while downplaying others, in the very composition of the image itself. The observer is not passive, but rather actively composes the image into something meaningful through image-processing practices. Learning to see and learning to draw are intertwined. This representational work of purposeful visual construal constitutes what Coulter and Parsons call the "praxiology of perception": "modes of perceptual orientation as forms of practical, social actions, capacities and achievements" (1991, 252).

The planetary scientists I spoke to in my research repeatedly explained to me that the point of their image manipulation was "to see new things," to make a hidden feature "pop out," to discriminate between different units that otherwise appeared the same in one filtered image but which might, upon combination with other filtered images, prove to have different spectral characteristics, pointing to a difference in composition. For example, a scientist I interviewed who was looking for sulfate content on Mars explained, "If you get a particular [filter] combination, the sulfates just jump out at you. It's like they turn green or blue or something." This does not imply a change in the underlying dataset: only a change in visual orientation or aspect due to the combination of filtered images. Another scientist explained, "The data is the same, the difference is in what you see." An engineer on the team echoed this statement: "The image never changes, but you can manipulate the image, and everyone sees something different." Another insisted that this ability to see something different with each click of the mouse is the key to his digital work with Pancam images: "If you were walking around with your rock on Mars without Pancam you might not even know that these [two layers of rock] were different! . . . The ability to discriminate between these units is the real power of Pancam . . ."

There is a clear resonance here with postphenomenology, in which instrumental embodiments mediate human perception and the formulation of scientific sight (Idhe 2009; see also Engström and Selinger 2009). But what I wish to draw our attention to is how the *seeing as* experiences that the Pancam produces through its "instrumental embodiment" are due to specific practices with visual materials: in this case, digital images and software suites. This work of purposeful image construal enables practitioners to see, discriminate and characterize, and enforce changes in aspect that, in turn, allow new elements in the image to be appreciated as foreground instead of background. Through the work of digital image processing, a skilled vision is crafted into the image from the outset. The resulting picture thus depicts what the object ought to be *seen as*. This purposeful image construal, this active representational disambiguation, is the practice of *drawing as*.

4   The Many Visions of Tyrone: *Drawing as* in Action

Despite its Wittgensteinian heritage and postphenomenological resonances, *drawing as* is not a philosophical construct: it is a practical one. It focuses our attention on what scientists *do* with images, their work of image construal that makes scientific objects (such as soil composition on Mars) legible to their peers. It reasserts our focus on the observable activities of scientists at their desks, in the field, or wherever they perform image work, examining these activities as constitutive of perception, while at the same time maintaining interest in perceptual change in aspect, scientific seeing as expert bias, and visual suggestibility. *Drawing as* also permits us to maintain our interest in actors' epistemologies as they are made visible and accountable through practical action, and how these practices of ways of knowing can produce objects and interactions too. Let us return to the opening example to see how this is accomplished as a function of representational work.

Susan Lee began her investigation of Tyrone by using it as a test object on which to practice different techniques of composing false-color images. The Rover had taken four filtered images of Tyrone, which she combined through red, green, and blue color channels in the Pancam's software to produce True Color and false color. In doing so, she first noticed a slight color distinction: that what looked like just a patch of white soil in the filtered images seemed to display as two slightly different colors in the True Color and false-color imagery (figure 2.1, bottom right). Facing a possible distinction in the terrain, Susan then attempted to isolate this distinction and amplify it, to make it "pop out." She turned to an image-processing technique called a decorrelation stretch, which accentuates the contrast between pixel values on one of the three contributing filtered images, but not equally across all three of the contributing images.[5] The result is a picture that bears resemblance to the color scheme of pop artist Andy Warhol, with the garish colors representing the disparity of pixel values across the composite filters, indicating a possible difference in mineralogical composition (figure 2.4).[6] This was Susan's first *drawing as* activity to disambiguate the visual data: purposefully composing the filtered frames into an image that depicts fine-grained spectral differences, making the two-toned soil "pop out."

Susan then moved from simply discriminating between colored materials to characterizing them in order to say something about their classification or origin. To do so, she turned to the numerical side of the image. First, she asked the computer to display all the image's pixel values at once on a graph (figure 2.1, bottom left). That is, she *drew* the pictorial data *as* a histogram: a graph in which individual pixel values are plotted together. This would help her to isolate the spectral properties of the two different kinds of soils and possibly make a determination about their composition. As she declared, "I'm not looking at pretty image, I use [a] histogram . . . if my purpose [is] to see if [it is] two different type[s of] material."[7] Construed in this way, the image data showed two distinct clusters of pixel values. Susan interpreted these two branches of the histogram as two different types of material, whose properties of light absorption were so different that they produced radically different pixel values.

But while her histogram showed that two different kinds of material were present in the image data, it did not show *where* that material was located or what properties that material had. She therefore used another Pancam tool to "separate [the two materials] spatially." When one branch of the histogram was colored in green, all the pixels plotted on that branch lit up in green on the picture version of the file (figure

Tyrone: **Decorrelation Stretch**

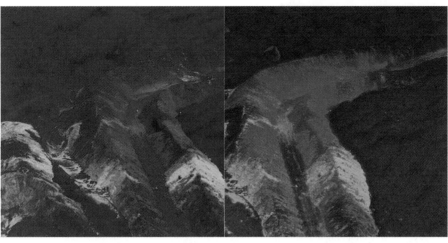

L5-L7-L2   R2-R3-R7

Figure 2.4
Susan Lee's decorrelation stretch of Tyrone uses color to demonstrate compositional difference in the two soils. End of Sol Presentation, 11 October 2006. With permission.

2.5). She could then see where that material was scattered. She proceeded to color the other branch of the histogram in yellow, lighting up a different patch of white soil. Thus two different kinds of soil with different spectral characteristics were confirmed. And because of where those different patches of soil lit up in the image in green and yellow—what Susan called "spatial correlation"—she could see that the yellow material was buried deeper in the wheel track than the green. She therefore suggested that *Spirit*'s recently broken wheel had turfed up a deeper layer of soil that was previously invisible to the team. Finally, having identified where those distinct materials were located, Susan used another Pancam tool to plot the pixel values for each type of soil across all the Pancam's filters, from visible light into the near infrared. As different minerals reflect different amounts of light in different wavelengths, the resulting graph revealed the mineralogical content of the two salts visible in the scene.

As Susan moved from false color, to decorrelation stretch, to histogram, to spatial correlation image, to pixel graph, her image-processing techniques revealed different aspects of Tyrone around which to organize her visual experience. Just as an ambiguous gestalt figure may resolve into the image of a duck or of a rabbit, so the image of Tyrone resolves anew at each click. Susan thus disambiguated the visual experience of Tyrone by isolating only a single aspect of it at a time, blinding or curtailing alternative

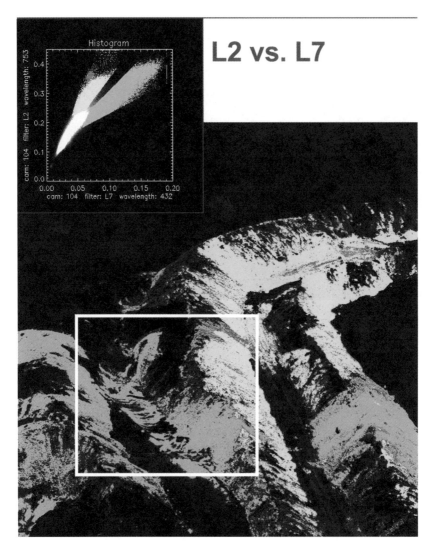

Figure 2.5
Coloring in a branch of the histogram in green lights up the location of those pixel values in the image. Coloring the other branch in yellow locates the second type of material. With permission.

aspects. With each transformation of Tyrone, she purposefully included those features that she considered salient and excluded or silenced other features, relegating them to the background. This highlights and even restricts the subsequent visual experience to that aspect, enforcing a situation of *seeing as*. Importantly, these classifying, sorting out, and discriminating practices of observing arise from and are recorded in the work of image making. And each of these transformations not only allowed Susan to *see* a distinction in the Martian terrain: bringing these aspects together in concert, she made a claim about the history of water in this region on Mars.[8]

5   The "Light-Soil Campaign": From *Drawing as* to *Seeing as*

Susan is adamant that the distinction and changes in the soil revealed by her image-processing techniques are based in fact. However, her use of color was important for "showing" this distinction both to herself and to others: "You decide the color you want to show, the color you want to use, but the data is there, it's not the color. . . . Because the existing data [images] contains this kind of information, you decide how you want to show [the data]."

Using green and yellow reconfigures the pictorial representation of the image such that this feature of the soil lights up. It also reveals "information" that is already *in* the image: it is not an interpretive annotation. But while the image "contains this kind of information" (e.g., the spectral properties of the soil), it is at Susan's discretion to "decide how to show" the data. That is, *drawing as* allowed her to first see and then characterize a distinction in the soil—and then to show her colleagues what to see in the soil. In this way, *drawing as* not only reveals new aspects of a visual dataset but also produces a *seeing as* experience for other viewers. Once the distinction has been made in one aspect, it cannot be unseen.[9]

This sort of operation is not limited to Susan's transformations of Tyrone, or to Pancam imagery alone. Across the mission, team members articulate the dawning of aspect when presented with digital images that are visually construed to present only particular properties. Expressions such as "now I see!" are frequently heard in meetings and at scientists' desks as they go through different image-processing routines or present their processed images to their colleagues. As one scientist examined an image he produced in his lab, he expressed, "It's efficient to have something like that [image] to communicate what you're showing, what your interpretation [is]." Even when operating a spectrometer, a team member explained that he had to "show other spectra to teach [my teammates] what to see," or that he took the approach of "I'm only gonna show you the part I want you to pay attention to." This is not hiding data that might be essential to interpretation, but rather limiting data to only those features that are relevant: an attempt to *draw as*, to delimit aspect in order to produce and reproduce a *seeing as* experience in the viewer. This presents implications for the results of visual interpretation.

Additional implications of *drawing as* also arose from Susan's representational work with Tyrone. Following her presentation in October 2006, Susan applied the same visual transformations to images taken at three other sites in the broader region around the Rover *Spirit*. Displaying these slides at the February meeting, she emphasized the "similar situation" between these locations, indicating that the phenomenon was widespread. Thus the Rover scientists who witnessed Susan's presentation acquired this same aspect, a particular vision of the surface that generated excitement about specific possibilities for further interaction and exploration. All present treated the existence of the two-toned soil and its distribution as observational fact: the question up for discussion was not whether the soil existed, but *why* it was there and how it got there. The result of this discussion was "the Light Soil Campaign": a series of follow-up observations, photographs, spectral readings, and drive directions that formed the basis of Rover operations on Mars for the following two weeks. These follow-up interactions in turn generated new suites of images to *draw as* Tyrone, thereby aiding in establishing, constituting, and working with new classes of objects on Mars.

Susan's *drawing as* practices—*drawing* Tyrone *as* composed of two distinct kinds of salty soils distributed in different layers, then applying that visual convention to other local sites to *draw* them *as* Tyrone—encouraged the rest of the team to *see* Tyrone *as* composed of those materials as she suggested, and to *see* other examples *as* cases of the same phenomenon. Following this practical work of drawing and seeing was a suite of Rover interactions and eventually a published paper bearing Susan's name along with those of her teammates.[10] Thus, contrasting Ian Hacking's (1983) point that how scientists intervene with an object informs how they represent it, a more entangled perspective is also true: practical interactions with an object (like Tyrone, or Mars) are also predicated on how that object is visually construed: how and what it is *drawn as*.

6   Conclusion

As Susan's way of drawing and seeing Mars began to take hold across the Rover team, the two-toned soil moved from being an individual vision of peripheral interest and idiosyncratic representation to one of the central questions of the mission, a shared way of representing Martian soil at *Spirit*'s landing site, and a cue for subsequent Rover activity. This vision of Mars did not come from "just seeing" the terrain. It was the result of purposeful practices of image construal using image-processing software that *drew* the soil *such that* the distinction could be seen. And as this visual framework was applied across the region, the scientists no longer *saw* the white soil *as* two-toned: they "just saw" the two-toned soil, and saw it everywhere. Lest this seem like a simple example of perceptual suggestibility, recall how this interpretation is *drawn into*, inscribed in, the very images that present the phenomenon, such that the phenomenon can be seen.

The practical work of digital image processing is not only a matter of revealing distinctions, patterns, or "information" in an imaged object. It also includes the work of producing and then inscribing a visual aspect into an image—the practical activity of *drawing* a natural object *as* an analytical object, such that subsequent viewers and image makers will see, represent, and interact with that same object the same way. As these images are deliberately constructed, manipulated, and pointed to with human hands and digital software suites, this practical work renders the pictured object meaningful and workable, even at a distance of millions of miles. Representation in scientific practice is therefore not a question of creating an ever more true or singular image of an object. Instead, we should note how practical work with images shuts down other ways of seeing in order to focus on one aspect, one set of salient relationships. It is the role of the analyst to identify each image's inscribed and implied aspect, to note the representational choices that produce and reproduce that aspect, and to remind us that it could always be seen otherwise.

This analytical approach applies to our representations of scientific practice as well. Throughout this chapter I have emphasized certain practices and deemphasized others. As I represent my field site, I also disambiguate my actors' many complex practices to produce a coherent account of their work. As Woolgar and Pawluch (1985) suggest in their discussion of "ontological gerrymandering," other scholarly perspectives might easily be brought to bear to illuminate these examples differently, to draw and redraw the contours of salient moments in the field. Our selective attention in our research, and our ability to make new features of interest "pop out" from prior research findings, are themselves another example of *drawing as*. Like the skilled work of image processing, I argue, such work is not problematic for our claims: rather, it produces insights. Grounded as it is in disciplinary modes of attention and depiction, *drawing as* does not produce untruths or dubious results, but rather focuses expert attention on one feature at a time in an otherwise complex experience. *Drawing as* therefore presents analytical opportunities, but only if we remain reflexively attuned to our own position. After all, it is always possible to produce a change in aspect. Looking for moments in which we can say, "my site has not changed, and yet I see it differently" can be a powerful methodological tool for developing understandings of subjects, objects, and analysts in practices of knowledge production.

Acknowledgments

My sincere thanks to Michael Lynch, as well as to Steve Straker, Alan Richardson, and Simon Schaffer, for their intellectual engagement with previous iterations of this work; to Steve Woolgar, Catelijne Coopmans, and the anonymous reviewers for their help in strengthening and clarifying the argument and its contributions; and to Steve Squyres, Jim Bell, and the Mars Exploration Rover mission members for permission to study their team. Versions of this paper were presented at the 2006 4S Annual Meeting in Vancouver, and at the University of Toronto's IHPSC center conference, 2007. Work was supported under NSF Doctoral Dissertation Improvement Grant #0645945.

Notes

1. Radder's (2006) work in the philosophy of science also attempts to reclaim "theory-laden representation" in this vein, not through an analysis of visual practice but rather by drawing on work in embodiment from the sociology of science and from phenomenology.

2. The Galileo literature is vast. Of particular interest here is Horst Bredekamp's encyclopedic *Galilei der Künstler* (2007) and Biagioli (2006, especially 105–111); on recent interpretations of Thomas Harriot's lunar images as cartography and not visual astronomy in the Galilean sense, see Pumfrey (2009). I thank Eileen Reeves for her suggestions with respect to Galilean scholarship.

3. It is outside the scope of this paper to properly consider the role of bodies in image work; I refer the reader to Alač and Prentice (this volume); Myers (2008); and Vertesi (2012).

4. For more on the Wittgensteinian concept of aspect, see Verdi (2010).

5. For the scientists I observed, it was critical to apply the same effect to all frames involved in a composition, with the exception of the decorrelation stretch. However, even in the latter case, the emphasis was always on repeatability, largely through the application of mathematical formulae to image frames. This was often described as a practice that made their image-processing work scientific instead of artistic. It is beyond the scope of this paper to examine the relationship between visual techniques, trust, and morality (Daston and Galison 2007), but see Vertesi (forthcoming).

6. The relationship between astronomical image processing and modern art is discussed in Lynch and Edgerton (1996).

7. Susan's rejection of the "pretty image" resonates with Anne Beaulieu's work on visual iconoclasm (2002). However, Susan's condescension here is not aimed at the images, which she uses equally alongside the numerical transformations to characterize Tyrone, so much as at the aesthetics of the image to the detriment of her scientific sight. See Lynch and Edgerton (1988) on this point.

8. Note that this processing requires a particular virtuosity in terms of both image-processing capability and familiarity with the visual conventions in the field (such as visual languages: see Rudwick 1976).

9. Also described in Shubin (2008).

10. See Wang et al. (2008).

References

Alač, M. 2008. Working with brain scans: Digital images and gestural interaction in fMRI laboratory. *Social Studies of Science* 38 (4):483–508.

Amann, K., and K. Knorr Cetina. 1990. The fixation of (visual) evidence. In *Representation in Scientific Practice*, ed. M. Lynch and S. Woolgar, 85–121. Cambridge, MA: MIT Press.

Barad, K. 2007. *Meeting the Universe Halfway: Quantum Physics and the Entanglement of Matter and Meaning*. Durham: Duke University Press.

Beaulieu, A. 2002. Images are not the (only) truth: Brain mapping, visual knowledge and iconoclasm. *Science, Technology and Human Values* 27 (1):53–86.

Bell, J. F. I., S. W. Squyres, K. E. Herkenhoff, J. N. Maki, H. M. Arneson, et al. 2003. Mars Exploration Rover Athena Panoramic Camera (Pancam) investigation. *Journal of Geophysical Research* 108:E1. doi:10.1029/2003JE002070.

Biagioli, M. 2006. *Galileo's Instruments of Credit*. Chicago: University of Chicago Press.

Bredekamp, H. 2007. *Galilei der Künstler*. Berlin: Akademie Verlag.

Coopmans, C. 2011. "Face value": New medical imaging software in commercial view. *Social Studies of Science* 41:155–176.

Coulter, J., and E. D. Parsons. 1991. The praxiology of perception: Visual orientations and practical action. *Inquiry* 33:251–272.

Daston, L., and P. Galison. 2007. *Objectivity*. New York: Zone Books.

Edgerton, S. Y. 1984. Galileo, Florentine "disegno," and the "strange spottednesse" of the moon. *Art Journal* 44 (3):225–232.

Engström, T., and E. Selinger. 2009. Reinventing sight: Theories and practices of imaging. In *Rethinking Theories and Practices of Imaging*, ed. Timothy Engström and Evan Selinger, 21–61. New York: Pallgrave Macmillan.

Goodwin, C. 1994. Professional vision. *American Anthropologist* 96 (3):606–633.

Goodwin, C. 1995. Seeing in depth. *Social Studies of Science* 25:237–274.

Hacking, I. 1983. *Representing and Intervening*. Cambridge: Cambridge University Press.

Hanson, N. R. 1958. *Patterns of Discovery: An Inquiry into the Conceptual Foundations of Science*. Cambridge: Cambridge University Press.

Haraway, D. 1991. *Simians, Cyborgs and Women: the Reinvention of Nature*. New York: Routledge.

Idhe, D. 2009. *Postphenomenology and Technoscience*. Albany: SUNY Press.

Kaiser, D. 2005. *Drawing Theories Apart: The Dispersion of Feynman Diagrams in Postwar Physics*. Chicago: University of Chicago Press.

Latour, B. 1990. Drawing things together. In *Representation in Scientific Practice*, ed. M. Lynch and S. Woolgar, 19–68. Cambridge, MA: MIT Press.

Latour, B. 1995. The "pedofil" of Boa Vista: A photo-philosophical montage. *Common Knowledge* 4 (1):144–187.

Law, J., and M. Lynch. 1990. Lists, field guides, and the descriptive organization of seeing: Bird-watching as an exemplary observational activity. In *Representation in Scientific Practice*, ed. M. Lynch and S. Woolgar, 267–299. Cambridge, MA: MIT Press.

Longino, H. 1990. *Science as Social Knowledge: Values and Objectivity in Scientific Inquiry*. Princeton: Princeton University Press.

Lynch, M. 1985. Discipline and the material form of images: An analysis of scientific visibility. *Social Studies of Science* 15:37–66.

Lynch, M. 1990. The externalised retina: Selection and mathematization in the visual documentation of objects in the life sciences. In *Representation in Scientific Practice*, ed. M. Lynch and S. Woolgar, 153–186. Cambridge, MA: MIT Press.

Lynch, M. 1991. Science in the age of mechanical reproduction: Moral and epistemic relations between diagrams and photographs. *Biology and Philosophy* 6:205–226.

Lynch, M., and S. Edgerton. 1988. Aesthetics and digital image processing: Representational craft in contemporary astronomy. In *Picturing Power: Visual Depiction and Social Relations*, ed. G. Fyfe and J. Law, 184–220. London: Routledge and Kegan Paul.

Lynch, M., and S. Edgerton. 1996. Abstract painting and astronomical image processing. In *The Elusive Synthesis: Aesthetics and Science*, ed. A. I. Tauber, 103–124. Netherlands: Kluwer Academic Publishers.

Mol, A. 2002. *The Body Multiple: Ontology in Medical Practice*. Durham: Duke University Press.

Myers, N. 2008. Molecular embodiments and the body-work of modeling in protein crystallography. *Social Studies of Science* 38 (2):163–199.

Pinch, T. 1985. Towards an analysis of scientific observation: The externality and evidential significance of observation reports in physics. *Social Studies of Science* 15:167–187.

Pumfrey, S. 2009. Harriot's maps of the moon: New interpretations. *Notes and Records of the Royal Society* 63:163–168.

Radder, H. 2006. *The World Observed/The World Conceived*. Pittsburgh: University of Pittsburgh Press.

Rudwick, M. 1976. The emergence of a visual language for geological science 1760–1840. *History of Science* 14:149–195.

Shubin, N. 2008. *Your Inner Fish: A Journey into the 3.5 Billion Year History of the Human Body*. New York: Vintage Press.

Squyres, S. W., R. E. Arvidson, S. Ruff, R. Gellert, R. V. Morris, D. W. Ming, et al. 2008. Detection of silica-rich deposits on Mars. *Science* 320 (5879):1063–1067.

Verdi, J. 2010. *Fat Wednesday: Wittgenstein on Aspects*. Philadelphia: Paul Dry Books.

Vertesi, J. 2012. Seeing like a Rover: Visualization, embodiment and interaction on the Mars Exploration Rover mission. *Social Studies of Science* 42:393–414.

Vertesi, J. Forthcoming. Documenting Mars: Lookiloo and constraints in 21st century planetary image processing. In *Documenting the World*, ed. K. Wilder and G. Mittman. Chicago: University of Chicago Press.

Wang, A., et al. 2008. Light-toned salty soils and co-existing Si-rich species discovered by the Mars Exploration Rover Spirit in Columbia Hills. *Journal of Geophysical Research* 113, E12S40, doi:10.1029/2008JE003126.

Winkler, M. G., and A. Van Helden. 1992. Representing the heavens: Galileo and visual astronomy. *Isis* 83:195–217.

Wittgenstein, L. 1953. *Philosophical Investigations*. Oxford: Blackwell Press.

Woolgar, S., and D. Pawluch. 1985. Ontological gerrymandering: The anatomy of social problems explanations. *Social Problems* 32:214–227.

3   Visual Analytics as Artful Revelation

Catelijne Coopmans

In February 2010, the *Economist* published a fourteen-page special report titled "The Data Deluge—And How to Handle It." The cover showed a male figure holding an inverted umbrella toward a sky from which streams of binary data rain down. With his umbrella, the man catches some of the streams, using the distillate to water a flower by his side. Other zeros and ones rain down without bothering the man or his plant. The umbrella device thus helps manage an otherwise overwhelming "torrent of information" (*Economist* 2010, 15), extracting beneficial elements from it. This idea is made salient inside the special report by a series of articles on how scientists, business analysts, and others engage with the expansive realm of digital data. One article emphasized the usefulness of data *visualization*. It stressed the "natural" affinity humans have for identifying visual patterns: "Looking through a numerical table takes a lot of mental effort, but information presented visually can be grasped in a few seconds" (*Economist* 2010, 13). Such grasping, according to the article, is now increasingly facilitated by software programs in which "computer science, statistics, artistic design and storytelling" are brought together, enabling specialists and nonspecialists to work with, and extract insights from, large datasets.

Data visualization, also sometimes called information visualization, visual analysis, or the term I shall adopt in this chapter, *visual analytics* (Lawton 2009), has found a wide range of uses and users. Traders in the financial industry rely on dynamically configured, color-coded graphs and scatter plots to monitor "the market" and to strike when they "see" an opportunity (Beunza and Stark 2004; Pryke 2010). Other domains known for investing in visual analytics include pharmaceutical research, supply chain management, manufacturing, and, in the United States, homeland security (Lawton 2009).[1] Meanwhile, free data visualization tools are proliferating on the Internet, encouraging users to experiment with visualizing their own or public datasets, and to share and comment on the results.

Like such visualization technologies as the microscope, the telescope, and the X-ray detector before it, software for visual analytics is associated with the capacity to reveal what has hitherto been hidden, to give access to what is otherwise inaccessible. In the

1980s and 1990s, before the onset of interactive, screen-based applications, Edward Tufte, a famous proponent of information design, already stipulated that a good data visualization must "reveal correlations not otherwise evident" (Grady 2006, 236).[2] With the computerized data displays that underpin contemporary visual analytics the emphasis on revelation has intensified, even as the job of selecting what to view and how to view it has become shared between designer and end user–analyst (cf. de Rijcke and Beaulieu, this volume). Charles Kostelnick, a specialist in the use of visual rhetoric in professional communication, characterizes a digital graphical display as "a mere surface chart that readers can metamorphose in a variety of ways, with a rich underlayer of possibilities waiting for the reader to discover" (2008, 123). Visual artist Richard Wright (2008, 81) writes that, by harnessing the power of computers together with that of human perception, data visualization may make a dataset "give up its secrets." Similar statements can be found in sociological writings on data visualization, particularly in finance. Michael Pryke (2010, 434) notes that "the software in many cases enables the visualization of overlooked or even undetectable market characteristics lost in previous representations of market action." Adding a metaphorical flourish that emphasizes enhanced vision, Daniel Beunza and David Stark (2004, 389) say that "traders put on the financial equivalent of infrared goggles that provide them with the trader's equivalent of night-vision."

What can we make of such claims that visual analytics makes hitherto obscure relations within a dataset amenable to perception? We could, in a move that will likely be familiar to readers of this volume, understand such language as shorthand for the practices that coax data into visual patterns, and for the situated ways of seeing that make sense out of such patterns. In this chapter, I want to go in a different direction—namely by attending to how *revealing* as an epistemological trope both animates and is animated by visual analytics as a way of working with data.

Ludmilla Jordanova's essay "Nature Unveiling before Science" provides a reminder of the long-standing fascination with the prospect of unveiling truths hidden "beneath a layer" (Jordanova 1989, 90)—something she locates in eighteenth- and nineteenth-century art and commentary on science and medicine. The rich connotations of veiling/unveiling, Jordanova suggests, prompt reflection on what is contained in "the very act of looking itself, an act that lies at the heart of our epistemology" (1989, 91). Visual analytics is a domain in which some of these connotations are presently reenacted. In this chapter, I examine how this is done in online software demonstrations that portray data as a source of value, and visualization as a means to unlock that value.

One aspect I will discuss in particular is the paradoxical manner in which "plain sight" is co-opted as a key ingredient for visual analytics—and the implications of this for how investment in data visualization practices is created and sustained. Indeed, *because* of its emerging nature as an influential technovisual arrangement for how people see, know, and engage with the world, visual analytics belongs in a volume that revisits representation in scientific practice. The amalgamation of vision, knowledge, discovery, and value that lies at the heart of visual analytics cuts across scientific and other forms of data analysis, especially insofar as it moves people to relate to data in a particular way.

My chapter discusses software demonstrations of visual analytics vendor Tableau Software, a company whose goal is "to help people see and understand data," and whose marketing is targeted at "a broad population of business users" (descriptions from its website). In particular, I will focus on its webinars, online "web seminars" that, by displaying the screen work of experienced users, provide prospective customers with a chance to witness visual analytics in action.[3] The rhetoric of showing and telling manifested in these webinars bolsters expectations that insights inhere in data and can be visually apprehended, at the same time as it renders such insights conditional and elusive. This showing and telling, I suggest, can be characterized as *artful revelation*.

1 Artfulness and Revelation

In science and technology studies (STS), to speak of representational practice as *artful* usually is to dispel any notion of a natural order ready to reveal itself through the tools, techniques, and methods of science. It draws attention to the work involved in making phenomena—such as the home ranges of lizards and the anatomy of axonal sprouting in the brain (Lynch 1985a), or the interaction between forest and savannah in the Amazon (Latour 1999)—tractable for scientific investigation through a progressive process of visualization and mathematization (see also Lynch 1990). Artful practice is also contained in "visual exegesis" or "the work of seeing what the data consist of" (Amann and Knorr Cetina 1990, 90), an important aspect of which is the situated accomplishment of the distinction between signal and noise, or findings and artifacts (Lynch 1985b). The work of making visible and that of reading visible outputs are brought together in *image processing*, which since early discussions in the 1980s has become an ever more central part of various fields of science. Rendering image data in such a way as to highlight "the observable and measurable properties of the data" (Lynch and Edgerton 1988, 208) can be characterized as artful revelation, in the sense that practitioners work to draw out those things that are properly contained within the data. In some circumstances, the art entailed in such practices makes them vulnerable to the charge of "unscientific" aestheticism, prompting scientific journal editors to campaign against misleading data displays (Frow, this volume) and scientists themselves to rhetorically demarcate proper science from the production of "pretty pictures" (Beaulieu 2002).

Like image processing, visual analytics involves the rendering of data in visual form. Such rendering, while software-supported, is not automatic: practitioners make judgments on how to do it. At the same time, visual analytics cultivates a neutral stance to what a visual display is supposed to show: "you do not know what you are looking for but will recognize it when you find it" (Beunza and Stark 2004, 373). Instead of presupposing a referent, visual analytics presupposes users' pattern recognition ability and makes this a central feature of the way it supports discovery. Richard Wright puts it this way:

The images frequently exhibit the continuous qualities of the familiar visual world despite the fact that they are utterly constructed. It is these implicit visual properties that are valued for their openness to perceptual inference—a continuous interplay of surface features rather than discrete graphic elements or symbols. At this end of the spectrum, visualization is nonrepresentational because it is speculatively mapped from the raw, isolated data or equations and not with respect to an already validated representational relation. A visualization is not a representation but a means to a representation. (Wright 2008, 81)

Unlike a graphical pattern that is read as the "visible traces *of* invisible reactions" (Knorr-Cetina and Amann 1990, 263, emphasis added), or as an unfolding indication *of* the nature and success (or failure) of an experiment (Woolgar 1990), Wright suggests that the graphical views that support visual analytics are composed, by software algorithms and by the user-analyst, *for* perceptual inference.[4] While sharing features with the active coaxing into view that Janet Vertesi in her chapter has termed *drawing as*, this kind of data visualization is invested less in a particular "something" to be seen, and more in sight itself as a detector of whatever stands out. A trading "opportunity" (Beunza and Stark 2004), or business "insight" (North 2006) is not treated by practitioners as a natural phenomenon like the soil on Mars, but at the same time the technology naturalizes *vision itself*, and thereby raises expectations that it can reveal such things before their eyes. This is not to say that what counts as a meaningful insight is determined at a glance (North 2006), but that the glance is given a key role in making previously inaccessible statistical relations accessible to anybody. The co-optation of plain sight gives visual analytics a public dimension: once relations within data are presented visually, anybody with normal eyesight is empowered to notice patterns and outliers and to use such noticing as a basis for exploring what is contained within the data.

Rather than dispel this appeal to plain sight and revelation as a fallacy, I seek to understand its role in establishing and maintaining a user community for visual analytics, one in which sight is harnessed to produce insights and knowing one's data is a premise for success.[5] Other work in STS has shown that purportedly naive understandings of "unmediated witnessing" and "direct access to an object of concern" remain a central and generative feature of the engagement with the products of science and technology. When visual evidence is presented in a court of law, plain sight vies for priority with a sight that is learned and practiced (Goodwin 1994; Jasanoff 1998). Technology demonstrations accommodate an audience's desire to see for itself, even as they are understood to be staged (Smith 2009). In related ways, visual analytics webinars provide occasions at which unmediated access is both affirmed and disavowed, and the fact that the very idea of visual analytics is premised on natural pattern recognition ability makes it especially pertinent to examine this dynamic here. Accordingly, the remainder of this chapter will address the multilayered artful revelation at work in visual analytics webinars.

2 Watching Webinars

Between July 2009 and August 2010, together with three student assistants, I analyzed a selection of webinars on visual analytics. These webinars were organized by Tableau Software, a Stanford University spinoff company that has achieved considerable recognition in the software industry since its inception in 2003 (it was also mentioned in the *Economist*'s special report on the data deluge).[6] Webinars are a relatively subtle form of marketing: they blend promotion of the technology with instruction in the way of seeing it affords.[7] Presenters tend not to assume that the audience is already highly familiar with the software, although the jargon they use suggests that they do assume a level of familiarity with the particular organizational practices discussed in the webinar, such as marketing, logistics, human resource management, or financial retail. My students and I examined how presenters articulated linkages between "data," "visualization," and "analysis" and how they promoted, as well as visually demonstrated, the value of visual analytics as a way of seeing-into-data. Where presenters offered case studies of organizational success in this regard, we were interested in how things were made to stand out visually, what significance was attributed to this, and how the ease and immediacy of "seeing" was affirmed (or denied) in talk and action.[8]

Webinars are "live" events that work like teleconferences. Anyone who is interested in the advertised topic can register, dial in, and participate through questions or comments. Webinars are not meant to achieve sales, but participants can expect their details be used for follow-up emails and offers of a free trial.[9] Anything that is audible and visible during the webinar—the voices of presenters, moderators, and (occasionally) audience members, the view of the software or of PowerPoint slides on the presenters' desktop—is captured in a recording of the event, which is subsequently made available on Tableau's website to find a larger audience. These recorded webinars appear alongside other resources such as case studies, white papers, training videos, blogs, and an online customer support center. While Tableau Software does organize face-to-face demonstrations, training events, and customer conferences, the Internet plays an important part in its engagement with prospective and existing customers.

Webinars put practitioners' screen work on display for online audiences. By focusing on how visual analytics is made available for witnessing in and through these webinars, we can begin to understand the amalgamation of vision and technology with notions of knowledge, discovery, and value that underpins the practice. This involves the multiple facets of visual analytics that I have already alluded to and will now examine in more detail: the emphases on surface views as beacons and shortcuts in the process of developing insight into data, on plain sight as the linchpin that draws human and machine capabilities together, and on the public nature of the access to statistical relations thereby facilitated.

3   From Sight to Insight: Showing Screen Configuration

This is how I drill

In the January 2009 webinar "Raising the Bar on Marketing Analysis at Wells Fargo," user-analyst Kyle Biehle recounts how his team upgraded from the use of Excel and SQL (a widely used database language for managing data in relational databases) to Tableau's visual analytics software coupled with a real-time, multimillion-row "data warehouse." Excel, according to Mr. Biehle, supports "charting" but not "true visual analysis." As he switches from PowerPoint slides to a desktop view for a live demonstration, Mr. Biehle tells the audience to expect a "simplified visual analysis," adding that "hopefully some things will pop and we can drill in and move around, and I'll show you guys how it is that I work with Tableau."

Popping, drilling, moving around: these visual and spatial metaphors are exhibited in the extract of Mr. Biehle's demonstration presented below (figures 3.1–3.4). The display is arranged into "small multiples" (Tufte 2001): a series of miniature graphical displays that only differ on some parameters, so that their combination supports analysis through visual comparison. The data are sales figures, plotted over time and disaggregated in multiple ways: There are five *regions* (A, B, C, F, and G), represented as columns in the display. There are four *measures to evaluate sales* (absolute volume count, percentage volume count, absolute balance, and percentage balance), represented as rows. Finally, there are four *customer segments* (C, D, H, and X), represented as graph colors.

The audience can follow Mr. Biehle's exploration of his data, because every next move is precipitated by what stands out among the visual patterns on the screen. In this particular sequence, the "double spike" is the lead motif. The audience witnesses how Mr. Biehle follows the double spike *into* the data until its "story" is revealed. High-level views that graphically summarize sales data provide the starting point for an analysis that traces a contrasting pattern to its most fine-grained form. In this tracing, the demo locates the advantage of visual analytics over spreadsheet technology. As one "drills down" to a lower level, patterns become specific enough to qualify as insights one can act on. Conversely, the importance of such insights and ensuing actions is established by keeping the connection to the bigger (or higher-level) "picture."

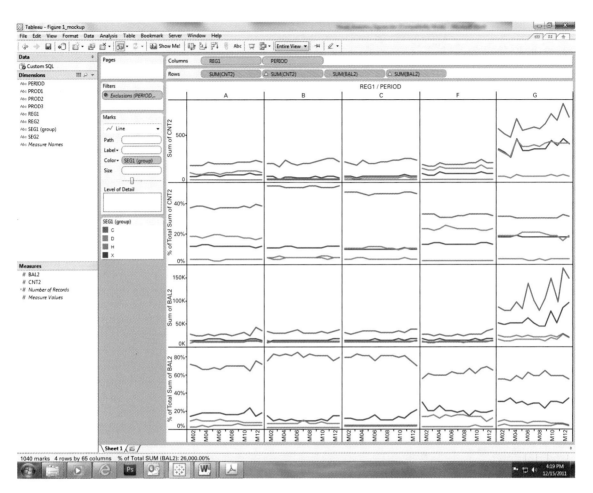

Mr. Biehle: *"So it looks like there's a lot going on in region G here. Region G has got, you know, a lot of volume. And this double-spike phenomenon seems only to be happening in region G. . . . I'm going to focus my analysis, drilling on just region G."*

A click on the column that represents region G makes a new view appear (figure 3.2).

Figures 3.1–3.4

Credits: Adapted from screen shots from "Raising the Bar on Marketing Analysis at Wells Fargo" (2009), with Kevin Brown from Tableau and Kyle Biehle from Wells Fargo, http://www .tableausoftware.com/resources/webinars/visual-analytics-raising-bar-marketing-analysis-wells -fargo. Graphic design by Alfons van Stiphout, Amsterdam.

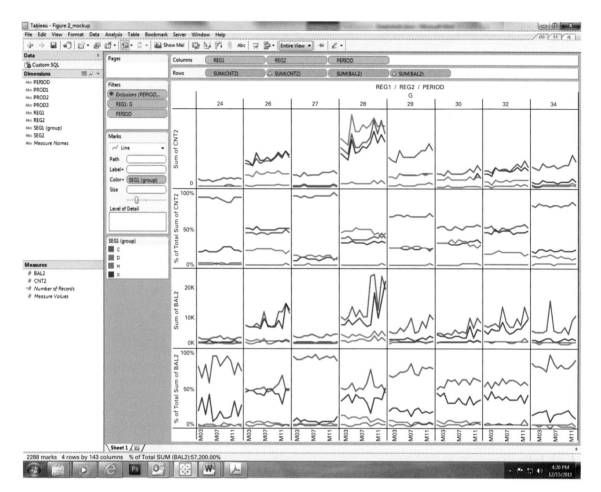

The screen shows eleven subregions (G24–34), which jointly but unevenly account for the spiky pattern in region G. Mr. Biehle points out that the double spike is prominent in subregions 26, 28, and 32 but not in the others. Subregion 28 is particularly noteworthy to him because, in the top row that represents sales volume, the line graphs for 28 reach a higher point than those in any other subregion.

"This looks like region 28 has a lot of the volume. . . . So I want to drill in a little more on that. What's going on with this double spike? What products maybe are causing that?"

Mr. Biehle clicks on G28 and for a moment we only see one big column. Then he drags a new dimension, "product categories," onto the column shelf at the top of the screen, and the column splits again (figure 3.3).

Figures 3.1–3.4
(continued)

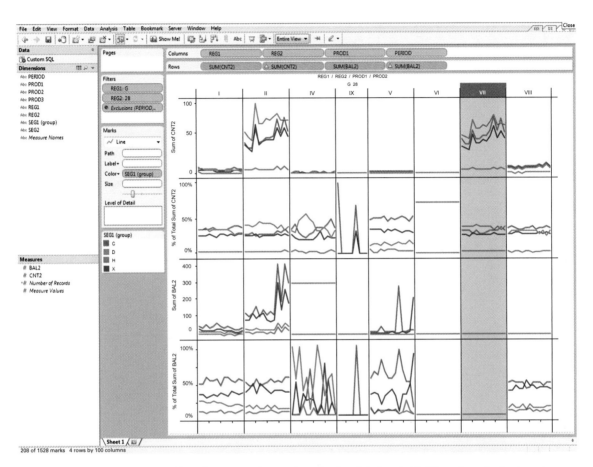

Now the screen shows plotted sales data for eight different product categories within region G28: I, III, IV, IX, V, VI, VII, and VIII. As Mr. Biehle moves the cursor across, different columns or rows are highlighted (here column VII).

Mr. Biehle: *"It looks like a lot of the volume, the account volume, is happening with this product three, and product seven. But the balances are really moving around in product three, and down here in product five. But five doesn't have a lot of the account volume, so I'm not that interested in that. Looks like there's some other noise going on here, maybe some kind of data anomalies that are happening. But I want to focus on product three and see what's going on there."*

To "see what's going on," Mr. Biehle clicks on the column that represents product III (figure 3.4).

Figures 3.1–3.4
(continued)

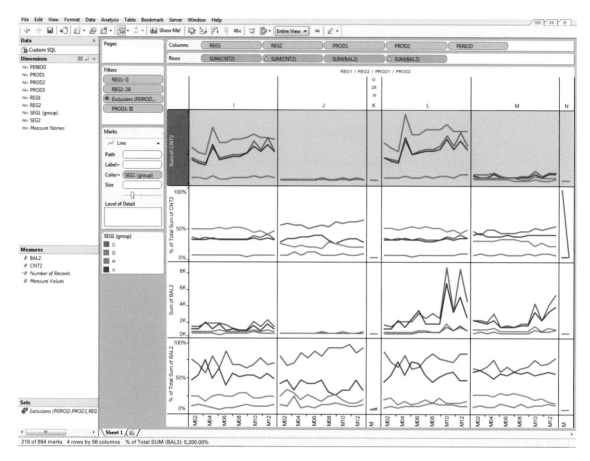

Mr. Biehle: *"[H]ere is . . . the story. We can see in accounts, it looks like this segment D is driving a lot of the account volume. However, the spikes are being driven in the account by these two other segments—segment C and segment X. And the actual balances are largely coming from this product L. You see the spikes repeated. And also product M has a lot of balances as well, but not a lot of accounts."*

The screen shows six product subcategories within product III (I–N). The top row of minigraphs (representing "account volume") has been highlighted with the cursor. Mr. Biehle draws attention to the orange lines that dominate several of the graphs (customer segment D), as well as the blue and red lines (segments C and X) which in recent months have been showing "spikes." In the quote above, he comments on the difference between products by comparing patterns across columns (product L versus product M) and rows ("balances" versus "accounts").

Mr. Biehle: *"So I could keep going further if I wanted, I could keep going down to product level three and see if there's any more detail. But I thought I'd stop there. I just wanted to give you guys a demonstration of kind of how it works and how I drill. And I wanted to give you a pretty good sense of how you can do this, this visual exploration, with Tableau so easily."*

Figures 3.1–3.4
(continued)

By pursuing what stands out visually, Mr. Biehle models visual analytics as a practice propelled by pattern recognition. It seems that, once in possession of the software, all the analyst has to do is add plain sight! At the same time, the demo contains features that work against attributing the progressive revelation of the "story" to the software alone. Not everything that visually stands out merits further exploration: Mr. Biehle swiftly dismisses some spikes (figure 3.3, column IV) as "noise," while affirming others (such as the patterns for product III) as worth drilling into. Throughout the process, judgments are being made regarding which patterns are interesting, how they can be disaggregated into subcomponents, and when to declare the "story" complete. Anticipating that some may feel intimated by the level of skill on display, webinar host Kevin Brown reassures the audience after the demo: "This wasn't meant to be a training session, so don't worry." The showing and telling exhibited in the webinar let the audience partake in the process while affirming the role of the practitioner: it is the analyst who makes the revelation happen.

The power of Tableau

In the January 2010 webinar "Visual Spend Analytics—Take Your Spend Analysis to the Next Level," John Simon from the professional services firm Alvarez and Marsal shares his desktop with the online audience. He is going to demonstrate how he helped a client identify opportunities to save money. Mr. Simon brings up a scatter plot (figure 3.5A) that shows the dollar amount his client spent on materials plotted against the percentage of savings achieved in a particular time period. The color of the plot marks signifies a category of supplies (e.g., green represents packaging/labels), and the shape signifies the business unit responsible for purchasing. Mr. Simon continues:

I'm going to show you the power of Tableau where you can analyze thousands of data points quickly and rapidly. So, I'm going to drag in a couple different measures and dimensions into the "level of detail" [refers to a field on the screen]. I'm going to drag in the "item," and now you'll see a lot more marks, and I'm going to drag in the "vendor" as well. And I know what you're thinking: that *this is going to be hard to look at and analyze*.

The number of plot marks has expanded from fifteen to 532, and most of them are concentrated in the same area (figure 3.5B). Mr. Simon concedes that this "is going to be hard to look at," before demonstrating how the dense clutter he created can be disentangled with the help of the software's zoom function and filters. He shows how these tools let him configure the view in multiple ways, revealing savings opportunities at the intersection of all the measures and dimensions dragged into view. As in the previous case, the demo models software-enabled seeing at the same time as it draws attention to the analyst's skill. The latter is exhibited by the rapid pace at which Mr. Simon creates and reconfigures graphs and plots: the visual display changes every few seconds and the cursor moves constantly to highlight changes and guide the gaze.

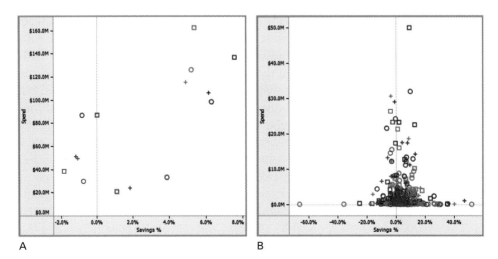

A   B

Figure 3.5
John Simon demonstrates the ability to move back and forth between an organized and a cluttered field of vision, dragging in more dimensions than the eye can handle as a step toward further insight. Credits: Fragment of screen shot from "Visual Spend Analytics—Take Your Spend Analysis to the Next Level" (2010), with Paul Teague from Purchasing Magazine and Ken Tsai and John Simon from Alvarez and Marsal, http://www.tableausoftware.com/resources/webinars/visual-spend-analytics. Images courtesy of John Simon.

Show Me

In the demonstrations by Mr. Biehle and Mr. Simon, visual analytics entails knowing how to create and navigate visual configurations of data on the screen. At the same time, the analytical sequences of drilling and disaggregating are pushed by the software's internal logic, a logic expressed in visual contrasts designed for anyone to see. This oscillation between skilled vision and plain sight is also encapsulated in a particular feature known as the "Show Me" button (Mackinlay et al. 2007). John Simon calls it "one of the best features . . . in the tool." He elaborates:

I had no formal training with Tableau, all I did is watch several videos on Tableau's site. . . . It's so easy after you have the data set up to create any chart you want. With this Show Me feature, that's where you really learn: you just select the dimensions and then hit Show Me, and it's going to recommend what chart to look at, you can click on that one, if it doesn't really show you what you want to see or it's difficult to read, just hit Show Me again and you can select another one.

The "Show Me" button provides the user-analyst with a default option for visualizing data. Based on "the best practices of graphic design" (Mackinlay et al. 2007, 1137), the software chooses the display that is most suitable for the measures and dimensions of the data. For example, time series data are displayed as a line graph, while quantitative data, such as sales results, are displayed as a bar graph. Mr. Simon's commentary hints at a paradox: "Show Me" is the best way of displaying data for visual perception, but it may not "show you want you want to see." Indeed, in his own demonstration Mr. Simon decides to override the default:

I'm going to override, you can see the blue box, this is the one that they're recommending but I'm going to select a heat map instead. Just my opinion, this is probably the one that I would use to analyze [this].

Overriding the software's automatic visualization recommendation is something that the expert user-analysts in other webinars also do. Mr. Biehle, for example, explains that "Tableau's guess for this quantitative comparison is to make it . . . bars. I happen to prefer to look at lines." The two presenters appeal to personal preferences when explaining their reasons for not following "Show Me." In the quest to make accessing data as seamless as possible, designers of visual analytics systems increasingly seek to reconcile such personal user preferences with research-based notions about how humans best see trends and other patterns (Kostelnick 2008; Pretorius and van Wijk 2009). The paradox, and the way in which "Show Me" exhibits artful revelation, is that on the one hand the user relies on the software to reveal, at the click of a button, patterns and relations in otherwise hard-to-explore data, while on the other hand users have to cultivate the art of knowing what they do or do not want to see. Mr. Biehle and Mr. Simon sustain this paradox by overruling the default while also crediting the software with the ability to make salient features "pop" on the strength of its visual organization of data.

So far, the artful revelation that characterizes visual analytics has been discussed by juxtaposing skilled vision with plain sight: a "seeing" as work, as *achieved*, with a "seeing" as non-work, as *immediate*. Insight is attributed to the "power" of the software to make data relations visually available, as well as to the skill of the user-analyst in determining how to configure the display, which visual patterns warrant further drilling and which can be discarded as "anomalies," how to create and disentangle visual mess, and at what point the "story" can end, at least for the time being. The manifestation of this dual rhetoric in the webinar demonstrations allows an audience to "see for itself" and yet to be aware of its reliance on the analyst. Access to the insights data hold is thereby presented as available to anyone, but also as contingent upon artful practice.

4   What You See Is (Not) What You Get

Access to Excel is ubiquitous for many people with office jobs, and free data visualization tools can be found on the Internet (Tableau launched its own version in February 2010). So what makes visual analytics an attractive *commercial* proposition? The answer to this question is rooted in arguments about the growing importance of "data" as a strategic resource. Tableau's marketing communications advance this view, and they are by no means unique—as illustrated by the *Economist* special report mentioned at the start of this chapter. Business leaders are told that they can scarcely afford not to compete on analytics (Davenport and Harris 2007), and, now that every forward-looking organization needs one, "data scientist" has been hailed as "the sexiest job of the 21st century" (Davenport and Patil 2012).

Webinars convey that it is worth investing in visual analytics through an intricate play of sharing and withholding. As a genre, webinars are structured to go beyond simple sales pitches; they exhibit openness, candor, and even a degree of intimacy.[10] Attendees are encouraged to ask questions, which presenters frequently go out of their way to answer. John Simon, in his webinar on visual spend analytics, provides several short demonstrations in response to questions from the audience. Wade Tibke from Tableau, following a demo of his own use of the software for classifying marketing leads, goes further in emphasizing his readiness to engage beyond the webinar:

OK, so that is the end of my presentation. Again, this is the URL. We'll send you guys out an email. We are going to put up a white paper based on this. I will put up the [slide] deck, and of course you can always reach out to me [points cursor to contact information on the screen]. I'm always available to chat. I love . . . love marketers, I love talking about this stuff. I'm particularly interested in marketing technology, and . . . please feel free to reach out to me. (from "Visual Scoring—the 360° View")

Such willingness to explain, to expand, and to be contacted, on the part of those who have already achieved cost savings, organizational efficiency, or increased sales through visual analytics, evokes a sense of community, one in which aspiring users can take part not just to see data but also to see value.

Tableau's webinars also and simultaneously indicate limits to such access, making viewers aware that what is shown cannot be taken at face value. This happens through the surfacing of missing "hows" and missing "whats." The first of these refers to the back-end work needed to enable visual analytics; that is, the infrastructure to get the data organized for visualization. Prompted by audience questions, presenters acknowledge, and to a lesser degree describe, the hard grind and (occasional) frustrations entailed in this back-end work. This does not become visible in the screen work on display in the webinar, but the fact that it is a recurrent theme in Q & A sessions highlights that claims about the software's capabilities cannot be divorced from the organizational circumstances of its use.

The missing "whats" come into play when presenters make announcements that deny or take away from the "reality" of the data used in the demonstration. Such announcements are common in the webinars my research assistants and I reviewed. Kyle Biehle refers to his as "dummy data." John Simon says: "the data, we made it up for this demo today." Some other presenters, such as Bonnie Elliot from Providence Health and Services, include a formal disclaimer on one of their presentation slides: "The data provided in this presentation have been altered for confidentiality." Such announcements tend not to specify the sense in which, or the extent to which, data have been altered,[11] their point being simply to caution the audience not to put too much stock in what is shown.

In the June 2009 webinar "Visual Data Mining," data scientist guest presenter Stephen McDaniel talks about the way his team helped REI, a company that sells products for outdoor activities, optimize its marketing expenses. The focus of his presentation is "coupon sales"—sales that result after discount coupons are given to customers. There is no live demonstration, but Mr. McDaniel has summarized the visual analytics for several projects in a series of PowerPoint slides. Before he starts, there is a slide with a disclaimer that reads (emphasis added):

About the example slides

• All the data are from actual projects and views used by the team.
• The data and results in this presentation *have been randomized and "scrambled" to hide actual outcomes.*
• However, these are great examples of results that REI has achieved with Tableau in just six months.

The audience is thus informed that what will be shown is *not* what the team really discovered in the data; it merely represents the *type* of visual analysis undertaken at REI. The audience can witness how Mr. McDaniel was able to get meaningful insights from data, but cannot become privy to what the actual insights were. This simultaneous showing and hiding is communicated through periodic reminders about what can and cannot be inferred from the data displays presented on the screen. For example, when Tableau host Elissa Fink asks what the *y*-axis represents in one of the data displays, Mr. McDaniel responds:

Yeah, the *y*-axis is total sales. Obviously it isn't our *real* total sales for coupons. It's a very different number, but each one is showing the amount of sales that involve that coupon within that quarter.

The audience is thus reminded *not* to see the display as though it represented actual insights obtained by REI, while at the same time it is assured of the display's close connection to those actual insights. Later in the presentation, Mr. McDaniel talks about golf products in relation to seasonal and recession-affected cycles of sale. Pointing to a slide that shows various upward- and downward-sloping graphs, he explains sales figures per quarter in the recession year that spanned 2008 and 2009: "we can see a lot of weakness in this golf category—you know golf is an expensive game, and I guess it's not too surprising." The colored patterns on the screen direct the viewer's gaze, in particular, to the subcategory of golf accessories, which appears to be showing a downward trend. But while Mr. McDaniel singles out golf in his demonstration of data-driven insights, and highlights characteristics of golf—such as its expense and seasonal character—as essential to appreciating what is conveyed by the visual display, the audience knows that these patterns are divested of reality. This is because, earlier,

Mr. McDaniel had told viewers that although "you're going to see things in here like golf and polo . . . we don't sell golf and polo products."

Announcements that deny the reality of what is seen complicate the notion that these webinars help prospective users understand what it takes to get value from visual analytics. Showing almost-real results with altered data stands in the way of audience members' ability to see for themselves the benefits a visual approach has conferred upon REI and other Tableau clients. At the same time, the announcement that the display has been altered in order to become *less* revealing raises expectations of the success that Mr. McDaniel's team must have had, while the trouble taken to "randomize and scramble" data suggests that what is now obscured would otherwise be (too) readily available for all to see.[12] In this way, the concealment entailed in the use of altered data further affirms the "revealing" nature of software-based visual analytics and the power invested in plain sight. By putting actual insights beyond the audience's purview, this play of sharing and withholding bolsters the expectation that insights inhere in data and can be unlocked by sight.

5   Conclusion

The immanence/conditionality of visual insight

This analysis of visual analytics as *artful revelation* has explored how art and revelation are concurrently asserted in the showing and telling that goes on in webinars. The tension contained in the phrase thus takes on a different character than it would in STS ethnographies of representational work in which "revealing what the data show" is treated as the upshot of artful practice. Instead, it is the *concurrence* of artfulness and revelation that animates the visual analytics enterprise. Through this concurrent enactment—and this is how the rhetoric of visual analytics gains force—the insights afforded by a visual approach are portrayed as both immanent and conditional.

Immanence is expressed in the enduring emphasis on data as full of potentially significant correlations that (only) a visual approach can reveal. The audience can follow how a dataset is made to "give up its secrets" as webinar presenters make things "pop" (the demonstrations by Mr. Biehle and Mr. McDaniel), oscillate between complexity and simplicity (Mr. Simon's demonstration), and display data relations in a form that best suits the nature of the data (the "Show Me" button). Watching webinars, one may readily imagine oneself following in the footsteps of these analysts (even if it would take time to reach their level of skill), arranging data in patterns amenable to visual perception, exploring these patterns for their potential significance, and configuring and reconfiguring the view until a "story" that evidences such significance has been located. Visual analytics as shown in the webinars suggests that insights are there for the taking and that the ability to unveil, with a mouse click, ever more sales, marketing, or cost-saving opportunities is vital to business success.

Even the alteration or substitution of data contributes to this understanding, because it suggests that salient data relations might otherwise be left for all to see (see also Coopmans 2011).

Conditionality is manifested when insight is presented in the form of abstract stories (in Mr. Biehle's demonstration), as likely to result from the data configuration maneuvers being demonstrated (Mr. Simon's entangling and disentangling, and his use of the "Show Me" button), or as achieved in private but not disclosed in public (Mr. McDaniel's data alteration). In each case, there is something missing, something that *in its absence* promotes the confluence of vision, knowledge, discovery, and value that animates the very idea of visual analytics.[13] Members of the webinar audience *would* be able to see the software's true value *if* they had the skill of Mr. Biehle or Mr. Simon, or the unscrambled data of Mr. McDaniel. The fact that they do not have such skill means that judgment on the value of the software cannot be made here and now; it will have to be deferred to a future moment when the conditions for achieving insight from visual analytics are properly satisfied. In every situation, data will have to be organized and analytical skill developed, making users accountable for their own success. The incompleteness of the success stories demonstrated by webinar presenters may thus help spur investments of time and money in the practice of visual analytics, until such practice brings about a future state in which "sight" is properly mobilized to yield valuable insights.

While the demonstrations of real-life analytics practice in Tableau's webinars complicate the notion that visual access to data is granted simply by purchasing the software and pressing a button, they also sustain beliefs in data visualization as a form of revealing, beliefs that reanchor "the very act of looking . . . at the heart of our epistemology" (Jordanova 1989, 91). The rhetoric of showing and telling conveys that promises of insight can only be fulfilled in and through *practice*, at the same time as it fuels expectations that visual analytics works like magic, opening up to sight what might otherwise remain hidden within a dataset.

A little bit of mystery goes a long way

In a 2009 article, Tableau Software's CEO Christian Chabot sets out to dispel four stereotypes about visual analytics. One of these is the notion that visual analytics is primarily aimed at revealing hidden insights. Chabot grants that "[a] visual analysis session *might* unearth a hidden gem" (2009, 87, emphasis added), but argues that this actually is not the point:

The problem with the belief that these "aha" moments are the crux of visual analytics, however, is that they aren't representative of the analysis process. Not for experts. Not for everyday people. Most of the time that people spend with data is in exploring it, cleaning it, gaining confidence in it, summarizing it, pursuing inconclusive paths, confirming facts, and presenting findings. None of these steps necessarily has anything to do with finding a hidden insight. (Chabot 2009, 87)

I bring up Chabot's bid to correct what he sees as an unhelpful stereotype in order to address a possible criticism of my approach, namely that it *mystifies*, rather than helps to understand, the actual ways in which visual analytics is ordinarily practiced. Attending to how user-analysts work with data visualization software is certainly important, but we should also recognize how such practices are mediated by and complexly intertwined with the promise of sight as the key to knowledge. Chabot, in suggesting that stereotypes about visual analytics do not accurately represent analytical practices, evokes a distinction between discourse and practice that is deeply ingrained in Anglo-Saxon ways of thinking (critically discussed by Woolgar 1986). I wish to make a case for a more integrated study of visual practice and visual rhetoric, one that allows examination of the continuing importance and renewed manifestations of revealing as an epistemological trope in current arrangements for working with data.

This means holding on to the mystery of visual analytics as a feature of how it works. The rhetorical form of the webinars exhibits a generative tension between immediate revealing and skillful accomplishment, and between disclosing and obscuring. I call this tension "generative" because it makes it possible to believe in the revealing power of visual analytics at the same time as one experiences it in the rather more mundane way described by Chabot. As analysts of the technovisual arrangements through which we probe the world, we should be careful not to overplay the dichotomy between belief and skepticism and in the process ignore what makes such arrangements compelling.[14]

Enthusiastic investor response after Tableau Software became a publicly traded company in May 2013 shows confidence that data visualization is something many people want and need. While the amalgamation of vision, knowledge, and value that lies at the heart of visual analytics reanimates a long-standing epistemological trope, it does so in a way that may prove especially influential in and characteristic for the twenty-first century. This is because claims about the power of sight are accompanied by emerging imaginations of "data" as abundant and as containing latent value— imaginations increasingly at play in scientific, business, and other practices geared toward data-driven discovery.

Acknowledgments

Research for this chapter was supported by a start-up grant from the Faculty of Arts and Social Sciences at the National University of Singapore. Thank you to my coeditors and two anonymous reviewers for helpful comments; Brian Rappert for inspiring me to do work on revelation and concealment; my former students Yap Xiong, Soon Chuan Sheng, and Dina Delias for their contributions to and beyond our joint webinar analysis sessions; and Elena Simakova, Alfred Montoya, Ingmar Lippert, and above all Antonio Alvarez for input and support at important moments. Kyle Biehle and John Simon, as well as Elissa Fink and Doreen Jarman from Tableau Software, generously helped me assemble the visual materials for this chapter, and I am grateful to Alfons van Stiphout for his meticulous graphic design.

Notes

1. Increased attention to visual analytics at the start of the twenty-first century was in large part due to American concerns about homeland security. In the wake of the terrorist attacks on the World Trade Center in 2001, the US government supported the emergence of a subfield of computer science called "visual analytics" by establishing the National Visualization and Analytics Center at the Pacific Northwest National Laboratory in Richland, Washington state, in 2004 (Thomas and Cook 2005). Since 2006, visual analytics has had its own annual academic meeting, the IEEE Symposium on Visual Analytics Science and Technology (VAST). Since 2008, Europe has had a virtual Visual Analytics Network that focuses on, among other things, urban planning and transport research. In this chapter, the focus is on commercial visual analytics applications; these often have roots in academia or policy-related research, but that aspect will not be further discussed here. For a brief introduction to the history and heritage of visual analytics, see Lawton (2009).

2. Tufte's "rules" for displaying data—such as which chart type to use (and avoid) for which kinds of data—have found their way into software packages for visual analytics (see the section on "Show Me" in this chapter). Tufte himself was influenced by earlier work on the display of statistical information, including that by William Playfair (1759–1823), the inventor of the line graph, bar chart, and pie chart.

3. While correlations, trends, and outliers are made *calculable* by database technology and statistical software, it is a visual interface that makes them available to the user-analyst in the form of bar or line graphs, bubble charts, scatter plots, heat maps, and combinations of these. It is the latter rather than the former that is being analyzed here.

4. This leads Wright to argue that "a visualization is not a representation but a means to a representation." Knorr Cetina and Bruegger (2002) make a similar point when they argue that screens *appresent* data for financial analysis.

5. It may seem strange to speak of "plain sight," given that there is so much excellent scholarship on how sight, including scientific ways of seeing, is learned and practiced. It should be clear that what I am concerned with is not the truth (or otherwise) of sight as natural or immediate, but the way in which representations of visual analytics mobilize sight as the key to knowledge.

6. The company has won several awards, including "Best Overall in Data Visualization" by *DM Review*, "Best of 2005 for Data Analysis" by *PC Magazine*, and "2008 Best Business Intelligence Solution" (CODiE award, based on peer recognition within the software industry) by the Software and Information Industry Association.

7. In the case of instruments for observation, this confluence of instruction and sales has a long-standing history: early modern European shops in which telescopes were sold played "a vital role in disseminat[ing] a form of natural philosophical practice to a broad public" (Bennett 2002, 389).

8. Between 22 February 2010 and 15 July 2010, we accessed the recordings of twenty webinars via Tableau's website, watched them, took notes, and selected six for detailed analysis. This involved the production of transcripts as well as a series of joint webinar analysis sessions.

9. At the time, the software cost $999 per license for the personal edition and $1,800 for the professional edition.

10. According to the advice of one online consultant: "You need to be respectful of both the time and money that people spend to attend your webinars. That means providing them with truly useful information and not dragging the presentation on for an overly long period of time. Even if your webinar focuses on a product or service your company offers, be sure to include information that attendees can use without buying anything from you. Do not worry about giving away your company's secrets to your potential clients. It is actually good for business for clients to understand why your approach to business works and to see the reason why your company can benefit them." (http://ezinearticles.com/?Use-Webinars-to-Educate-Customers-and-Market-Your-Products&id=3858918, accessed 27 December 2012.)

11. Sometimes presenters choose to mask data in a way that is less ambiguous, namely by blurring the descriptions of data categories, such as names of customers, employees, and products. In one of the webinars we watched, this was explained in the following way: "Your eyes aren't tricking you, that data is blurred. Most of the information that we deal with is proprietary information, and it's blurred just to protect the customers involved in the information" (John Hoover, "Transforming Disparate Data").

12. Presenters say that they alter data to protect "customer privacy," but the hiding of "actual outcomes" (in Mr. McDaniel's presentation) inevitably raises expectations as to the competitive value of the insights obtained through the analysis.

13. For the productive role that "absence" can play in technology demonstrations, see also Simakova (2010).

14. These last few remarks are directly indebted to Michael Taussig's (2003) work on beliefs in shamanism and magic healing, and his critique of anthropological studies that account for the endurance of such beliefs by associating them with a naivete that evidences lack of uptake of a more "modern" empirical skepticism. Taussig shows that belief and skepticism are eminently compatible—indeed, he argues that skepticism harbors belief, an insight he suggests animates not only shamanism but also Western epistemic practices, prompting reflection on the assumptions underpinning many ethnographic studies.

References

Amann, Klaus, and Karin Knorr Cetina. 1990. The fixation of (visual) evidence. In *Representation in Scientific Practice*, ed. Michael Lynch and Steve Woolgar, 85–121. Cambridge, MA: MIT Press.

Beaulieu, Anne. 2002. Images are not the (only) truth: Brain mapping, visual knowledge, and iconoclasm. *Science, Technology and Human Values* 27:53–86.

Bennett, James A. 2002. Shopping for instruments in Paris and London. In *Merchants and Marvels: Commerce, Science and Art in Early Modern Europe*, ed. Pamela H. Smith and Paula Findlen, 370–395. New York: Routledge.

Beunza, Daniel, and David Stark. 2004. Tools of the trade: The socio-technology of arbitrage in a Wall Street trading room. *Industrial and Corporate Change* 13:369–400.

Chabot, Christian. 2009. Demystifying visual analytics. *IEEE Computer Graphics and Applications* 29:84–87.

Coopmans, Catelijne. 2011. "Face value": New medical imaging software in commercial view. *Social Studies of Science* 41:155–176.

Davenport, Thomas H., and Jeanne G. Harris. 2007. *Competing on Analytics: The New Science of Winning*. Boston: Harvard Business School Press.

Davenport, Thomas H., and D. J. Patil. 2012. Data scientist: The sexiest job of the 21st century. *Harvard Business Review* (October 2012): 70–76.

Economist. 2010. The data deluge—and how to handle it: A 14-page special report. *Economist* (February 25).

Goodwin, C. 1994. Professional vision. *American Anthropologist* 96:606–633.

Grady, John. 2006. Edward Tufte and the promise of a visual social science. In *Visual Cultures of Science: Rethinking Representational Practices in Knowledge Building and Science Communication*, ed. Luc Pauwels, 222–265. Hanover, NH: Dartmouth College Press.

Jasanoff, Sheila. 1998. The eye of everyman: Witnessing DNA in the Simpson trial. *Social Studies of Science* 28:713–740.

Jordanova, Ludmilla J. 1989. Nature unveiling before science. In Jordanova, *Sexual Visions: Images of Gender in Science and Medicine between the Eighteenth and Twentieth Centuries*, 87–110. Madison: University of Wisconsin Press.

Knorr-Cetina, Karin, and Klaus Amann. 1990. Image dissection in natural scientific inquiry. *Science, Technology and Human Values* 15:259–283.

Knorr Cetina, Karin, and Urs Bruegger. 2002. Global microstructures: The interaction practices of financial markets. *American Journal of Sociology* 107:905–950.

Kostelnick, Charles. 2008. The visual rhetoric of data displays: The conundrum of clarity. *IEEE Transactions on Professional Communication* 51:116–130.

Latour, Bruno. 1999. Circulating reference: Sampling the soil in the Amazon forest. In Latour, *Pandora's Hope: Essays on the Reality of Science Studies*, 24–79. Cambridge, MA: Harvard University Press.

Lawton, George. 2009. Users take a close look at visual analytics. *IEEE Computer* (February), 19–22.

Lynch, Michael. 1985a. Discipline and the material form of images: An analysis of scientific visibility. *Social Studies of Science* 15:37–66.

Lynch, Michael. 1985b. *Art and Artifact in Laboratory Science: A Study of Shop Work and Shop Talk in a Research Laboratory*. London: Routledge and Kegan Paul.

Lynch, Michael. 1990. The externalized retina: Selection and mathematization in the visual documentation of objects in the life sciences. In *Representation in Scientific Practice*, ed. Michael Lynch and Steve Woolgar, 153–186. Cambridge, MA: MIT Press.

Lynch, Michael, and Samuel Y. Edgerton. 1988. Aesthetics and digital image processing: Representational craft in contemporary astronomy. In *Picturing Power: Visual Depiction and Social Relations*, ed. Gordon Fyfe and John Law, 184–220. London: Routledge.

Mackinlay, Jock D., Pat Hanrahan, and Chris Stolte. 2007. Show me: Automatic presentation for visual analysis. *IEEE Transactions on Visualization and Computer Graphics* 13:1137–1144.

North, Chris. 2006. Toward measuring visualization insight. *IEEE Computer Graphics and Applications* 26:6–9.

Pretorius, A. Johannes, and Jarke J. van Wijk. 2009. What does the user want to see? What do the data want to be? *Information Visualization* 8:153–166.

Pryke, Michael D. 2010. Money's eyes: The visual preparation of financial markets. *Economy and Society* 39:427–459.

Simakova, Elena. 2010. RFID "Theatre of the proof": Product launch and technology demonstration as corporate practices. *Social Studies of Science* 40:549–576.

Smith, Wally. 2009. Theatre of use: A frame analysis of information technology demonstrations. *Social Studies of Science* 39:449–480.

Taussig, Michael. 2003. Viscerality, faith, and skepticism: Another theory of magic. In *Magic and Modernity: Interfaces of Revelation and Concealment*, ed. Birgit Meyer and Peter Pels, 272–306. Stanford: Stanford University Press.

Thomas, James J., and Kristin A. Cook. 2005. *Illuminating the Path: The Research and Development Agenda for Visual Analytics*. National Visualization and Analytics Center.

Tufte, Edward R. 2001. *The Visual Display of Quantitative Information*. 2nd ed. Cheshire, CT: Graphics Press.

Woolgar, Steve. 1986. On the alleged distinction between discourse and praxis. *Social Studies of Science* 16:309–317.

Woolgar, Steve. 1990. Time and documents in researcher interaction: Some ways of making out what is happening in experimental science. In *Representation in Scientific Practice*, ed. Michael Lynch and Steve Woolgar, 123–152. Cambridge, MA: MIT Press.

Wright, Richard. 2008. Data visualization. In *Software Studies: A Lexicon*, ed. Matthew Fuller, 78–87. Cambridge, MA: MIT Press.

Webinars

Providence Health & Services Case Study: Using Data to Understand and Manage Your Workforce (2010), with Elissa Fink from Tableau and Bonnie Elliot from Providence Health & Services. Available at: http://www.fiercehealthcare.com/webinars/providence-health-case-study-using -data-understand-and-manage-your-workforce (accessed 27 December 2012).

Raising the Bar on Marketing Analysis at Wells Fargo (2009), with Kevin Brown from Tableau and Kyle Biehle from Wells Fargo. Available at: http://www.tableausoftware.com/resources/webinars/ visual-analytics-raising-bar-marketing-analysis-wells-fargo (accessed 27 December 2012).

Transforming Disparate Data into Actionable Analysis (2010), with John Hoover from Norfolk Southern and Elissa Fink from Tableau. Available at: http://www.tableausoftware.com/resources/ webinars/transforming-geographic-and-disparate-data-actionable-analysis (accessed 27 December 2012).

Visual Data Mining: Finding Patterns and Key Insights in Online Marketing Data (2009), with Elissa Fink from Tableau and Stephen McDaniel from REI and Freakalytics. Available at: http:// www.tableausoftware.com/resources/webinars/visual-data-mining-finding-patterns-online-marketing-data (accessed 27 December 2012).

Visual Scoring—the 360° View (2009), with Wade Tibke from Tableau. Available at: http://www .tableausoftware.com/resources/webinars/visual-scoring-360 (accessed 27 December 2012).

Visual Spend Analytics—Take Your Spend Analysis to the Next Level (2010), with Paul Teague from Purchasing Magazine and Ken Tsai and John Simon from Alvarez and Marsal. Available at: http://www.tableausoftware.com/resources/webinars/visual-spend-analytics (accessed 27 December 2012).

4 Digital Scientific Visuals as Fields for Interaction

Morana Alač

As visual renderings in sciences are becoming increasingly entangled with computers and computational formats, their digital materiality calls for a distinct approach. To tackle the digitality of scientific visuals,[1] attention turns to how they are engaged as a part of "local, interactionally produced, recognized, and understood embodied practices" (Garfinkel et al. 1981, 135). When scientific visuals are analyzed in a published format, their digitality is, obviously, not directly accessible. This may even be the case when the analysis is based on interviews and classical ethnographic observations. But, if doing and making are considered, the digital materiality of those renderings comes to the fore. The details of how computer screens are manipulated, and how they are coordinated with ongoing talk as well as gesturing bodies in a scientific setting, become relevant.

This turn to the dynamic bond between computers and scientists' working and gesturing bodies has consequences for the understanding of those visuals, as it calls for an examination of their representational status and their boundaries. The move, furthermore, allows us to recover aspects of phenomena that scientists engage with in the laboratory. In other words, by turning toward the scientific visuals as an interface between digital screens and lived bodies, we attend to the objects of knowledge as they are enacted in the midst of the everyday work of science.

1 Studying the Digitality of Scientific Visuals

This paper focuses on renderings of the human brain generated by functional magnetic resonance imaging (fMRI) technology and employed in research laboratories of cognitive neuroscience. Together with its forerunner MRI, fMRI is a key digital imaging technology used for medical and scientific purposes. The goal of MRI is to provide detailed static renderings of the anatomical structure of internal body parts, such as the brain. This technique uses radiofrequency, magnetic fields, and computers to create visual renderings based on the varying local environments of water molecules in the body. To obtain such renderings, a person (or, in fMRI practitioners' jargon, an *experimental subject* or a *subject*) is scanned. During an MRI brain scanning session, hydrogen protons in brain tissues are magnetically induced to emit signals that are detected by the computer. Such signals, represented as numerical data, are then converted into renderings of the brain anatomy of the experimental subject. The mapping of human brain *function* by use of fMRI represents a newer dimension in the acquisition of physiological and biochemical information. The technique is used to observe dynamic processes in the brain that are demonstrated by visualization of the local changes in magnetic field properties occurring as a result of changes in blood oxygenation. The role of fMRI, thus, is to display the degree of activity in various areas of the brain: if the experimental data are obtained while a subject is engaged in a particular cognitive task, the visual can indicate which parts of the brain are most active during that task.

To think about the character of fMRI brain visuals, I work with the ideas from interpretative semiotics of Charles Sanders Peirce (relying on his *Collected Papers* [Peirce 1934–1958]). I start from Peirce's *icon-index-symbol* typology to argue that in research laboratories in the field of cognitive neuroscience fMRI visuals function as iconic signs. I suggest, more precisely, that these are *diagrammatic* signs (a kind of iconic sign), because their specificity lies in how they are engaged in practice. To describe this engagement I draw from ethnomethodology (Garfinkel 1984; Garfinkel 2002) and conversation analysis (Jefferson 2004; Sacks 1992; Sacks et al. 1974).[2] Such an approach allows me to enter into dialogue with the literature on scientific visuals that relies on Peirce's categories (see for example Gross 2008; Rheinberger 1997) while highlighting how scientists treat those visuals in practice.

My descriptions of scientists' everyday practices are further grounded in a recent research trend that aims at recovering fine details of the *multimodal* interactional organization (e.g., Goodwin 1994, 2000a; Heath and Hindmarsh 2002; Koschman et al. 2007; Mondada 2007; Ochs et al. 1996; Streeck 2009; Suchman 2000; see also Goffman 1976). Similar to ordinary-language philosophy (Austin 1962; Wittgenstein 1953) and the approaches of semiotics (Benveniste 1971; Peirce 1934–1958), these studies point out that talk, as well as bodily conduct that engages with material elements in the setting, participates in the practical accomplishment of social activities. To highlight the importance of going beyond an analysis of communication focused exclusively on ongoing talk, Charles Goodwin (e.g., 1994, 2000a) describes how interactants coordinate *multimodal semiotic means*—talk, gestures, bodily conduct, prosody, visual orientation, facial expressions, and the material elements of the setting—to accomplish meaning through actions. These actions are *situated* (Suchman 1987)—always realized, moment by moment, with respect to the environments in which they are lodged, while they constitute the local context.

The actions I discuss draw from an ethnographic study I conducted between 2002 and 2005 in three university laboratories in cognitive neuroscience that used fMRI. During that study, I videotaped work and training sessions to trace how fMRI practitioners used their computers and interacted with one another. In particular, my attention was directed toward the gestural and embodied semiotic[3] engagement with the digital renderings. As the videotapes indicate, how fMRI practitioners gestured in front of and touched computer screens while working and communicating with their colleagues afforded both themselves and the ethnographic analyst access to the interface between the body and technology (Ihde 2002). Video recordings are still not widely used in science and technology studies (STS) as a methodological tool to study scientific practices, even in studies of visual imaging. However, if we accept that scientists accomplish their work through tactile interaction with technology and visible semiotic comportment, we shall examine these acts. My use of video recordings to examine scientists at work (see also Goodwin 1995; Ochs et al. 1996) supplements and enhances earlier STS studies based on audio recordings (see, for example, Amann and Knorr Cetina 1990; Garfinkel et al. 1981; Lynch 1985; Woolgar 1990).

As we go beyond linguistic aspects of communication (such as talk and writing), our intuition and memory are not reliable sources with which to document the complexities that characterize multimodal interaction. Furthermore, people are often unable to provide sufficiently detailed a posteriori accounts of their own conduct for an effort to recover the organization of multimodal communication. For example, gestural articulation in an environment of practice cannot be fully reported in an interview or accurately remembered by an observer. A scientist who was involved in the work when a gesture took place may have only tacit knowledge of how it was done, and thus be unable to articulate it, while an ethnographer who saw the gesture is at a loss when trying to represent its temporal unfolding and coordination with other elements of the semiotic action. The problem lies not only with the gesture taken in isolation but also with its fine embeddedness in the complexities of the moment-to-moment practice. The gesture is contingent upon the local and spatial organization of the setting, and is produced in relation to the ongoing talk and the action of the coparticipants. To access multimodal semiotic aspects of working hands in the laboratory, video recordings, with all their insufficiencies and their inevitably incomplete output, are currently the best way to record the dynamic actions in a setting in which work and multimodal interaction take place (Goodwin 2000b). That is not to say, of course, that every discussion of embodiment in STS should rely on video recordings; it is to say, however, that there are multimodal and temporally unfolding phenomena constitutive of scientific practice that may go unnoticed if video recordings are excluded as a component of observational approach.

In order to aid my description of gestural and embodied engagement, I coordinated written transcripts of interaction with line drawings. Following Goodwin's (2000b) technique of transcribing visual phenomena, I turned still photographs (retrieved from the video) into such drawings. Using software programs, I delineated the contours of practitioners' bodies and relevant elements of the setting by working directly on the photographs. My goal was to let the reader see as much as I saw while indicating elements of the practice that the scientists were treating as relevant in their work and interaction. In that sense I positioned myself in parallel with what I described. Just like the fMRI scientists who engaged their hands with digital visual fields, I explored the affordances of digital technology by repeatedly inspecting my records, transcribing the chosen moments, and working with the stills to delineate the contours of the bodies and spatial arrangements so that I could make them visible.

2   fMRI Brain Visuals as Diagrams

Observing the gestures, practical engagement of hands, and the overall orientations of the scientists' bodies as they worked with fMRI brain visuals problematizes the semiotic character of those visuals. Such brain visuals are instances of Peircean *signs* in a straightforward sense: cognitive neuroscientists examine and make sense of the brain processes by consulting their digital renderings. But what kinds of signs are they? In other words, how do such renderings function in the laboratory? The aim of this question is to recover aspects of phenomena with which scientists engage in their work of fMRI.

In his famous characterization of the sign's relationship to its object, Peirce distinguishes between icon, index, and symbol (Peirce 1934–1958, 4:531).[4] For Peirce, whereas a symbol (most closely related to the Saussurian language-like sign) is a conventional sign grounded in a rule, iconic and indexical signs are characterized by their materiality and embodiment.[5] An index is a sign that functions on account of being physically or causally connected with what it stands for, and an icon is a sign that shares characteristics with an object that is perceived as having some similarity with it. As an iconic sign is defined in terms of *likeness*, the central characteristic of that sign is its capacity to affect us in a way that has similarities with how we would be affected by the object the sign stands for—the two would "excite analogous sensations" (Peirce 1894, §7).

Among iconic signs, Peirce further distinguishes between *images* and *diagrams*. "Whereas images are the iconic signs that have the same simple quality as their objects, diagrams are the signs whose parts have analogous relations to those of their objects" (Peirce 1934–1958, 2:277). Thus, a portrait would be an example of an image, and a map would be an example of a diagram. Peirce points out that diagrammatic signs "allow for experimentation and generate insight as they can be used to draw new conclusions about the relations existing in the world" (1934–1958, 2:279, 4:531).[6]

Alan Gross (2008, 381) has suggested that the character of fMRI brain visuals should be understood in terms of indexical signs. According to Gross, fMRI brain visuals are indexical signs as they are causally linked to the external world. I claim, instead, that fMRI digital visuals are iconic signs. I agree with Gross that they cannot be productively conceived as signs that *depict* (or images). Instead, I see them as the kind of iconic signs that Peirce calls diagrams. Cognitive neuroscientists use fMRI technology to distinguish between and locate the brain activations in the cerebral cortex; in other words, their enterprise importantly concerns spatiality. The scientists, thus, treat fMRI visuals as maps or diagrams rather than images. This is evident not only in their theorizing about "function localization," but also in the details of their everyday, embodied practice in the laboratory. In the laboratory, fMRI practitioners engage with digital scans and treat them as places for problem solving. They manipulate the scans by using computer commands as they involve their gesturing bodies in the work of neuroscience. When talking about maps as examples of diagrams, Peirce has pointed out that diagrams allow for engagement; this engagement can be accomplished in the imagination but also by direct involvement (1934–1958, 4:530).

To maintain that fMRI brain visuals are iconic and diagrammatic signs, however, does not imply a denial of their indexical and symbolic character. Peirce is clear in pointing out that every sign is a "mixture of likeness, indices, and symbols. We cannot dispense with any of them" (Peirce 1894, §9). In fact, fMRI brain visuals, as iconic and diagrammatic signs, do not function alone. As they go through processes of "spatial normalization," and inscribe debates such as the one over "maps versus modules," they have a strong conventional and symbolic component (see Alač 2011, chapters 2 and 7). Similarly, in their published format, the visuals are always encountered together with a variety of indexical signs: they are accompanied with labels that indicate geographical locations, annotations of statistical values, graphically provided units of distance, scale, time, etc. (ibid.). Furthermore, as part of laboratory practice, these icons are always grounded in specific circumstances of making and doing, and hence cannot be divorced from their indexicality. When Peirce explains that icons (not differently from symbols) cannot convey information apart from indexical signs,[7] he singles out their capacity for experience:

It is true that a map is very useful in designating a place; and a map is a sort of picture. But unless the map carries a mark of a known locality, and the scale of miles, and the points of the compass, it no more shows where a place is than the map in *Gulliver's Travels* shows the location of Brobdingnag. It is true that if a new island were found, say, in the Arctic Seas, its location could be approximately shown on a map which should have no lettering, meridians, nor parallels; for the familiar outlines of Iceland, Nova Zemla, Greenland, etc., serve to indicate the position. In such a case, we should avail ourselves of our knowledge that there is no second place that any being on this earth is likely to make a map of which has outlines like those of the Arctic shores. This experience of the world we live in renders the map something more than a mere icon and confers upon it the added characters of an *index*. Thus, it is true that one and the same sign may be at once a likeness and an indication. Still, the offices of these orders of signs are totally different. It may be objected that likenesses as much as indices are founded on experience, that an image of red is meaningless to the color blind, as is that of erotic passion to the child. But these are truly objections which help the distinction; for it is *not* experience, but the *capacity* for experience, which they show is requisite for a likeness; and this is requisite, not in order that the likeness should be interpreted, but in order that it should at all be presented to the sense. (Peirce 1894, §5)

In laboratories of cognitive neuroscience, this capacity for experience that character-izes iconic signs concerns fMRI brain visuals. There, the function of the visuals is to exhibit the character of the brain and its processes, rather than to foreground the causal or physical link with the brain processes, assuring the scientists of their reality (as indexes would do). These visuals, instead of depicting, allow for engagement and suggest new spatial relationships among the supposed states of things.

To discuss how fMRI scans as diagrammatic signs exhibit the character of the brain and its process in laboratories of cognitive neuroscience, I examine the everyday meth-ods that scientists employ when they work with digital visuals to show that fMRI scans are understood through an active visual inspection and embodied engagement. In this sense, while being importantly visual, they are not images or pictures to be passively looked at, but material to be experimented with. When coordinated with other semi-otic phenomena, they are spaces to be engaged by human bodies. In activities such as training apprentices and analyzing data, fMRI visuals are malleable fields. They are malleable because they are digital (practitioners use keyboard and mouse commands to alternate visual displays), but also because they can be transformed in interaction through the involvement of gesturing bodies. By virtue of being both visual and highly malleable, the scans thus function as centers of action *with* which (not only *on* which) the work is performed. They allow the practitioners to deal with experimental data in a way analogous to their engagements with physical objects and with one another.

This attention to visual renderings as they are bound with the world implies a turn toward *ontologies* and *what* scientists know (Mol 2002; Daston 2008). To see fMRI brain visuals as primarily iconic (rather than indexical) lets certain kinds of objects into the laboratory. During an fMRI brain scanning session (see, for example, Alač 2011, chap-ter 3), the fMRI researchers do not look at the brain and its processes in the way that a photographer looks at the scene she is photographing. Instead, the brain processes they inspect become visible consequently when the researchers work on their comput-ers in the laboratory. In this sense, as fMRI visuals *exhibit* the brain's character and its processes as a part of embodied work and interaction (rather than functioning by virtue of being physically or causally connected to the scanned brain), they are *fields for interaction* (Alač 2011).

As such, fMRI brain visuals are not thoroughly "constructed" by the scientists. Rather, as they afford practitioners the ability to skillfully turn their bodies to their experimental data, the visuals allow for an *intertwining with reality* (Merleau-Ponty 1968): the coordination of the visuals with the practitioners' bodies *enacts* what these visuals target. In parallel with an argument made by Lorraine Daston, in fMRI laborato-ries these enacted objects are sustained by scientific observations (see also Alač 2011):

It is observation, grounded in trained, collective, cultivated habit, that fuses these bits and pieces into a picture—often a literal picture crafted by techniques of scientific visualization. And it is the picture, seized at a glance, all at once, that guarantees the sturdy existence of a world. This is not quite the vision of angels, who, according to Bonaventure and Aquinas, saw only universal forms, not individual particulars. It is not a metaphysics at all, not a God's eye point of view, but only an ontology for humans, with their eyes wide open. (Daston 2008, 105)

When the material that scientific observations are rooted in is digital, nevertheless, this "eyes wide open" means the multimodal engagement of working, experiencing, and semiotic bodies. To make sense of the fMRI brain visuals and to decipher what they say, practitioners engage their eyes, as well as their hands and entire bodies, in their everyday work with computers.

3 In the Laboratory

To consider fMRI visuals from the perspective of real-time, practical engagement, we join a training session in a laboratory of cognitive neuroscience. Recent efforts in STS have highlighted the pedagogical dimension of scientific practice and the mutual reliance of training and research (e.g., Kaiser 2005; Mody and Kaiser 2008; Nersessian et al. 2003). Here I focus on instructions that take place as a part of an analysis session where practitioners work with previously collected experimental data. The goal for the session is to accomplish the analysis while allowing a new laboratory member to acquire certain *habits* (Peirce 1934–1958, 5:397, 5:400) and regularities in her interaction with the world within the context of *language games* (Wittgenstein 1953, §23) that the laboratory practitioners and the larger cognitive neuroscience community share (Crocker 1998). To attain these habits, and eventually proceed on her own, the new laboratory member centers her work on brain renderings. She has to learn how to recognize the visual aspects of the brain scans that her more senior laboratory colleagues notice when they engage with those renderings.

As a part of this training session we shall focus on moments when the researchers, as they prepare the experimental data for statistical examination, have to assess whether the brain scans can be aligned. Over the course of the experiment, a series of scans are recorded, each scan standing for a slice of the brain. The assessment of the alignment among the scans is achieved by viewing slices in the axial, sagittal, and coronal views[8] shown on the computer screen over the course of the experiment (figure 4.1). Figure 4.2 shows an example of two sagittal-view visuals in a series. To identify the nonaligned visuals, the practitioners do not inspect the visuals one by one, but use mouse commands to alternate the views of individual scans on the computer screen. For example, in regard to figure 4.2, the practitioners first see the brain visual on the left-hand side, then the visual on the right-hand side.

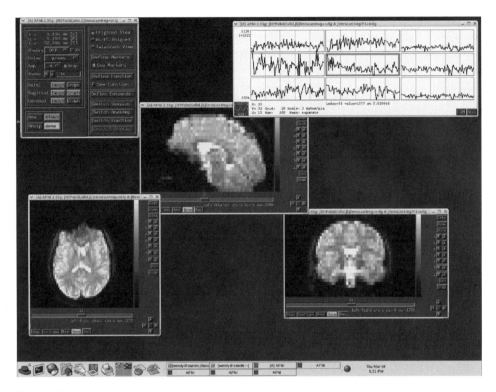

Figure 4.1
The computer screen displays brain slices over the time course of the experiment.

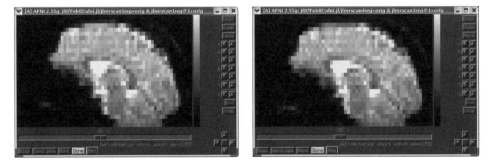

Figure 4.2
Two sagittal-view brain visuals in a series.

The practitioners' engagement with the computer thus has a salient temporal dimension. When the visuals are not perfectly aligned—as exemplified in figure 4.2—their rapid alteration generates the appearance of motion. The phenomenon of "apparent motion" has been studied since early efforts in experimental psychology, and is investigated today by neuroscientists themselves.[9] The trick of apparent motion has also been explored by a variety of devices and techniques, from optical toys (such as the eighteenth-century zoetrope) to contemporary cinema, television, and computer animation. The specificity of the appearance of motion in the fMRI laboratory has to do with the involvement of the viewer's hands; by engaging computer interface devices, the practitioner dictates the rhythm of the succession of visuals that produces the illusion of motion.

Laura Mulvey's classic "Visual Pleasure and Narrative Cinema" (1975) provides an image of the movie theater in which an immobile, individual spectator (invisible to others) is buried in the darkness while only the dim cone of the projector beam and the surface of the screen are visible. In contrast, in the fMRI laboratory, due to the digitality of the material, the viewer's hands are directly involved in creating the apparent motion.[10] In this regard, figure 4.2 is an inaccurate rendering of the laboratory phenomena. Constrained by the paper medium, the figure shows two static and discrete visuals; yet, in the laboratory, the practitioners understand discrepancies among the visuals as a unified whole. In fact, the difference between the two visuals is rather difficult to identify when displayed in a stationary manner, as in figure 4.2. However, when the same visuals are engaged in the digital format, their nonalignment becomes readily available.

In this simple example, where the fMRI practitioners explore the capacity of the digital matter by swiftly changing the computer display, of primary interest is digitality as directly manifested to the viewers: the fMRI brain renderings are present to the researchers as visual phenomena that can be manipulated via computer interaction devices (mouse and computer keyboard). Obviously, these properties of the brain renderings are grounded in the underlying level of discrete, numerical values and software code. Even though the practitioners do not have to constantly access this level, and more often delegate it to the machine, their work cannot be divorced from the specific properties of the material they engage with.[11] Yet what primarily matters for solving the problem at hand—identifying the nonaligned visuals—are manipulability and the visual character of the data. The practitioners' manual engagement with the visuals presented on the screen in the three views and ordered in the series allows them to eliminate visible defects in the scans (such as blurry splotches). This activity of "cleaning the data" of motion *artifacts* is a routine procedure in the laboratory.[12]

To account for the artifact and make sense of the nonalignment, the practitioners explain that it is due to the movement of the scanned body. Therefore, their immediate task is to figure out the morphology of the subject's movement that could have caused the nonalignment. Whereas they understand the subject's brain and its function as natural phenomena to be examined, they consider the potentially unintentional movements of the subject's body to be the cause of an intrusion in the visibility of fMRI scans. They thus seek to identify the character of the aberrant movement that the experimental subject performed while inside the magnet so that they can remove distortions from the data set. However, they did not actually see the movement they believe caused the nonalignment, since the minute movements of the experimental subject lying in the scanner are not available to direct inspection.

Rather than relying on direct experience of the scanned body movements, the practitioners attempt to evoke laboratory knowledge and employ multimodal semiotic means, such as talk, gestures, visual orientation, bodily conduct, and facial expressions. These resources are dynamically coordinated with the brain visuals so that the practitioners can *see* the *movement of the subject in the brain visuals*. As they identify the movement in the misaligned consecutive visuals (by exploiting the visual and digital character of the scans), they attempt to understand that movement by publicly enacting it and feeling its effect in their own bodies. The movement is not simply seen in the visuals, but is enacted through a coordination of bodies and technology inscribed with cultural knowledge distributed through the laboratory.

The excerpt we are studying reports on 20 seconds of interaction between a graduate student who is an *old-timer* in the laboratory (Olga) and an undergraduate student who is a *newcomer* (Nina).[13] Olga is an easygoing but ambitious and hard-working Ph.D. candidate in psychology who has worked in the laboratory for several years. Because of her familiarity with the laboratory procedures, Olga has been asked by the director of the laboratory to guide the new member in acquiring laboratory skills so that she can get involved with the current research projects. Nina, who is majoring in cognitive science, has just started her internship in the laboratory with the goal of gaining practical knowledge in experimental methods. During the observed interaction, Olga is introducing Nina to the data analysis procedure.

As Olga (on the left in figure 4.3) guides Nina through an actual data analysis procedure, the teaching session is also an instance of work practice. The practitioners are seated in front of a computer in one of the laboratory's rooms where, while engaged in the work, they do not seem to pay much attention to the other events taking place in the room (laboratory members entering and exiting the room, and the ethnographer videotaping the scene). The atmosphere between the two practitioners is friendly but strictly work-oriented. Olga is patient and clear in her explanations, and Nina—softspoken and sometimes shy—shows her eagerness to become a competent member of the laboratory by intently engaging in work and interaction.

During the practice of motion correction, Nina controls the computer, using the mouse to engage the computer display. While she does so, her actions are closely coordinated with the old-timer's. In our excerpt, Olga notices a misalignment among visuals and directs the newcomer's attention toward the computer screen. Olga then performs a gestural reenactment of the noticed misalignment, to which Nina responds, quickly picking up the technique. After searching through the data, Nina also indicates a misaligned visual. She points toward the screen and then gestures by employing her hands, upper body, shoulders, and neck, enacting the movement of the subject in the brain visuals.[14]

The excerpt is rendered according to the transcription style of conversation analysis (Sacks 1992). To indicate the intricate ways in which interlocutors coordinate with each other, the transcription adopts the following conventions (Sacks et al. 1974; Jefferson 2004):

= Equal signs indicate no interval between the end of a prior and start of a next piece of talk,

(0.0) Numbers in brackets indicate elapsed time in tenths of seconds,

(.) A dot in parentheses indicates a brief interval within or between utterances,

() Parentheses indicate that transcriber is not sure about the words contained therein,

(()) Double parentheses contain transcriber's descriptions,

°°° Degree signs are used to indicate that the talk they encompass is spoken noticeably quieter than the surrounding talk,

// The double oblique indicates the point at which a current speaker's talk is overlapped by the talk of another,

: The colon indicates that the prior syllable is prolonged,

___ Underscoring indicates stressing,

.,? Punctuation markers are used to indicate "the usual" intonation:

. Dot is used for falling intonation,

? Question mark is used for rising intonation,

, Comma is used for rising and falling intonation.

The action line, charted above the talk, follows transcription conventions from Schegloff (1984):

o Indicates onset of movement that ends up as gesture,

a Indicates acme of gesture, or point of maximum extension,

h Indicates previously noted occurrence held,

t Indicates thrust or peak of energy animating gesture,

hm Indicates that the limb involved in gesture reaches "home position" or position from which it departed for gesture,

p Indicates point,

.... Dots indicate extension in time of previously marked action.

The transcript and line drawings based on the video image (Goodwin 2000b) are combined with arrows to capture body movement in static form.

The excerpt first indicates how fMRI brain scans acquire their meaning through an active coordination with the practitioners' working and gesturing bodies. To help the newcomer learn to see the specific feature in the stream of visuals while she directs the changes on the computer screen,[15] the old-timer keenly involves her gesturing body. In line 1, as she notices that there is a misalignment between two consecutive brain visuals in the sagittal view that appear, one by one, on the computer screen, Olga extends her left hand toward the computer (figure 4.3) to draw Nina's attention to the misaligned visuals while confidently saying: "That's definite I can see her in this plane." Thereafter (line 2), Olga sweeps her gesturing arm in a downward motion (figure 4.4) and halts at a certain point (figure 4.5) while briefly looking at Nina to check on her comprehension.

The gesture not only allows Olga to position herself in the activity as an experienced member of the laboratory, but it also provides *scaffolding* (Vygotsky 1978) for the newcomer's examination of the brain visuals. As such, the gesture organizes the seeing in terms of a three-dimensional horizontal movement. The horizontal position of the gesturing hand suggests that Olga's semiotic act, rather than merely standing for the brain scans (and depicting what is already present on the screen), also evokes the head of the experimental subject moving in the scanner. The practitioner's acts, therefore, aren't simply a matter of pointing to a referent, or even interpretively seeing what is on the screen, but concern a way of seeing, saying, and showing multiple and mutually elaborating articulations of the experimental data so that their coordination will generate comprehension. This editing of the scans—which takes place both with the brain renderings on the screen and with the embodied interaction around them—allows the newcomer to understand the fMRI data in terms of a human head that moves, despite the fact that none of the visuals taken in isolation (or even placed in a series) would indicate such a movement to an untrained eye.

The excerpt also shows how the newcomer indicates her comprehension of what her colleague is demonstrating, and how she learns to see by coordinating her own body with the digital renderings of the brain. She does so by using mouse commands to manipulate the visuals and by aligning her embodied semiotic engagement with them. After Olga has performed the gestures in lines 1 and 2, the two practitioners comment while carefully observing the changes on the computer display (lines 3–6), organized by Nina's manual engagement with the computer. In line 3, the old-timer's "Aaaa" turns the attention toward the brain scan, followed, in line 4, by the newcomer's similar vocalization combined with the utterance, "This is a good one." While the newcomer clicks on the mouse to rapidly display the visuals ordered in a series, she spots, indicates, and characterizes the nonaligned visual as a "good one," so that what needs to be discarded is not only indicated but exposed and framed. Whereas the action provides the old-timer with a chance to monitor the trainee's understanding (displayed through her gestures and talk, as well as her work on the computer), it

1 O: That's definite I can see her in this plane=

o... p.........

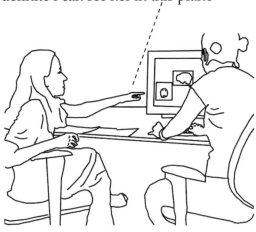

Figure 4.3

2   going from ₁ here to here.

.................a..........t......hm

Figure 4 4 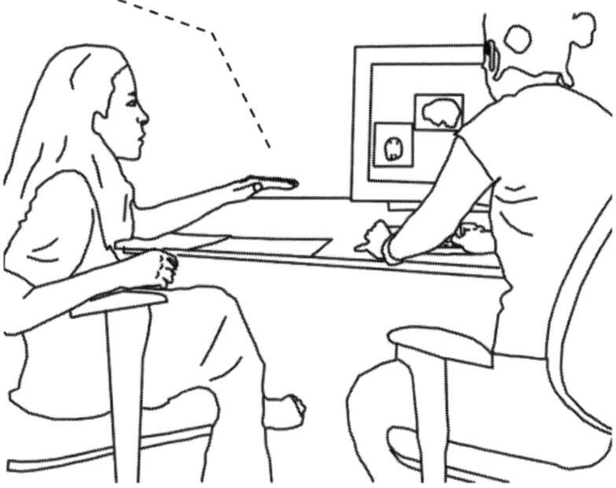   Figure 4.5

₁ ((Briefly looks toward N))

3   Aaaa ₁

₁ ((Disapproves of what she sees))

                o.p............hm
4   N: (Aaaa, this ₁ is a good one)

                 ₁((Points to the sagittal view of the brain slice on the screen))

5   O: Yeah. Slice, edit ₁

                 ₁((Instructs N how to select commands on the computer screen))

6   N: (Oh, here we are)

7   O: ₁ Is she moving any more in thirty?

                 ₁((Moves closer to the screen))

                o..a
8   N: She's (0.1) ₁ moving – =

Figure 4.6    Figure 4.7

        ₁((O turns toward N))

o..p.........
9   Like this ₁one

Figure 4.8

₁((Points to the axial view of the brain slice))

.... a.......t..hm
10   Is going down. Hhhhh ₁

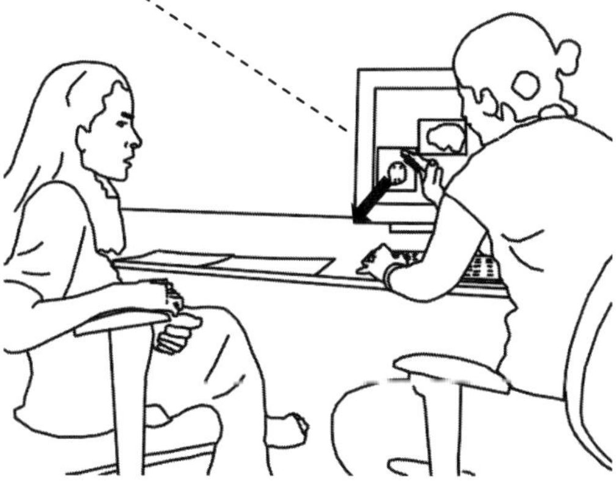

Figure 4.9

₁((Laughs))

allows the newcomer to indicate her ability to identify the misalignment and see the brain renderings as an fMRI practitioner. Once again, this seeing does not concern any one visual taken in isolation. Nina indicates and characterizes specific visuals as "good ones," but she sees them as such only in relation to the others: those of the same brain slice in different views, and those in the same view organized in the series (whose appearance and disappearance on the screen she directs).[16]

In fact, as the interaction proceeds, Nina reports on another find: "Oh, here we are" (line 6). At this point (line 7), Olga moves toward the computer screen as if she is unable to see what her colleague is indicating, and asks: "Is she moving any more in thirty?" In response, Nina takes the floor. She illustrates her full command of what she sees by performing the subject's movement in terms of a complex action that coordinates the brain scan with a series of embodied semiotic acts (lines 8–10). In line 8, while her hands are engaged with the keyboard, Nina hunches (figure 4.7). In line 9, she points toward the computer screen (figure 4.8). Finally, in line 10, she executes a hand gesture in which she swings with the right hand downward (figure 4.9).

In line 8, after saying "She's moving," Nina abruptly bends her shoulders and neck toward the screen (figures 4.6 and 4.7). The hunching movement is accompanied by a chuckle, as she displays her negative judgment toward what she points at (figure 4.8). By physically enacting the process of change attributed to the brain visuals, Nina employs her torso to create the hypothesized movement of the experimental subject. Her semiotic enactment is not redundant in reference to the axial view that she points to. First, the visual by itself does not indicate the movement performed by the practitioner. Second, even though the appearance of motion is generated on the screen, the screen does not show the movement that the practitioners attribute to the experimental subject. That movement becomes available through the performance of their bodies coordinated with Nina's work on the computer. While Nina performs the hunch, Olga turns her attention from the computer screen toward her (figure 4.7), treating the hunch as a response to the question: "Is she moving any more in thirty?"

The hunching gesture is also a means for Nina to learn about the trouble in the experimental data (not only a way to respond to her colleague). In addition to understanding the nonalignment by comparing the visual features on the computer screen (or by observing the old-timer's action), the newcomer learns about brain visuals through an embodied experience. As even more clearly indicated by the excerpt that follows, the old-timer does not simply ask the newcomer to work with the computer and look at the screen. Instead, Olga—by example—encourages Nina to comprehend the artifact by enacting the movement attributed to the experimental subject. Apparently, Olga has at least two reasons to do so. First, what is seen in the comparison between the brain renderings significantly differs from observing a person who moves (to see it requires training). Second, the practitioners never saw and cannot see or experience the subject's movement during the scanning.[17] But they can coordinate their work on the screen and their lived bodies to make sense of the trouble in the data. They can understand the cause of the defect by gesturing (as seen in Olga's hand gesture in line 2, and Nina's hand gesture in lines 9–10, for example), and by experiencing it in the flesh. Because they, as it seems, operate under the assumption—somewhat akin to Alfred Schutz's *reciprocity of perspectives* (e.g., Schutz 1990, 11–13)—that they would have the same experience as the subject had if they were scanned, they resort to their own embodied engagement as a mean of understanding the artifact. Nina's hunching movement generates first-person understanding for its performer, allowing the trainee to experience in her own body a movement similar to the one that the experimental subject may have produced in the scanner: Nina reads the computer screen by feeling a feature of the experimental data in her own body.

This experiential and semiotic action is further elaborated as Nina points to an exemplary brain visual on the screen ("like this one," line 9) and performs another gesture (lines 9–10). She first points to the visual (figure 4.8), and then enacts the movement by sweeping her gesturing hand downward (figure 4.9).[18] The articulation of the gesture is rather similar to Olga's gesture enacted at the outset of the excerpt (lines 1 and 2). Yet, by mirroring the old-timer's action, Nina demonstrates her competence in seeing what her colleague didn't notice at first. She progressively acquires the habit of seeing like a professional researcher by manipulating the visuals via computer and by aligning what she notices on the screen (and what the old-timer points out) with the gestures and experiences of her semiotic body.

4   The Imaged Body in the Brain Visuals

Science studies scholars (e.g., Knorr-Cetina 1981; Latour and Woolgar 1979; Star 1989) have discussed how final accounts of scientific work (i.e., in the form of scientific articles, conference talks, or newspaper reports) standardize, while averaging and filtering out local contingency or individual differences: "One of the mandates of science is to create generalizable results, which are meant to be universal, and this mandate is often conflated with the deletion of local contingency" (Star 1989, 93). Method sections of fMRI articles list repeatable and well-defined steps of the experimental procedure, as they—with the exception of single-subject studies—report on experimental participants in terms of their number, gender, and social group, thus making individual bodies utterly invisible.

This view of deletion and simplification (of reduction of initial data) has been challenged by Michael Lynch's early work on visualization (e.g., Lynch 1990), where he showed that the succession of renderings in laboratory specimens (and field specimens) doesn't just reduce but also adds and highlights significant features, reads them, codes them, essentializes them. An important point is that contingency and context do not simply drop out; they are reconfigured as renderings are called into play in different circumstances, in texts, etc. Not dissimilarly, the analysis performed by Nina and Olga highlights the individual and the particular. During their engagement with the digital data, the experimental subject, because of her movement, becomes visible. Rather than simply noticing the difference between the two brain scans, Olga "sees" the experimental subject moving: "I see her in this plane" (line 1). When in lines 4 and 6 Nina says "Aaaa, this is a good one," and "Oh, here we are," Olga follows by reintroducing the personal pronoun, as she asks Nina to elaborate: "Is *she* moving any more in thirty?" (line 7). Nina answers by saying: "She is moving" (line 8).

Certainly, this resurfacing of the experimental subject is recognized during the data analysis so that it can be canceled. The perceived consequences of "her" undesired behavior have to be removed so that the general character of the data can be regained. Similarly, if the data do not show any particular problems, practitioners tend to treat them as anonymous and general: as long as the subject's body and behavior are seen as yielding themselves to the prescribed procedure, the experimental subject is promptly translated into background. Such anonymous and general data are what the scientists look for, as their quest is to describe the functioning of the brain, not the idiosyncrasies regarding an individual brain.[19] Nevertheless, this moment of passage—where the brain visuals still belong to a single individual whose individuality is being noticed so that it can be eradicated from the data—is relevant as it allows us to encounter things and bodies that populate fMRI practice, and, thus, to describe how digital visuals function in the laboratory.

Undoubtedly, the imaged body does not show itself in the laboratory as a somehow "natural" and independent thing. During the scanning session, the body is controlled by the requirements of the scanning procedure while being constrained by the theoretical expectations in cognitive neuroscience (see Alač 2011, chapters 2 and 3); in this manner, only certain features of the lived body, relevant for the research activity, are transposed into the visuals, and only some among the many features that can be accessed through the visuals are desirable. Also, while the experimental subject's body was in the scanner, the practitioners were not able to witness the subtle movement that they are now reconstructing. Instead, the subject's movement becomes retrospectively witnessable when it is enacted in the laboratory: the movement is seen in the differences between the scans that the practitioners' working hands manipulate, and in their semiotic bodies that they coordinate with those scans.

In their discussion of the observation of the first optical pulsar, Harold Garfinkel, Michael Lynch, and Eric Livingston (1981) use an analogy of the pulsar and a potter's object in the process of being formed:

The analogy to the oscilloscopically displayed pulse is the developingly observable object of the potter, where the pulse takes "shape" in and as of the way it is worked, and *from* a place-to-start *to* an increasingly definite thing. . . . Our policy, and the point: We want to examine the pulsar for the way it is *in hand* at all times in the inquiry. We want to see the way it is "performatively" objective. We did *not* examine and we want *not* to examine the end-point object for its correspondence to an original plan. We want to disregard, we want *not* to take seriously, how closely or how badly the object corresponds to some original design—particularly to some cognitive expectancy or to some theoretical model—that is independent of their embodied work's *particular occasions* as of which the object's production—the *object*—consists, only and entirely. (Garfinkel et al. 1981, 137)

At the same time, this appearance of the imaged body places conditions on the visuals and limitations on what goes on in the laboratory. As the practitioners understand the morphology of the subject's movements by coordinating their trained and experiential bodies with the manipulation of the computer screen, their movements are not constructed intentionally and defined a priori (Garfinkel et al. 1981; Suchman 1987). Instead, as they relate to the "what" they help make evident and communicable, the movements are formed through the practitioners' moment-by-moment coordination with each other and with the computer. In this sense, the subject's body—seen as the cause of the disturbance in the experimental data—articulates the movements of the practitioners. Even though the practitioners openly disapprove of what the subject did in the scanner, to understand and prepare their data for analysis they have to coordinate with brain visuals to enact the imaged body's movements.

These acts of coordination with the reality indicate the contours of the laboratory objects. In the fMRI laboratory, the observed body-in-the-scan takes shape as it is engaged by the researchers who gesture and work with the computer: the movements of the practitioners' bodies progressively become the subject's movements, as the subject's movements enter the laboratory through the researchers' engagement shaped by the affordances of the technology. The reality that poses resistances is enacted through this encounter.[20]

This suggests that fMRI visuals neither function by virtue of their physical or causal connection with their objects, as indexical signs may do, nor are iconic signs because they "look like" what they "stand for." Instead, as iconic signs, the brain scans are bound up with the world through the embodied engagement in the laboratory. They are bound with the lived bodies of fMRI practitioners as well as with the imaged body. This intertwining of the practitioners and the object of their study is realized through the digital materiality that characterizes these iconic signs. As the two practitioners move their semiotic bodies in coordination, the trainee acquires the habit of spotting the artifact as she swiftly compares the visuals across the three views and taps on the mouse to create an impression of movement between the visuals ordered in the series. Manipulability combined with the visual character of the experimental data is crucial in these step-by-step acts of subtle coordination.

5 Digital and Visual as Interdependent

The claim that digital scientific visuals are diagrams, which must be understood with respect to how they are worked with, semiotically engaged, and experienced, has consequences for how the apparent dichotomy between the numerical and visual character of brain scans can be understood. When investigating social aspects of MRI, fMRI, and PET, researchers have been busy discussing this distinction (e.g., Beaulieu 2002; Dumit 2004; Joyce 2005, 2008). Anne Beaulieu (2002), for example, found out that the neuroscientists she interviewed highlighted the potential of brain imaging measurement to render spatial components and anatomic referents while, at the same time, they downplayed the visual form this information took to emphasize the quantitative information it represented. Beaulieu understood this negation of the importance of visual knowledge in brain-mapping research as related to the way evidence is evaluated in modern Western science. She argued that, because visual evidence has been regarded as appealing first to the senses as opposed to reason, and hence is seen as lacking a solid relationship to the truth, visual evidence is judged as not having a particularly high position in the hierarchy of types of scientific evidence. The interviewees claim that those most interested in the visual aspects of brain-mapping techniques are usually clinicians, not scientists, suggesting a hierarchy in which the visual is associated with the lower echelon of applied research.

Kelly Joyce (2005, 2008), who studied the use of MRI in clinical settings, agrees with the claim made by Beaulieu's interviewees. When introducing the history of MRI, Joyce describes how Paul Lauterbur, an American chemist credited with being the first person to use MRI to generate visuals of human anatomy, talked about those renderings in terms of maps, rather than images and pictures, defining them as a "mathematical representation of spatial information" (Joyce 2008, 32). Joyce, in contrast, points out that clinical practitioners prototypically talk about *pictures* of the human body, as their language reflects the centrality of the visual and visible in contemporary life:

Today, language that highlights the relation of the image to pictures of the anatomical body are often used in clinical practice, while language that calls attention to maps and spatiality is less common. . . . This linguistic difference occurs in part because of the broader recognition of the centrality of images to contemporary life as visualizing technologies such as cameras, computers, video games, and picture-producing cell phones become more common. (Joyce 2008, 32)

Discussions of the tension between the visual and the numerical, and the decision to talk about "pictures" and "images" when referring to MRI, fMRI, and PET visuals, are certainly relevant. On the one hand, they document how practitioners rationalize and talk about their work (these categories are useful to the practitioners in making claims about the novelty of what they do, and in distinguishing themselves from what others do by insisting on the categorical differences). On the other hand, they highlight the pervasiveness of the current focus on scientific texts, larger communities, and societal phenomena in social studies of science and technology. Yet once we turn our attention to the real-time practical work in neuroscience, and we adopt the understanding of brain visuals in terms of iconic signs, this dichotomy disappears.

Rather than associating the visual character of fMRI visuals with transparency while also coupling their digitality with mediation, interpretation, and choice, the analysis of laboratory work indicates that visual and digital aspects are codependent. In fact, the centrality of the embodied engagement concerns not only the digital but also the visual character of fMRI brain scans. When scientists work with fMRI scans, their vision is accomplished through a coordination of eyes with the action of the hands and the workings of fMRI technologies. Consequently, fMRI visuals involve an active, distributed engagement, with scientists engaging the visuals as if dealing with the everyday physical objects. In this respect, the contribution of each constituent is indispensable. In fact, the relationship with the digital screens would not be possible (at least not to this extent) if the manipulable data with which Nina and Olga worked were not also visual.

My analysis thus follow Michael Lynch's discussion of the two orders of laboratory "space," *opticism* and *digitality*: "The paradigm for the former is the lensed instrument and the scrutinizing eye, while the latter is embodied by the play of fingers (digits) on a keyboard instrument" (Lynch 1991, 56). Lynch points out that digitality does not displace opticism; rather, the two orders coexist and overlap with each other across historical periods. As seen in the interaction between the fMRI practitioners, the digital brain scans, engaged with hands, bodies, and eyes, are neither only visual nor only digital. Instead, because of their diagrammatic character, they are at the same time visual and digital.

6   Extending the Boundaries of Scientific Visuals

This move toward engagement when discussing scientific visuals calls for extending the boundaries of what those visuals include and implicate. Digital scientific visuals gain meaning in the laboratory not in isolation but when organized in a series, manipulated via computer commands, and embodied by the practitioners. This means that the practitioners' gestures, their touching, and their alterations of the scientific visuals (directly enacted or evoked) are the constitutive elements of those visuals. The visuals do not end at the borders of the computer screens, but stretch out into the world of the imaged and working bodies.

Even though these extensions are tamed in the laboratory, and largely erased once the visuals have been translated into paper format (Amann and Knorr Cetina 1990, 116), they are crucial in understanding how scientific evidence is generated in the age of computers. In this text my aim has been to indicate how they may allow us to recover aspects of phenomena and objects of knowledge as they are enacted in the actual moments of scientific practice. What matter then are the efforts in documenting the coordination across multiple embodied and social agents, technology, and communicative actions.

Notes

1. I use the term "visuals" rather than "images" to avoid some of the connotations implied in the word "image," as explained later in the text.

2. Ethnomethodology is known for its principle of avoiding any "generically theorized representations" (e.g., Garfinkel 2002, 136) and accounting for observed events in strictly "local" terms. Thus, my bringing together of Peirce's semiotics and ethnomethodological approaches could be seen as problematic, as the use of Peirce's categories may imply a description of scientific practice in terms that are not naturally available to scientists as a matter of their practice. Yet even though the scientists may not use the term "iconic" when reflecting on the character of fMRI brain renderings, they do highlight the spatial character of those visuals. What is more, my analysis and employment of Peirce's terms are intended to render what is observably done by the scientists when they engage with the fMRI visuals.

3. "Semiotic" should not be reduced to "symbolic." As proposed by Peirce, semiotics has to do with phenomenological aspects of communication and interaction. According to his (1867) phenomenology (see Peirce 1934–1958, 1:545–559), every experienceable entity possesses the properties of firstness (as a phenomenal entity in itself), secondness (as it stands in dyadic relationships with other entities), and thirdness (as it stands in triadic relationships with other entities) (also see, for example, Ransdell 1989; Rosensohn 1974). As Peirce builds his semiotics based on this distinction, his understanding of meaning has a distinctly pragmatic character. Peirce's signs gain their meanings through their concatenation or *semiosis*, which is a time-bound, context-sensitive, interpreter-dependent, and materially extended dynamic process (Queiroz and Merrell 2006). I adopt Peirce's semiotics in this paper to talk about "embodied" and "multimodal" interaction, and to signal that not only language tokens but also gestures, nonlinguistic vocalizations, visual orientations and movements participate in accomplishing actions in specific practical circumstances.

4. Based on the triadic idea of the sign, Peirce generated multiple typologies of signs (in his 1903 account of semiotics he suggested ten classes of signs, while announcing sixty-six classes of signs in his final typology), the distinction between icon, index, and symbol being the best known.

5. Another well-known reference to Peirce's symbol-icon-index distinction, as a part of the STS discussion on representation, can be found in Hans-Jörg Rheinberger's *Toward a History of Epistemic Things* (1997, 103). Rheinberger relies on Derrida's conceptualization of the *trace*: "Engaging in the production of epistemic things means engaging in the potentially endless production of traces, where the place of the referent is always already occupied by another trace. To use a terminology familiar from linguistics, there is a permanent gliding replacement of any presumed 'sig-

nified' by another 'signifier'" (1997, 104). In this paper, however, in contrast to the Saussurian model of the sign that *brackets the referent* (excluding from the domain of interest any reference to objects existing in the world), I consider that Peirce's conception of the sign includes what the sign stands for as its necessary part. I find this conception, with its concern for the materiality of the world, to be valuable when multimodal sign systems and the digitality of the visuals are of interest. For a further discussion of the problem concerning the infinite regress of signifiers and the *limits of interpretation*, see Eco (1991).

6. For a further discussion of this point, see Alač (2011), 41–43.

7. Note that Peirce's view of indexical signs is not incompatible with how ethnomethodologists and conversation analysts conceive of indexicality. In their discussion of indexicality in the history of philosophy, Garfinkel and Sacks (1970, 348) refer to Peirce (together with the Wittgenstein of *Philosophical Investigations*).

8. The axial sections are vertical sections made from the front to the back of the brain. The sagittal sections are vertical sections ordered from the center of the brain out to the side. The coronal or horizontal sections are displayed from the top to the bottom of the brain.

9. The classical experimental work of Max Wertheimer (1912) and Adolf Korte (1915) has pointed out that the subject will sometimes report seeing motion between still images flashed in succession at specific temporal and spatial distances, and contemporary work in fMRI brain imaging (e.g., Muckli et al. 2002) has showed that the perception of apparent motion can be correlated with brain activations in hMT+(V5), the human "motion complex."

10. Because the seeing of each individual member has to be aligned with the knowledge of the scientific community, that seeing also often (at least in its early, learning stages) takes place in pairs or larger groups of practitioners.

11. The fMRI practitioners are certainly aware of this underlying level. Several of the practitioners that I observed write their own computer programs, and others are skillful users of off-the-shelf fMRI packages. The laboratory discussed in this chapter uses an AFNI set of programs to process, analyze, and display their data.

12. For a discussion of how scientists deal with artifacts in the laboratory, see, for example, Lynch (1985) and Latour and Woolgar (1979).

13. I use the terms "newcomer" and "old-timer," adopted from Jean Lave and Etienne Wenger's (1991) discussion of communities of practice (where people learn together by participating in a common endeavor), to refer to recently joined laboratory members and those who joined the community at a more distant time.

14. We shall notice how the practitioners designate this compound phenomenon with the pronoun "she."

15. See figures 4.1 and 4.2, as *inaccurate* examples of the scenic features on the screen.

16. Note that if I were to interlace screen shots with the transcript or provide another way of representing the visual display that the practitioners are confronting (see, for example, Woolgar

1990, 127–130), I would reduce the digital renderings to static ones (images), and thus misrepresent what the practitioners see (see figure 4.2 and my discussion of that representation).

17. This is not to say that the defective visuals do not have any indexical characteristics. They do: the practitioners attribute the cause of the nonalignment to the subject's movement. Yet how they deal with the visuals suggests that the prevalent attribute of those visuals is the capacity to be engaged and, thus, to allow the practitioners to identify, through this engagement, the character of the subject's movement.

18. It is worth noting that at this point in the practice the brain visual assumes a strong indexical character, practically achieved through the moment-to-moment embodied interaction. Importantly, however, during the practice through which this indexicality is achieved, the fMRI visuals function as prominently iconic signs: they allow the practitioners the possibility of experience and are engaged in an embodied manner.

19. In clinical work, it would probably be a very different situation, with an orientation to the particular case. Even there, however, there would be concerns with artifacts, since they would be seen as interfering with the diagnosis of the individual's intrinsic condition and not as momentary fluctuations in the way the brain is visualized.

20. My argument is akin to what Steve Woolgar (1990, 137–140) calls the *constitutive* position: "the phenomenon is constituted in and through descriptive work, and importantly, this work includes such practices as the assignation of alternative versions, the invocation of relevant mediating circumstances, and so on" (137). However, because Woolgar refers to the semiotics of Ferdinand de Saussure while I bring up Peirce, I evoke the object as a constitutive element of the sign (which is deemphasized in the linguistically based sign of Saussure's structuralist semiotics).

References

Alač, Morana. 2011. *Handling Digital Brains: A Laboratory Study of Multimodal Semiotic Interaction in the Age of Computers*. Cambridge, MA: MIT Press.

Amann, Klaus, and Karin Knorr Cetina. 1990.The fixation of (visual) evidence. In *Representation in Scientific Practice*, ed. Michael Lynch and Steve Woolgar, 85–121. Cambridge, MA: MIT Press.

Austin, John L. 1962. *How to Do Things with Words*. Oxford: Clarendon.

Beaulieu, Anne. 2002. Images are not the (only) truth: Brain mapping, visual knowledge, and iconoclasm. *Science, Technology and Human Values* 27:53–86.

Benveniste, Emile. 1971. *Problems in General Linguistics*. Coral Gables, FL: University of Miami Press.

Crocker, Thomas. 1998. Wittgenstein's practices and Peirce's habits: Agreement in human activity. *History of Philosophy Quarterly* 15 (4):475–493.

Daston, Lorraine. 2008. On scientific observation. *Isis* 99 (1):97–110.

Dumit, Joseph. 2004. *Picturing Personhood: Brain Scans and Biomedical Identity*. Princeton: Princeton University Press.

Eco, Umberto. 1991. *The Limits of Interpretation*. Bloomington: Indiana University Press.

Garfinkel, Harold. 1984 [1967]. *Studies in Ethnomethodology*. Cambridge, UK: Polity Press.

Garfinkel, Harold. 2002. *Ethomethodology's Program: Working out Durkheim's Aphorism*. Lanham, MD: Rowman and Littlefield.

Garfinkel, Harold, Michael Lynch, and Eric Livingston. 1981. The work of a discovering science construed with materials from the optically discovered pulsar. *Philosophy of the Social Sciences* 11(2): 131–158.

Garfinkel, Harold, and Harvey Sacks. 1970. On formal structures of practical actions. In *Theoretical Sociology*, ed. John C. McKinney and Edward A. Tiryakian, 338–366. New York: Appleton-Century-Crofts.

Goffman, Erving. 1976. Replies and responses. *Language in Society* 5:257–313.

Goodwin, Charles. 1994. Professional vision. *American Anthropologist* 96 (3):606–633.

Goodwin, Charles. 1995. Seeing in depth. *Social Studies of Science* 25:237–274.

Goodwin, Charles. 2000a. Action and embodiment within situated human interaction. *Journal of Pragmatics* 32 (10):1489–1522.

Goodwin, Charles. 2000b. Practices of seeing, visual analysis: An ethnomethodological approach. In *Handbook of Visual Analysis*, ed. T. van Leeuwen and C. Jewitt, 157–182. London: Sage.

Gross, Alan. 2008. The brains in *Brain*: The coevolution of localization and its images. *Journal of the History of the Neurosciences* 17 (3):380–392.

Heath, Christian, and Jon Hindmarsh. 2002. Analyzing interaction video ethnography and situated conduct. In *Qualitative Research in Action*, ed. Tim May, 99–121. London: Sage.

Ihde, Don. 2002. *Bodies in Technology*. Minneapolis: University of Minnesota Press.

Jefferson, Gail. 2004. Glossary of transcript symbols with an introduction. In *Conversation Analysis: Studies from the First Generation*, ed. G. H. Lerner, 13–31. New York: John Benjamins.

Joyce, Kelly. 2005. Appealing images: Magnetic resonance imaging and the production of authoritative knowledge. *Social Studies of Science* 35 (3):437–462.

Joyce, Kelly. 2008. *Magnetic Appeal: MRI and the Myth of Transparency*. Ithaca: Cornell University Press.

Kaiser, David, ed. 2005. *Pedagogy and the Practice of Science: Historical and Contemporary Perspectives*. Cambridge, MA: MIT Press.

Knorr-Cetina, Karin. 1981. *The Manufacture of Knowledge: An Essay on the Constructivist and Contextual Nature of Science*. Oxford: Pergamon Press.

Korte, Adolf. 1915. Kinematoskopishe Untersuchungen. *Zeitschrift für Psychologie mit Zeitschrift für angewandte Psychologie* 72:194–296.

Koschman, Timothy, Curtis LeBaron, Charles Goodwin, Alan Zemel, and Gary Dunnington. 2007. Formulating the triangle of doom. *Gesture* 7:97–118.

Latour, Bruno, and Steve Woolgar. 1979. *Laboratory Life: The Social Construction of Scientific Facts.* London: Sage.

Lave, Jean, and Etienne Wenger. 1991. *Situated Learning: Legitimate Peripheral Participation.* Cambridge: Cambridge University Press.

Lynch, Michael. 1985. *Art and Artifact in Laboratory Science: A Study of Shop Work and Shop Talk in a Research Laboratory.* London: Routledge and Kegan Paul.

Lynch, Michael. 1990. The externalized retina: Selection and mathematization in the visual documentation of objects in the life sciences. In *Representation in Scientific Practice*, ed. Michael Lynch and Steve Woolgar, 153–186. Cambridge, MA: MIT Press.

Lynch, Michael. 1991. Laboratory space and the technological complex: An investigation of topical contextures. *Science in Context* 4 (1):81–109.

Merleau-Ponty, Maurice. 1968. *The Visible and the Invisible.* Evanston: Northwestern University Press.

Mody, Cyrus, and David Kaiser. 2008. Scientific training and the creation of scientific knowledge. In *The Handbook of Science and Technology Studies*, 3rd ed., ed. Edward J. Hackett, Olga Amsterdamska, Michael Lynch, and Judy Wajcman, 377–402. Cambridge, MA: MIT Press.

Mol, Annemarie. 2002. *The Body Multiple: Ontology in Medical Practice.* Durham: Duke University Press.

Mondada, Lorenza. 2007. Multimodal resources for turn-taking: Pointing and the emergence of possible next speaker. *Discourse Studies* 9 (2):194–225.

Muckli, Lars, Nikolaus Kriegeskorte, Heinrich Lanfermann, Friedhelm E. Zanella, Wolf Singer, and Reiner Goebel. 2002. Apparent motion: Event-related functional magnetic resonance imaging of perceptual switches and states. *Journal of Neuroscience* 22: RC219.

Mulvey, Laura. 1975. Visual pleasure and narrative cinema. *Screen* 16 (3):6–18.

Nersessian, Nancy, Elke Kurz-Milcke, Wendy Newstetter, and Jim Davies. 2003. Research laboratories as evolving distributed cognitive systems. In *Proceedings of the Cognitive Science Society* 25, ed. Robert Alterman and David Kirsh, 857–862. Hillsdale, NJ: Lawrence Erlbaum.

Ochs, Elinor, Patrick Gonzales, and Sally Jacoby. 1996. "When I come down I'm in a domain state": Talk, gesture, and graphic representation in the interpretative activity of physicists. In *Interaction and Grammar*, ed. Elinor Ochs, Emanuel Schegloff, and Sandra Thomson, 328–369. Cambridge: Cambridge University Press.

Peirce, Charles Sanders. 1894. What is a sign? In *The Essential Peirce*, vol. 2, 4–10. Bloomington: Indiana University Press. http://www.iupui.edu/~peirce/ep/ep2/ep2book/ch02/ep2ch2.htm

Peirce, Charles Sanders. 1934–1958. *Collected Papers.* Cambridge, MA: Harvard University Press.

Queiroz, João, and Floyd Merrell. 2006. Semiosis and pragmatism: Toward a dynamic concept of meaning. *Sign System Studies* 34 (1):37–66.

Ransdell, Joseph. 1989. Peirce est-il un phénoménologue? *Études Phénoménologiques* 9–10: 51–75.

Rheinberger, Hans-Jörg. 1997. *Toward a History of Epistemic Things: Synthesizing Proteins in the Test Tube*. Stanford: Stanford University Press.

Rosensohn, William. 1974. *The Phenomenology of Charles S. Peirce*. Amsterdam: B. R. Grüner.

Sacks, Harvey. 1992. *Lectures on Conversation*. 2 vols. Oxford: Blackwell.

Sacks, Harvey, Emanuel A. Schegloff, and Gail Jefferson. 1974. A simplest systematics for the organization of turn-taking for conversation. *Language* 50:696–735.

Schegloff, Emanuel A. 1984. On some gestures relation to talk. In *Structures of Social Action: Studies in Conversation Analysis*, ed. J. Maxwell Atkinson and John Heritage, 266–296. Cambridge: Cambridge University Press.

Schutz, Alfred. 1990. *The Problem of Social Reality. Collected Papers*, vol. 1. Vienna: Springer.

Star, Susan Leigh. 1989. *Regions of the Mind: Brain Research and the Quest for Scientific Certainty*. Stanford: Stanford University Press.

Streeck, Jürgen. 2009. *Gesturecraft: The Manu-Facture of Meaning*. Amsterdam: John Benjamins.

Suchman, Lucy. 1987. *Plans and Situated Actions: The Problem of Human-Machine Communication*. New York: Cambridge University Press.

Suchman, Lucy. 2000. Embodied practices of engineering work. *Mind, Culture, and Activity* 7 (1–2):4–18.

Vygotsky, Lev. 1978. *Mind and Society: The Development of Higher Psychological Processes*. Cambridge, MA: Harvard University Press.

Wertheimer, Max. 1912. Experimentelle Studien über das Sehen von Bewegung. *Zeitschrift für Psychologie mit Zeitschrift für angewandte Psychologie* 61:161–265.

Wittgenstein, Ludwig. 1953. *Philosophical Investigations*. Oxford: Blackwell.

Woolgar, Steve. 1990. Time and documents in researcher interaction: Some ways of making out what is happening in experimental science. In *Representation in Scientific Practice*, ed. Michael Lynch and Steve Woolgar, 123–152. Cambridge, MA: MIT Press.

5   Swimming in the Joint

Rachel Prentice

With whose blood were my eyes crafted?
—Donna Haraway (1990)

A retired gynecologist I know, citing nineteenth-century surgeon William Halsted, said that anything he can see, he can operate on. The statement appears to be self-evident. Indeed, "exposure" is a surgeon's term for interventions that make injury or pathology available to sight and action. But what happens when new technologies reconfigure the relationship of hands, eyes, tools, and patient body? This chapter examines the technical and perceptual skills surgeons deploy as they work to see and to act upon patients' bodies in the operating room. I interrogate examples of open and of remotely mediated surgeries to show how action produces and shapes a surgeon's embodiment, the clinical perception and techniques unique to surgeons. These surgeries exemplify moments when the relationships between action and embodiment come into view, revealing how technology can lead to new perceptual experiences, but also how those experiences emerge from the broad cultivation of a surgeon's craft.

Learning to see in surgery involves the crafting of much more than eyes. Surgical sight emerges from a link between seeing and acting that is so tight that seeing should not slip into the representational language of a medical gaze or disembodied cognition. Rather, sight comes into being with the embodied work surgeons perform when they interact with tools and other bodies in surgery. Sight and touch are intertwined; that is, they "belong to the same world" in each individual's body and "yet they do not merge into one" (Merleau-Ponty 1969, 134). Put more simply, most people can sense what something might feel like when they see it and most can sense what something might look like when they touch it (for example, most of us can sense the roughness of a tree's bark before we touch it). All senses come into play during sensory interactions in ways that typically are taken for granted.

Medical ethnographers writing about anatomical and surgical dissection have described the visual (Good 1994) and representational (Hirschauer 1991) aspects of dissection. Surgeons, fully aware of the physical aspects of their work, often default to the language of mental models. This emphasis on visual and mental aspects of anatomical and surgical dissection downplays the significant ways in which surgeons engage all of their perceptual faculties. Focusing on embodiment allows me to open up how surgeons and trainees at various levels come to acquire surgical means of perceiving and acting, especially perceiving and acting with technological mediation. I show how a surgeon's body must be crafted from practitioners' social and technical actions as they unfold in the specific situation of the operating room.

Anthropologist Charles Goodwin has written about the construction of "professional vision," means of visual knowing unique to individual professions, among archaeologists, police officers, and to a lesser extent anthropologists themselves (Goodwin 1994). Science studies scholars have exhibited increasing interest in scientific, technological, and medical visualizing technologies (Daston and Galison 2007; Dumit 2004; Hacking 1983; Lynch and Woolgar 1990). But they have exhibited less interest in the construction of visual skills unique to scientific, technological, and medical professions. Further, the role of the practitioner's body—its trained senses, its movements—remains understudied as a tool of knowing (though see Myers 2007). This chapter shows how the surgeon's body plays a central role in the construction of the operative site as a three-dimensional space suitable for surgical intervention, regardless of whether the site is open or mediated by surgical visualizing technologies. Treating the body as joining condensed social and physical practice with improvised action in the present allows me to consider continuity and change in embodiment.

My work builds on eighteen months of fieldwork at three academic medical centers in North America, where, among other activities, I observed surgical procedures, interviewed surgeons, and worked with a group that was engaged in building simulation technologies for teaching surgery. I first became aware of the extent of physicians' perceptual training during an anatomy course I took at the start of my ethnographic fieldwork in 2001. My work locates perceptual responses to new technologies within preexisting training and cultural regimes, while simultaneously giving technologies that extend human senses some agency in shaping those perceptual responses. This chapter draws upon the insights of Maurice Merleau-Ponty, whose work explores the perceptual roots of human consciousness. Merleau-Ponty argues that the effects of history and culture upon an individual are real and that they overrun our full awareness of them (see Merleau-Ponty 1964, 1969, 2002). His arguments are deeply consonant with anthropological efforts to unearth taken-for-granted understandings of particular social and cultural traditions. By examining visualizing technologies, their structuring effects, and the perceptual skills needed to use them, I contribute to studies that explore technologies for representing and manipulating scientific and medical objects (Daston and Galison 2007; Hacking 1983).

The examples I recount make clear that minimally invasive technologies foster new relationships between practitioners' and patients' bodies. They also represent a new form of surgical embodiment. I examine two essential aspects of these phenomena: first, how embodied skills and experiences shape surgical perceptions; second, how mediating technologies interact with embodied skills to create new perceptions. While observing surgeries in 2001 and again in 2006, I saw surgeons and trainees put their three-dimensional spatial sense to work during both traditional surgeries and minimally invasive surgeries. The languages of visualization and representation clearly were inadequate for describing such interactions. Neither an open operative site nor an operative site depicted on a monitor is an image the surgeon views. Rather, the interplay of perception and action that "takes place" in the circuit created by the patient's body, the surgeon's body, and the monitor reconstruct the screen as a three-dimensional space the surgeon inhabits.

1 Sensing and Acting

I begin with two examples of open surgeries that reveal the broad embodiment at work in the operating theater. These two moments occurred during the same surgery. The patient was a middle-aged man with a tumor called a Klatskin's tumor at the top of his bile duct. When I arrived in the operating room, Dr. Marcos Alexander, the surgical fellow, and Dr. Jill English, the chief resident, had made a long incision across the abdomen and had retracted ribs and reflected muscles and intestines to reveal the liver.[1] I stood behind the anesthesiologist's drape and looked over at the operative site. While Marcos and Jill worked to expose the tumor, the patient started to bleed heavily into his abdomen. The operative team kept working silently, looking for the source of the bleeding. Jill accidentally rubbed her head against the handle on the overhead lamp and a nurse started to swap out the handle to maintain sterility. "This is not a good time," Jill told the nurse in a monotone. "We've got some bleeding. We need the lamp now." The surgeons had nicked the patient's vena cava, the largest vein in the body, which returns all blood from the body to the heart. Dr. Nick Perrotta, the attending surgeon, told the anesthesiologist to call his chief, saying with typical surgical understatement, "We've got a little bit of a problem."

The vena cava travels between the liver and the rear wall of the abdomen. The large vessel runs deep, at the back of a curved abdominal space that cradles the liver. The upper abdomen was rapidly filling with blood. The surgeons could not clearly see the enormous vein's path to the patient's heart. Nick reached into the cavity, spent a moment exploring the space, then removed his hand and showed Marcos how he believed the vena cava ran. He held his palm upward and pushed his curved hand up and to the left, as though he was following the vessel's path as it ascended into the chest. He instructed the surgical fellow to reach in and feel it. Marcos mimicked him, also tracing a curve to the left. They repeated the same movement of the palm until they agreed that this was indeed the curve the vessel followed. Nick said several times that the curved instrument would have to move to the left and not straight upward, where it would cause damage. Having traced the vena cava virtually, Marcos blindly slipped an instrument under the liver to lift it. Nick stitched the opening and, a very few moments later, they closed the hole and continued the procedure.

These surgeons could not see exactly what they were doing in this space. They could feel the path the retractor would have to take and they could demonstrate using hand gestures that they had felt it. The demonstration had two purposes: to communicate the venous anatomy's path to each other and to rehearse the movements needed to slip the instrument under the liver. The gesture drew upon and captured both surgeons' experience working in this abdominal space. For both surgeons, many years of working within similar abdominal spaces allowed them to sketch this particular patient's abdomen in the air. Nick and Marcos used touch and gesture to make the abdominal space virtually present to their bodies before they literally inhabited the abdominal space to repair the injury. After this gestural practice, they lifted the liver and stitched the cut vessel. The movement involved simultaneously imagining and practicing, learning with their hands. In this case, knowing was based on accumulated practice and gestural signals. Both surgeons understood the vena cava's path with their bodies, but they could not, in the strictest sense of the word, "see" it.

The second moment came later during the same surgery. At a critical moment of this difficult bile duct resection, the anesthesiologist's machines broke down. Blood pressure and other anesthesia monitors extend the patient's body by making it emit signs that speak for the patient (Hirschauer 1991, 290). Particularly during long, difficult operations like this one, monitors tell surgeons and anesthesiologists alike whether the patient's body has destabilized. Low blood pressure is the ideal state for this type of operation, so the anesthesiologist must pharmacologically keep the pressure down while watching to ensure that it does not dip too low. If the pressure drops, the surgeons must step away to give anesthesiologists time to raise it. Late in this long operation, the blood pressure readout plunged. Glancing at it, Jill, the chief resident, asked, "Is this a real number?" The anesthesiologist insisted that the numbers reflected a problem with their machines, not with the patient's body. The team proceeded. Nick asked the anesthesiologists repeatedly if everything was OK. Each time he asked, he placed his hand inside the abdominal cavity and lifted his eyes to the monitors. The anesthesiologists insisted that everything was fine with the patient, while they rushed around trying to get their machines to work. They used a manual backup to ensure that blood pressure was adequate, but the surgeons could not see the numbers. The team completed the last steps of the resection, as well as the rest of the operation, without benefit of a monitor the surgeons could see.

After the operation, I asked Nick for his account of what happened. "I could feel the aorta beating under my hand," he said. Each time he placed his hand inside the abdomen, the strong pulse from the aorta told him that the patient's heart was pumping blood through his body adequately, defying the numbers on the screen. The machine and the patient's aorta told him two different things. "I would have preferred the numbers," Nick said, wanting quantitative proof of what he could feel with his hand. The monitor also could measure the patient's blood pressure with a precision that Nick's hand could not match. Without a working monitor, Nick used his hand and its ability to understand the pulse beating through the aorta, rather than the information provided by the machine. Touch, bolstered by the anesthesiologists' reassurance, told him what he needed to know to continue the operation.

In both moments during this surgery, the surgeons continued to work effectively, despite the loss of direct visual perception. Both surgeons had long since embodied the relevant abdominal anatomy and intervention techniques. Thus, both surgeons' bodies had already synthesized the look, feel, and motions of this region of the body. Merleau-Ponty describes perception as taking advantage of "familiarity with the world born of habit, that implicit sedimentary body of knowledge" (2002, 277). With the implicit knowledge of the patient's body born of carefully honed and frequently practiced techniques of the body, the two surgeons were able to utilize touch, gesture, and language to overcome their inability to see. Seeing for both Nick and Marcos involved tracing the cava's path in gestures and rehearsing the correct movement until both surgeons were satisfied that Marcos could slide the instrument under the liver without doing any damage.

From Foucault onward, writers about medicine often have discussed the medical "gaze," an amalgamation of sensory cues and an organization of medical spaces, logics, and apparatuses of knowing that could tell physicians what they would see if they could open the patient up at autopsy (Foucault 1973). The concept captures the historical shift in the late eighteenth century from diagnosis based on nosologies of disease to diagnosis based on symptoms as presented within anatomical structures. Foucault's concept of the gaze (*le regard*) represents a broad, sociohistorical construction of perception that is quite close to the sense in which I discuss learning to see and act in surgery. But the gaze too easily comes to represent the slippage from vision to thought common to Western philosophical trends since Descartes and Locke (Rorty 1979). The physicians and many technology designers I encountered while doing fieldwork tended to sublimate bodily knowledge under the cognitivist label of "mental model." But, although these surgeons probably have an abstract understanding of the three-dimensional structures of venous anatomy (an understanding typically embodied and reinforced through regular interventions in this region of the body), shifting rapidly into visual or cognitive language elides other aspects of their embodied knowing, such as touch and gesture.

2   Inhabiting Minimally Invasive Space

Removal of a Klatskin's tumor is an open surgery: surgeons access the tumor through a large incision in the patient's abdomen. The remaining surgeries I examine all were done using minimally invasive surgical techniques. Minimally invasive surgery also is known as keyhole surgery, minimal access surgery and, depending on surgical specialty, arthroscopy, laparoscopy, or endoscopy. All minimally invasive techniques involve threading a camera into a natural or artificial hole in the patient's body and performing the work while watching instruments on a monitor. The dozens of techniques surgeons use today began to develop in the 1970s. To perform these techniques, the surgeon inserts a camera and instruments into "ports" or holes in the patient's body. The surgeon operates while looking at a monitor located somewhere nearby. Unlike in traditional open surgeries, the technology distances the surgeon's eyes and hands from the operative site. Putting the action on a monitor is the first, critical move toward surgical simulation, robotic surgery, and other kinds of remote surgical work (Satava 1997, 19).

The perceptual skills needed to work in minimally invasive space differ from those required during open surgery, leading to a new form of virtual embodiment that emerges from a new configuration of bodies and technologies during these surgeries. Surgeons have no direct manual contact with the insides of the patient's body. They cannot use their hands as probes, as Nick did when feeling the patient's aorta. They also have a less direct kinesthetic "feel" for the body as transmitted through the instrument. Further, they must continually extrapolate from a two-dimensional image to an interaction of bodies and instruments in three dimensions, sometimes "constructing" a three-dimensional space by verbally identifying anatomical structures as they appear on the screen.

The differences in embodied skills of beginners, competent practitioners, and experts become very clear, very quickly. Amal Nassif was an earnest first-year resident, just beginning his second month of residency. During the first surgery of the day, he needed careful supervision by nurses and surgeons to ensure that he maintained sterility, a clear indication that he was new to the operating room. Amal had also spent a few hours practicing with simulators, which gave him a basic feel for the instruments involved. Minimally invasive surgery requires surgeons to work over a fulcrum, or pivot point, the way a rower uses an oarlock as a fulcrum. Using instruments this way requires practice because the lever effect reverses the action; that is, one moves an instrument left to push its tip to the right, up to push the tip down, and so on.

While the team scrubbed for a second operation, a gall bladder removal, Dr. Tom Berg, the supervising surgeon, asked Amal if he would like to hold the camera. He eagerly said yes. Once the team had anesthetized and prepped the patient, Dr. Cory Nguyen, the surgical fellow, inserted the camera into a port that she had surgically incised in the patient's abdomen. She handed the camera to Amal, saying, "You keep the buttons up." The camera is a rigid stalk with a lens on one end that is inserted into the patients' body. At the other end is an easily gripped handle with buttons that allow surgeons to zoom in. The camera attaches to a large video monitor and recording deck. Because laparoscopic cameras often are angled, holding the buttons up provides important orienting information to surgeons, though it was unclear whether Amal understood this.

Cory told Amal how to direct the camera so she could see the abdomen from the inside and place several more ports. The "inside" view helped her ensure that she would not place a hole too close to a blood vessel or organ. Cory began to dissect the gall bladder's connective tissues. She told Amal several times to rotate the camera or pull it back to keep steam from the harmonic scalpel and loose material from clogging the lens. After a half hour of silently following directions, Amal said, "So 'up' is looking down?" "Yes," Cory said. Tom added, "From the top down. That's what it means." Amal's phrase is revealing: "up" meant moving the camera's base upward, so the camera tip inside the patient's body pointed downward: seeing down required a counterintuitive movement of hands and instrument. Amal was beginning to understand how the fulcrum effect applied to the camera.

Cory's repeated urging to pull the camera back to avoid clogging the lens indicated that Amal was not yet aware that he was working in a three-dimensional space. That is, he had yet to embody the two-dimensional space on the monitor as a three-dimensional space occupied by instruments under his control. Clearly, he did not know up from down, or near from far, or the rudiments of navigation in the three-dimensional abdomen depicted in two dimensions on the monitor. He had no feel for how the camera related to the inside of the patient's body. The abdomen for Amal was just a two-dimensional image on a monitor. As will become clear from the examples that follow, experienced surgeons do more than look at the image on a monitor. They treat these bodyscapes as three-dimensional spaces they work in, rather than as pictures they look at.

3 "Operating on Images"

My next example is of an experienced surgeon who started working with minimally invasive technology late in his career. Dr. Harry Beauregard, a now-retired gynecologist, began doing laparoscopic abdominal surgeries while peering through a microscopic eyepiece, an earlier generation of minimally invasive visualization. He found the transition to the monitor alienating. During an interview, Harry said:

It was the focus change from the patient to the monitor. That's where the action was and it was something I had to take into account. I mean I had to go there to do the work that the camera illustrated, allowed me to visualize. And so I would go there to work on the monitor. And so I was leaving the patient and looking up to a monitor where there I could do stuff. And with the tools of the minimal access, if I looked at the patient, I couldn't do anything. You see how absolute it was? And I could look at the handles, but couldn't see on the inside. It was totally useless. I had to go to the monitor to operate. And that's why I started saying I was operating on images, not on patients.

Harry described the operative site as though it had moved to the monitor, which, in terms of vision and action, it had. The position of his hands did not change much from the microscopic system. But he talked as though his entire body had moved with his eyes. He experienced himself as no longer working on patients but on images. Vision and action came together on the screen. His hands were outside the patient and the operative site was hidden inside. To see what he was doing, Harry had to look at the monitor. "I had to go there to work," he said.

To better understand this relationship between surgeon and remote technology, I asked Dr. Anna Wilson, an orthopedist, to watch videotapes of a shoulder arthroscopy she had performed and to explain to me what she was doing. Unlike Harry, Anna had performed minimally invasive surgeries since her residency. The tapes depicted the scope's view, the view from inside the patient's body. The patient had torn his biceps tendon years before. The destabilizing effect of the tear made his shoulder joint move improperly, wearing down protective cartilage and encouraging arthritic bone growth. Unlike the view of the abdomen through the scope, the view of the joint's interior was entirely unlike any anatomical view I have seen: the body looked incredibly abstract, like looking through a porthole at a red-and-white undersea floor with white tendrils undulating in the current. The view made me mildly seasick.

As the camera advanced into the shoulder, Anna said, "This is somebody with a terrible shoulder, a terrible shoulder." She described how she was running fluid through the joint, hence the sea floor effect.

Anna: That's the outflow. It's also the cannula for instrumentation in the front. This is not a good first one for you to look at. This is his humoral head and there's just a lot of arthritis, a lot of fibrillation.

RP: So arthritis, it's not like a neat bone growth, it's this messy crap?

Anna: It's messy crap. It's just bare, bare bone. So I'm coming from behind him and the glenoid is on our left and the big ball is on our right. So the camera is with me. It's kind of my view from chest level. So here I'm probing. I'm proving that he's got an arthritic shoulder. This is the remnant of his biceps tendon. This will definitely make you dizzy.

RP: You feel that or you see that?

Anna: I put through the cannula in front. I take the probe. That's my finger extender.

As this dialogue indicates, Anna had several ways of opening the operative site. The first was navigational. As the video advanced, she named anatomical structures, such as the ball of the humerus and the glenoid, or shoulder socket, as they came into view. One reason to share what she sees while operating, Anna said, is to establish common ground with the surgical team. Verbal navigation can be particularly important with minimally invasive surgery because the two-dimensional view can be deceptive and requires skill to read. Navigating a patient's anatomy this way was not something Anna did only while watching a video with an anthropologist. She also did this in the operating room with residents and fellows. Every surgeon I have watched does this with minimally invasive procedures. But I have never seen surgeons do as much narration during an open surgery.

The second method of establishing the operative site was through probing. Anna said the probe extended her finger. I often heard her tell trainees that instruments are extensions of her body. She sometimes struggled to describe in words exactly how she typically holds an instrument because, for her, the instrument becomes part of her body. Anna did not think about the probe. Rather, she used it as an extension of her finger, which was directed toward the arthritis. Probing the arthritis was the important action, not holding the probe. Merleau-Ponty uses the example of a blind man's cane to show how we use instruments to extend our senses—that is, our bodies and ourselves—toward objects in the world (Merleau-Ponty 2002, 165).[2] He argues that the cane extends the blind man's bodily consciousness into space. What remains unstated is that, like a surgeon, the blind man, too, needs years of practice to navigate in his world. Further, tapping does not resemble navigating either by sight or by unmediated touch. Both cane and probe have structuring effects on their respective users' embodiment.

Anna said something that suggests the broader embodiment at work. She said the image was her view "from chest level." This was an odd statement. We do not have eyes in our chests and, thus, have no view from chest level. But Anna located her body in relation to the patient's body. "The camera is with me," she said. She was standing behind the patient's right shoulder. Shoulder and scope were level with her chest. The technology gave a view from chest level. Just as the probe became her finger extender, the scope became her eye extender. Anna extended this technological eye from her body's position in space to the patient's body. Action began with her body and extended from there. It became a view from chest level. This was clearly not yet the case for Amal, the new resident. He could not quite connect what he saw with his eyes with what he did with the hands holding the camera.

I tried to get more detail about what Anna was doing:

RP: So then you feel the arthritis or you see it?
Anna: Yeah, both. It's very much a proprioceptive thing.

Anna proved that the man had an arthritic shoulder by sight and by feel. She verified the arthritis by probing the tissue's hardness. She said her identification of the arthritis was proprioceptive. Proprioception is our sense of our body in space, the sense that allows most of us to know, for example, where our left foot is without looking at it. Anna's statement that the visual and tactile confirmation of arthritis was proprioceptive appears to conflate vision with proprioception. Understood in terms of a narrow, objective definition this would be an incorrect use of the word. But I believe that a more subtle understanding was at work. Anna oriented her body and instruments so she could best see and operate on the patient's body. The connection between sight and action was so tight that vision and proprioception merged. Anna saw and probed the arthritis through her body. She extended herself—her senses and her being—into the operative site to make the diagnosis.

On the video screen, another right shoulder appeared. Anna used a probe to gently flick a small knob of flesh on the shoulder. She described this as "physical doodling." In other words, she was thinking about what she could do with this injury.

Actually what I'm doing also is, I'm externally rotating the shoulder to see the tension of the muscle. [She points.] That's the middle glenohumeral ligament. You can see the glenoid here. And you can see the humeral head over here. This is also a right shoulder. There's a lot of fibrillation coming down and actually this is probably going to be some of the rotator cuff falling in our face. The fibrillation is that gunky stuff. This is the rotator cuff tear.

Anna located us in relation to the anatomy and diagnosed the injury. She identified messy white tissue descending into the image frame as the rotator cuff tear and described the tissue as "falling in our face." This odd grammatical construction suggests several aspects of the embodiment at work. Tissue waving against the camera lens showed up on the monitor as tissue blocking our view. Anna, who sat next to me in a computer lab, placed our faces at the meeting place of tissue with camera, merging both our faces with the technological interface. This suggests that the apparatus of camera and monitor structured her perception: the monitor has just one "face," the camera lens inside the joint. Multiple people can look at the image depicted on the monitor, however, so it became "our face." In other words, Anna located our faces at the interface of the camera with the interior of the patient's shoulder joint. I have heard Anna make statements like this several times, but only while doing arthroscopy, never while doing open surgery. Unlike Harry, she did not experience this as alienating. She tells her residents that she becomes part of a joint when she does arthroscopy, using an analogy to Heisenberg's Uncertainty Principle to argue that her very presence in the joint causes significant changes. Instead of separating operative site from patient, Anna's body merged with the scope as it moved around inside the patient's body. The apparatus became part of her body. It also exerted its own agency in shaping her perception by allowing more than one person to be located in the space created by camera and monitor. Anna located our faces inside the patient's body and on the same scale as the magnified view of the shoulder's interior. As with her view from chest level, Anna placed herself where the technology was. Eyes and instruments merged at the operative site.[3]

Anna's experience of inhabiting the patient's body while doing minimally invasive surgery echoed that of other surgeons. I discussed a similar relationship with Dr. Ramesh Chanda, another orthopedist, who also trained using both open and minimally invasive techniques. What he said is worth quoting at length:

You have an image on the monitor. You have this thing in front of you, which is the actual patient, the patient's joint or whatever. In addition to this, there is also a third image, and that's the image which is in your head. And it's a combination of the two, the patient's image that you see and the stuff you see on the monitor, but also takes into consideration some cognitive aspects, some other issues, the haptic feedback you are getting from your hands. It's a mental model or image or whatever, and what I have felt as I have gone through my training is that I have tended to use that third image more and more, which in some ways draws upon what I am seeing on the screen, draws upon what I am faced with in front of me and am touching and holding and manipulating. So I am almost like, I almost imagine myself, almost routinely if I am doing an arthroscopy, sitting inside the joint. And I say, Oh, OK, I am looking up and I see the scaphoid or whatever, if I am in the wrist joint, for example. And of course the images on the screen are very important [for] guiding, in fact probably the most important. You certainly cannot do without that. But there are other pieces of information and that, in the end, becomes a guide.

Ramesh says he creates a composite bodily understanding of the patient's joint that unites the on-screen visual, the kinesthetic and tactile information coming from instruments and patient's body, as well as his experience and knowledge of anatomy. Anna and Ramesh both stumbled a bit verbally when trying to describe how they navigate patient bodies with minimally invasive technology. Anna said, "It's kind of my view from chest level." Ramesh said, "It's a mental model or image or whatever." Both surgeons are unusually articulate people, but these moments of imprecision reveal how perplexing some of these perceptual issues are. The two surgeons wrestled to describe experiences they have with their bodies. Anna's body merged with the scope; Ramesh dispensed with the technology and its limitations altogether. He inhabited the patient's body when he operated. He said he did not have this experience when doing open surgery.

I encouraged him to say more:

RP: It resonates very strongly with something [Anna] said, which is that when she is operating on the shoulder, she is part of the shoulder.
Ramesh: Yes, that's exactly how I feel.
RP: If you're thinking of yourself as inside the joint, do you actually position yourself, like my eyes are sitting on this piece of anatomy looking at whatever?
Ramesh: Yeah, and actually I would say I am sitting on that piece of anatomy, or rather that you are floating around, swimming around in the synovial fluid, so you can move around, look up, look down, look right, left. And actually the other thing is that you can also, in that mental model, come out of the joint. You can go in and out very easily, so you can visualize it from the outside. You can visualize it from the inside.

Ramesh located his entire body inside the patient's body. One could think of this as the disembodied gaze promised by writers about virtual reality (Balsamo 1996; Gibson 1986). But examining what Ramesh does while swimming in the joint suggests that this formulation is misleading. He looks at a monitor and, often, rotates the joint from the outside while using a probe to examine the internal effects of rotating. He says he experiences himself as sitting or floating inside the joint. Thus, his entire body engages with the joint when he operates. Further, he draws on a history of anatomical and surgical interventions in joints. Ramesh's actions condition his ability to place himself in synovial space. He described this as a synthesis of the view on the monitor with other sensory information, especially kinesthetic information. The perceptual synthesis that Ramesh described also gave him imaginary abilities that were technologically unavailable, such as the ability to move out of and back into the joint at unusual angles. Intriguingly, the perceptual tools he gave himself are exactly the kinds of technologies that he and other simulation and medical imaging experts wanted to develop. These technologies would allow a physician to glide through the patient's body, across membranes and through tissues, as though through water. His imaginary navigation of the body was also an imaginary of technology.

Multiple sensations are in play in minimally invasive surgery, including what the surgeon sees on the monitor, the tactile and kinesthetic sensations transmitted from the instruments to the surgeon's hands, and the surgeon's proprioceptive sense of his or her body in space. These sensations come together with the embodied skills developed in practice, as well as with understanding of human anatomy and surgical procedure. Harry, perhaps because he began using minimally invasive technology later in his career, experienced the patient as split in two when he used it: the image on the monitor and the actual patient's body. He repeatedly stated that he had to leave the patient's body to work on images. Anna and Ramesh, however, did not consider the monitor as such. The scale of their bodies was radically reduced, focused at the meeting place of scope and joint.

During a discussion over coffee, I asked the three surgeons together to speculate about the differences. Harry gave two possibilities. The first related to when in their careers—at the beginning or in the middle—they began working with a monitor; that is, when they began to train with the technology. Harry also suggested that this difference could relate to the size of joints versus abdomens. He said the abdomen is like a large room with darkened corners. It does not feel confined. Anna picked up his metaphor and began to play with it. "A shoulder is like a closet," she said. "Only it's like a California closet where everything should be neatly tucked away." The joint-as-closet analogy creates a strong sense of the joint's confined spaces and the disorder pathology creates. The patient's body itself contributes to these perceptual effects. Ramesh agreed with Harry and Anna and added that surgeons who work in the abdomen do not manipulate the body from the outside while viewing it from the inside. That is, they manipulate instruments, but do not rotate limbs to see how interior structures move as orthopedists do. This suggests that arthroscopy—minimally invasive surgery in joints—more tightly links the surgeon's body and the patient's body than laparoscopy.

Much later, after he read an early draft of this chapter, Harry found another explanation for the differences. He said he is a man who worked on women's bodies. These bodies were unlike his own. The intimacy the orthopedists experience would feel inappropriate, he said. As a gynecologist, he spent his career with hands, instruments, and eyes in intimate contact with women's bodies. But somehow, inhabiting a woman's body would have felt transgressive, suggesting a difference between putting his hands or eyes into the patient's body and putting himself into the patient's body. This suggests that these perceptual relationships also are shaped by gender and cultural experiences. In other words, he objectified his own body as it came into intimate contact with the patient's body.

Thomas Csordas (1990) argues that cultural experiences and habits shape perceptions, making a strong case for historical and cultural forces at work in constructing what counts as an object. All surgeons become habituated to seeking pathological objects in patient bodies. That does not mean that all surgeons experience patient bodies as identical (see Sacks 2003). Technologies help shape a surgeon's work space. Also, cultural concerns about how one relates to patient bodies can habituate the surgeon to experience particular kinds of relations with patient bodies. Similarly, the amount of time a surgeon has practiced in minimally invasive space can shape these experiences, as can the constraints of working in particular anatomical regions. Changes in surgeons' social, institutional, and technical worlds may impact their craft profoundly as they become incorporated into their embodied repertoire of skills.

4   Inhabiting Surgical Space

What do we make of these seemingly bizarre perceptual relationships? If Harry experienced his worksite as the image, is the image "just" a representation? And do these examples represent a complete departure from the bodily relations of open surgery?

To answer these questions, I consider the development of surgical embodiment as entailing a broad perceptual synthesis that builds from years of surgical interactions with patient bodies in the operating room to explain the alterations in a surgeon's location or scale that can occur with minimally invasive surgery. As Harry said, surgical seeing and acting are inextricable. Csordas's (1990) synthesis of Merleau-Ponty and Bourdieu can help explain sensory experiences as they are distributed by remote technologies. Csordas argues that embodiment is the existential ground of culture. He uses Merleau-Ponty's term "pre-objective" to show how cultural formation can shape perceptions before those perceptions coalesce into objects. By this logic, surgeons' medical training shapes how and what they come to perceive as objects. Thus, surgeons' perceptions—their "eyes" in Donna Haraway's shorthand—are crafted within a surgical environment (Haraway 1990, 192).

Sometimes described as primarily visual (Good 1994), anatomical dissection actually is embodied in complex sensory and affective interactions; trainees learn to identify and name parts by opening bodies and locating structures in the body's three-dimensional volume.[4] After taking the five-week summer course and after months of observing in anatomy laboratories and operating rooms, I became aware that my own sense of three-dimensional space had improved dramatically. Like dissection in the laboratory, surgical work has been described as the construction of an anatomical representation in the patient's living body (Hirschauer 1991). This characterization misses the ways in which surgeons intervene in patients' bodies to effect repairs and alterations. In other words, the purpose of surgical dissection is not primarily visual.

The first lessons in the operating room—including scrubbing, maintaining sterility, and obeying the staff—defamiliarize trainees with their own bodies, encouraging them to build a new, surgical stance toward patients and fellow practitioners. Repeated practice of the small actions of surgery, such as retracting and stitching, aggregate and condense to become bodily habits. Years of cultivation of surgical habits leads to surgical skill, a term surgeons use unflatteringly when qualified as technical proficiency alone, but which becomes high praise when incorporated with judgment and knowledge. According to Merleau-Ponty, "habit" is a "rearrangement and renewal of the body image" through which the body becomes "mediator of a world" (Merleau-Ponty 2002, 164). "Skill" connotes the effects of intentional training in ways "habit" does not and can be defined as purposeful habituation that leads to changes in body image. Building from this, "craft" becomes "skill, presence of mind and habit combined" (Mauss 2007, 58). The skilled body thus becomes the body habituated to particular kinds of intentional action through practice.

The accumulation of craft practices makes the body into a temporal joint that articulates past practices and present conditions. "Our body comprises as it were two distinct layers, the habit-body and that of the body at this moment" (Merleau-Ponty 2002, 95). The habit-body develops a generalizable capacity, the ability to do something or, more abstractly, the notion that one *can* do something (see also Dreyfus 1992). Drawing on the experience of amputees, Merleau-Ponty illustrates this with the example of a phantom limb, which joins a body habituated to having a limb to a present marked by its loss (Merleau-Ponty 2002, 88). In contrast, the habit-body that develops through practice in surgery becomes joined to a present in which variations in the milieu, such as new tools, unusual anatomy or pathology, or changes in team composition, generate improvisations that draw upon the general abilities of the surgeon's habit-body. The crafting of a surgeon's body also has a moral component: skill, judgment, and accumulated procedural techniques all qualify the surgeon to undertake this high-stakes activity.

Merleau-Ponty's habit-body corresponds roughly to Pierre Bourdieu's description of how bodily habits can reflect and create culturally conditioned dispositions to act according to "regulated improvisations" (rather than to follow preset rules) (Bourdieu 1977; see Csordas 1990, 1993). But Bourdieu's theme of *habitus* shows more clearly how bodily habits emerge from and shape social interactions: dispositions develop through practice in situations where symbolic, spatial, and social structures instill particular ways of perceiving and acting (Bourdieu 1977). Saba Mahmood makes a stronger case for practice as the embodying force, arguing that practices instill disposi-tions to act and believe in particular ways (Mahmood 2005). I argue that practices that take place within a structuring environment and guided by a respected teacher com-prise an extraordinarily powerful mode of subject formation, constructing an entire perceptual syntax that prepares surgeons to observe with all their senses and to inter-vene in human bodies.

Surgeons cultivate specific habits of perception and thought during years of train-ing. Obviously, a surgeon's visual and tactile perceptions are highly trained. They must learn to identify anatomy in indistinct flesh. They must also distinguish normal from pathological tissues. And, unlike anatomists, they must learn to repair the patient's body while doing as little extraneous damage to it as possible. While surgeons learn their craft in the operating room, the symbolic, spatial, and social structures of the operating room instill particular ways of perceiving and practicing upon the patient's body. Thus, surgeons bring to bear habits accumulated over years to handle clinical problems in the moment. Though these habits partly build upon an accumulation of specific techniques, they also build upon generalized dispositions that allow surgeons to improvise carefully crafted solutions to clinical problems. Technologies such as min-imally invasive cameras and tools may challenge a surgeon's skills, not least because they alter the relations among senses used in surgery. But with practice, surgeons learn to utilize accumulated perceptual skills to adjust to altered perceptual circumstances.

Surgeons learn the body even as they create the body from which they learn. That is, a surgeon learns by repeated crafting of anatomy in a patient's body. When blood, tissue, anatomy, or technological failure disrupt sight, the surgeon draws upon other senses to engage with the patient's body. When technology distributes the patient's insides and outsides, the equally distributed surgeon reunites it through the circuit of his or her own body. Harry, Ramesh, and Anna demonstrated this. Harry talked as though his hands followed his eyes to the monitor. Anna talked about arthritis falling in our collective face. And Ramesh described himself swimming in the joint. These surgeons' entire bodies were focused on the operative site, where seeing and acting come together.

So why do these surgeons describe such strange changes of location and scale of their bodies when doing minimally invasive surgery? Merleau-Ponty writes, "My body is where there is something to be done" (Merleau-Ponty 2002, 291). In surgery, the operative site is where the action is. The body's sense of space is brought into being through action. This is not a spatiality of position, but a spatiality of situation (Merleau-Ponty 2002, 115). When spatial perception is radically altered by technology, such as an apparatus that distorts vision, making the world appear at a 45-degree angle, the body creates a perceptually altered "virtual body" that, according to Merleau-Ponty, "ousts the real one to such an extent that the subject no longer has the feeling of being in the world where he actually is . . . he inhabits the spectacle" (Merleau-Ponty 2002,

291). For Merleau-Ponty, the "spectacle" is the world as it presents itself to our perceptions, which differs from the world as it might be described objectively. For a surgeon, the operative site as it presents itself to the surgeon's senses and interventions becomes the spectacle. Thus, the operative site, whether it is experienced on a screen or in the flesh, becomes a space the surgeon inhabits.[5]

The situation of surgery is to see and to act at the operative site. From an external, objective point of view, the surgeons' perceptions of themselves operating on images, or waving arthritic tissues out of their faces, or swimming inside the joint appear bizarre. And the surgeons themselves, if asked about the actual positions of their bodies in the operating room, would not describe their locations in this way. Instead, they would objectify their actions, describing themselves as an observer would. But these statements become clearer if we imagine surgical embodiment as developing from lengthy residence in a surgical culture dedicated to the art of seeing to intervene and intervening to see. The distributed bodies, instruments, sensations, and knowledges all focus on a single event: opening the operative site so the surgeon can work there. The surgeon's body unites sight, action, and technology. Harry located himself at the monitor so he could see enough to work. But he experienced his attention as divided between patient and monitor. Anna and Ramesh took a more radical step. They located their bodies in the one place where a person could see and operate without being divided—at the actual operative site, inside the joint.

Notes

This chapter is abridged from Rachel Prentice, *Bodies in Formation: An Ethnography of Anatomy and Surgery Education* (Durham: Duke University Press, 2013).

1. In this chapter, names of individuals and institutions have been replaced by pseudonyms.

2. The blind man's cane has a long history in philosophy that begins with Descartes and includes Heidegger and Polanyi. According to Descartes in his *Discourse on Optics*, the cane has one set of properties (long, thin) when described as an object, but becomes something quite different when used as an extension of the blind man's body. This distinction has been used to explore the difference between objective description of an object and subjective experience and use of an object.

3. For a fascinating discussion of perceptual changes that occur with technological mediation, see Reeves and Nass (1996).

4. For more information on the emotional training that takes place in the anatomy laboratory, see Prentice (2013).

5. In his important chapter on "space," Merleau-Ponty describes several experiments in which subjects' perceptions were altered fundamentally, for example by putting on glasses that turn retinal vision "right side up" (images received on the retina are "upside down" and the brain "corrects" them later in the process of seeing). Over the course of a week, subjects' perceptions

adjusted, normalizing the perceptual change, especially when the subject was in motion. These experiments reveal the body's orientation to motion (action) and its ability to construct "the virtual body" Merleau-Ponty describes, allowing the subject to "inhabit the spectacle" (2002, 291) as if it were the objectively presented world.

References

Balsamo, A. 1996. *Technologies of the Gendered Body: Reading Cyborg Women*. Durham: Duke University Press.

Bourdieu, P. 1977. *Outline of a Theory of Practice*. Cambridge: Cambridge University Press.

Csordas, T. J. 1990. Embodiment as a paradigm for anthropology. *Ethos* 18:5–47.

Csordas, T. J. 1993. Somatic modes of attention. *Cultural Anthropology* 8:135–156.

Daston, L., and P. Galison. 2007. *Objectivity*. New York: Zone Books.

Dreyfus, H. 1992. *What Computers Still Can't Do: A Critique of Artificial Reason*. Cambridge, MA: MIT Press.

Dumit, J. 2004. *Picturing Personhood: Brain Scans and Biomedical Identity*. Princeton: Princeton University Press.

Foucault, M. 1973. *The Birth of the Clinic: An Archaeology of Medical Perception*. New York: Vintage Books.

Gibson, W. 1986. *Neuromancer*. New York: Ace.

Good, B. J. 1994. *Medicine, Rationality, and Experience: An Anthropological Perspective*. Cambridge: Cambridge University Press.

Goodwin, C. 1994. Professional vision. *American Anthropologist* 96:606–633.

Hacking, I. 1983. *Representing and Intervening: Introductory Topics in the Philosophy of Natural Science*. Cambridge: Cambridge University Press.

Haraway, D. 1990. *Simians, Cyborgs, and Women: The Reinvention of Nature*. New York: Routledge.

Hirschauer, S. 1991. The manufacture of bodies in surgery. *Social Studies of Science* 21:279–319.

Lynch, M., and S. Woolgar. 1990. *Representations in Scientific Practice*. Cambridge, MA: MIT Press.

Mahmood, S. 2005. *The Politics of Piety: The Islamic Revival and the Feminist Subject*. Princeton: Princeton University Press.

Mauss, M. 2007 [1935]. *Techniques of the Body*. Ed. M. Lock and J. Farquhar. Durham: Duke University Press.

Merleau-Ponty, M. 1964. *The Primacy of Perception, and Other Essays on Phenomenological Psychology, the Philosophy of Art, History and Politics*. Evanston: Northwestern University Press.

Merleau-Ponty, M. 1969. *The Visible and the Invisible*. Evanston: Northwestern University Press.

Merleau-Ponty, M. 2002. *Phenomenology of Perception*. London: Routledge.

Myers, N. 2007. Modeling proteins, making scientists: An ethnography of pedagogy and visual cultures in contemporary structural biology. Ph.D. diss., History and Anthropology of Science and Technology, MIT.

Prentice, R. 2013. *Bodies in Formation: An Ethnography of Anatomy and Surgery Education*. Durham: Duke University Press.

Reeves, B., and C. Nass. 1996. *The Media Equation: How People Treat Computers, Television, and New Media Like Real People and Places*. Cambridge: Cambridge University Press.

Rorty, R. 1979. *Philosophy and the Mirror of Nature*. Princeton: Princeton University Press.

Sacks, O. 2003. The mind's eye: A neurologist's notebook. *New Yorker* 79:48.

Satava, R. M. 1997. Virtual reality for medical application. Information Technology Applications in Biomedicine, 1997. ITAB '97. Proceedings of the IEEE Engineering in Medicine and Biology Society Region 8 International Conference, 19–20.

6   Chalk: Materials and Concepts in Mathematics Research

Michael J. Barany and Donald MacKenzie

1   Chalk in Hand

Chalk in hand, his formulas expressed themselves, it seems, more easily on the board than they were able to with pen in his notebooks, for in his listeners' presence his fecund genius found again a new zeal, and a ray of joy illuminated the lines of his face when the proof he sought to render understandable struck his audience as obvious.[1]

So recounts an admiring biographer the pedagogical exploits of Augustin-Louis Cauchy, a towering figure of early nineteenth-century mathematics. Cauchy was trained and then became professor of analysis at the prestigious Ecole Polytechnique, a school for military engineers that not long before Cauchy's matriculation had become one of the first to make systematic use of a new mode of advanced mathematical instruction: lessons at a blackboard. Today, chalk and blackboards are ubiquitous in mathematics education and research. Chalk figures prominently in the imaginations and daily routines of most mathematicians.

For Cauchy's biographer, there was an organic link between chalk, genius, audiences, mathematical proof, obviousness, and understanding. This link persists to this day. There is, we contend, an essential relationship between the supposedly abstract concepts and methods of advanced mathematics and the material substituents and practices that constitute them. This process operates even in the rarefied realm of mathematical research, where the pretense of dealing purely in abstract, ideal, logical entities does not liberate mathematicians from their dependence on materially circumscribed forms of representation. That this self-effacing materiality is often unnoticed (unlike the visible and controversial materiality of computerized mathematical proof analyzed by MacKenzie 2001) makes the case of research mathematics all the more important to the social study of theoretical representations. Indeed, the very appearance of scholarly mathematics as a realm apart is a social achievement of practices that produce mathematical ideas using material surrogates.

This chapter reports a series of ethnographic findings centered on the theme of chalk and blackboards as a way of illustrating the distinctive modes of inscription underlying mathematical research. Chalk, here, functions both as a metaphor and as a literal device in the construction and circulation of new concepts. We begin, after a brief review of extant literature, by describing the quotidian contexts of such work. We then explore the blackboard as a site of mathematical practice, before finally expanding on its metaphorical and allusive significance in other forms of research.

Our observations have a dual character. On the one hand, we describe the supposedly distinctive realm of mathematics in a way that should appear consonant with other scholarly disciplines that one might imagine to be rather different from it. Observations that would be "old news" about other sciences or unsurprising to those acquainted with mathematical practice are nevertheless significant in a context where so few investigations of the sort we report here have been undertaken. On the other hand, we aim to account in some small way for the distinctiveness of mathematics, both as a field of study with its own characteristic objects and practices and as a domain that succeeds in appearing far more distinctive than we would suggest is actually the case.

In our account, the formal rigor at the heart of mathematical order becomes indissociable from the "chalk in hand" character of routine mathematical work. We call attention to the vast labor of decoding, translating, and transmaterializing official texts without which advanced mathematics could not proceed. More than that, we suggest that these putatively passive substrates of mathematical knowledge and practice instead embody potent resources and constraints that combine to shape mathematical research in innumerable ways.

2 Prior Accounts

This conclusion, developed through Barany's recent ethnographic study of university mathematics researchers,[2] builds on related literatures in the sociology and history of logic and the natural sciences, the history of mathematics, and the sociology of settled mathematics. Closest in methods and analytical orientation is a range of historical and ethnographic accounts of university researchers in "thinking sciences" such as theoretical physics,[3] artificial intelligence,[4] and symbolic logic.[5] These accounts collectively demonstrate how intersubjective resources are mobilized and disputed in the production of abstract accounts of physical, social, or logical entities. Their concern for the connection between theories and their means of articulation draws from early laboratory studies that documented the practical achievement of circulable data and principles of scientific knowledge through the use of instruments and other means of "inscription" or "rendering" that tame and transform unruly specimens.[6]

Two bodies of scholarship help us to adapt the foregoing insights to our study of mathematics. Historians and some empirically minded philosophers have used a variety of frameworks to trace the elaboration of specific mathematical theories and techniques.[7] Sociologists, meanwhile, have described mathematical pedagogy at many levels,[8] elementary proofs and examples,[9] and (less often) advanced theorems,[10] detailing in each case the modes and means of making already-established mathematical ideas intelligible. Some take an explicitly cognitive approach[11] and stress the mental and corporeal structures that ground mathematical thinking.

Most users of advanced mathematics, and indeed most mathematicians themselves, spend most of their time dealing with settled mathematics. This is the mathematics of teaching and of many forms of problem solving, even when these require deploying accepted results and methods in new ways, and it has generally proven amenable to social and historical analysis. Due to the obfuscations of temporal distance and conceptual difficulty, however, historians and sociologists of mathematics have struggled to account for the ongoing achievement of original knowledge in a research context, such as has been ventured for laboratory sciences. At present, those wishing to understand the core activity in most mathematicians' aspirations and self-identity must rely on accounts by mathematicians themselves or philosophically oriented treatises on the subject.[12] While we cannot pretend to fill this lacuna, our study offers a model for how such an account might proceed.

3   Mathematics in Action

On Mondays during term, members of the Analysis Group return from lunch and assemble to hear a local or invited colleague's hour-long presentation on the fruits and conundrums of his or her[13] recent and ongoing scholarship. These lectures are marked by a shared specialized vocabulary and expertise and sometimes-spirited outbursts of discussion over technical details. One gets the impression, however, that the specific mathematics of the presentation is of at best marginal interest to most of the gathered audience. Some jot notes or furrow their brows, but one is just as likely to see someone nodding off to sleep as nodding in agreement. Most audience members regard the speaker with a brand of reserved attentiveness that is easily mistaken for comprehension.

Lurking in the seminar's subtext and between the lines of multiple interviews was the open secret that mathematicians—even those in the same field, working on the same topics, or veterans of multiple mutual collaborations—tend to have comparatively little idea of what each other does.[14] Mathematics is a staggeringly fragmented discipline whose practitioners must master the art of communicating without co-understanding. Indeed, mathematicians seem persistently preoccupied with sharing their work with each other, boldly blinding themselves to the petty incommensurabilities of their studies in order to join, on scales ranging from meetings with collaborators to international congresses, in mutual mathematical activity.

Seminar performances are conditioned on a form of understanding whose pervasive presence and role in mathematical education and research stands in stark contrast to its minor role in extant accounts of mathematical proof and cognition. Most in the seminar audience do not aim for a detailed working knowledge of the results being presented—this can take years to acquire (after which the talk would not have much to offer)—but rather comprehend the talk in the sense of *following the argument*, engaging with the talk's conceptual narrative and technical and heuristic manipulations.

This "following" mode is reflected in how both speakers and participants prepare for the seminar—which is to say, in large part, how they do not prepare. Audience members do not typically study for upcoming talks by looking into the speaker's topic or previous work. Seminar goers are easily bored and prone to distraction, said one informant, adding that they rarely care in any event about the details behind the speaker's findings. Speakers indicated that their preparations, depending on the formality and importance of the occasion, ranged from "exactly four minutes" (an underestimate, but not a wholly misleading one) to a week of sporadic effort. For a chalk lecture, a single draft of highly condensed notes suffices.

Nearly all of the speaker's words and a varying but typically large portion of what is written on the blackboard during a seminar are produced extemporaneously. Speakers are expected to produce written and oral expositions with limited reference to notes, which serve primarily to help the speaker to recall precise formulations of nuanced or complex theorems or definitions. One result of the speaker's lack of premeditation regarding inscriptions is the frequent need to adjust notations mid-lecture—notations which do not necessarily correspond to the ones used in the limited paper notes the speaker had prepared.

Talks are not, of course, pulled from thin air. Rather, they rely on mathematicians' skill, honed through years of teaching, presenting, and interacting with colleagues, in constructing an argument at the board from a collection of principles and conventions. These arguments are built out of shared rhetorical scripts and graphical representations, practiced over many years and in many settings, that govern how commonly used ideas and methods are described and inscribed in mathematical discourse.[15] Those conventions also connect chalk writing to speaking, so that those who make a record of the talk tend only to transcribe text from the board, making comparatively few notes from the spoken component of the presentation.

Seminars are thus conditioned on a great deal of shared training in discursive and conceptual norms. Typically, however, the speaker's and audience's expertise and interests align only superficially. As one speaker put it: "it's not clear that there's anything in the intersection of what this person's thinking of and what I know how to do." But the seminar is far from pointless. "It's a bit like a beehive," the same speaker volunteered a few days before his talk: "Collecting nectar and pollen doesn't benefit the specific bee so well, but it's important for the community."

Indeed, seminar attendance is among the chief manifestations of the Analysis Group as a community. During the lecture, speakers constitute other communities as well by framing their research in terms of recognizable problems and approaches. These larger communities, organized around particular lines of expertise, structure the kinds of reasoning that can be implicated in a seminar's "following" activity. Researchers develop expectations about arguments, so that, as one explained, "If the argument is sort of well-established, . . . it can be the case that people know where it's going to break if it's going to break." Specialisms also supply canonical terms and arguments, dictating what claims can be made (and how) without further justification.

Specialized (and sometimes specialism-specific) ways of describing objects and rendering them on blackboards and other media are enculturated through attending and presenting lectures: "you somehow learn how to talk," explained an experienced speaker. Seminar presenters pepper their talks with remarks about "what everybody calls" certain objects, or citations of "some standard assumptions," and note standard approaches even when not using them. Speakers cite historical authorities in relevant subject areas and refer to colleagues (including some present at the seminar) to personalize these allusions. These references to people and concepts work to dissolve temporal as well as professional boundaries. In an interview, one junior researcher spoke undifferentiatedly of insights from a senior colleague gleaned, respectively, from a conversation the previous week and from a body of that colleague's work from more than two decades prior. So, too, do old and new theorems and approaches coexist in a seamless technical matrix on the seminar blackboard, thereby enacting an epistemology of mathematics that actively looks past the context-specificities of its concepts.

Like the neuroscientists studied by Lynch (1985), subjects for this study organized and narrated their research activity according to various projects.[16] Subjects typically maintained three active projects concurrently, often with many more investigations "on the shelf." Projects were distinguished by their set of collaborators, their animating questions, and the "tools" or methods they employed. Their progress was marked in researchers' accounts by the gradual reification and conquest or circumvention of barriers they classified as conceptual or (less often) technical. Projects rarely end decisively, but can be disrupted by the relocation of a collaborator, stymied or made obsolete by other researchers' results, or stalled in the face of particularly stark conceptual barriers. When a suitable partial result is obtained and researchers are confident in the theoretical soundness of their work, they transition to "writing up." Only then do most of the formalisms associated with official mathematics emerge, often with frustrating difficulty. Every researcher interviewed had stories about conclusions that either had come apart in the attempt to formalize them or had been found in error even after the paper had been drafted, submitted, or accepted. Most saw writing up as a process of verification as much as of presentation, even though they viewed the mathematical effort of writing up as predominantly "technical," and thus implicitly not an obstacle to the result's ultimate correctness or insightfulness.

Seminars have a special place in the temporal organization of mathematics research. For presenters, presentations can drive the writing-up process by forcing the speaker to cast recent results into a narrative that can be used in both talks and papers, one that mobilizes both program and project to construct an intelligible account of their work (cf. Ochs and Jacoby 1997). Preparing a piece of work for public consumption requires the imparting of an explanatory public logic in which ideas develop according to concrete and recognizable methods. Seminars force researchers to articulate their thinking in terms of a series of significant steps, unavoidably changing the order and character of that thinking in the process by forcing it to conform to a publicly viable model or heuristic. Finally, the members of the seminar audience join—through the facts and circumstances of their presence—in the constitution of a shared public logic that frames their own projects in turn.

Thus, the "following" that takes place in the seminar and extends to other areas of mathematical communication consists of more than a mere sequential comprehension of inscriptions and allusions. "Following" structures the production and intelligibility of entire programs of mathematical research, as well as of the communities that engage in those programs. These entities are built along figures of time and topic that underwrite the directed pursuit of new mathematics.

4   An Ostentatious Medium

We have just depicted a seminar room subtly suffused with concepts and allusions, but these invisible entities arise as little more than facile shorthands for what takes place in the seminar. Rather than treating mathematical communication as a trading zone for airy intellections, we aim to describe it in terms of the pointings, tappings, rubbings, and writings that more manifestly pervade our subjects' work.[17] In the seminar, these material constituents of mathematics are concentrated around the person of the speaker and the physicality of the blackboard.

There is nothing about the blackboard that is strictly necessary for a mathematician. There are other means of writing equations for personal or public display; other tokens on which to hang one's disciplinary hat. Outside of the seminar room, blackboards play a relatively limited (which is not to say insignificant) role in most mathematicians' daily work. The stereotype of the chalk-encrusted mathematician is nearly as misbegotten as that of the mathematician lost in his own mental world.

Nevertheless, mathematicians return to the blackboard. Introduced in its present form as a large surface for pedagogic chalk writing near the turn of the nineteenth century, its status as an iconic signifier for the discipline is no accident.[18] Blackboards dominate mathematics in two crucial spheres: the classroom and the seminar. It is with blackboards that young mathematicians learn the ins and outs of their art, and it is on blackboards that established scholars publicly ply their newly minted findings. These twin settings enshrine blackboard mathematics as an exemplary model that pervades all of mathematical practice, despite the blackboard's ever-growing appearance of obsolescence. Mathematics marked in dust, ink, and electrical circuits alike owes key features of its form and content to the venerable blackboard.[19] It matters little that the full measure of the blackboard's glory is confined to the narrow environs of its two principal uses. In the pregnant space between chalk and slate there reposes a germ of the bursts of inspiration, triumphs of logic, and leaps of intuition that dominate mind-centered accounts of mathematics.[20]

As components of the mathematics department's physical infrastructure, blackboards are most prominent in seminar rooms and lecture theaters. There, multiple boards are typically arranged to span the front of the room, sometimes in sliding columns that allow the speaker to move the boards up or down for writing and display (figure 6.1). Blackboards are also found in the tea room used by faculty and graduate students and in individual professors' and shared student offices.

Even as blank slates, blackboards are laden with meaning. As topical surfaces of potential inscription, they define the spatial outlay of lectures and tutorials, guiding

Figure 6.1
Blackboards, like these at the front of the Analysis Group's seminar room, offer mathematicians what we call an "ostentatious medium"—a robust, visible, and copresent substrate that orients the communal exposition and sedimentation of mathematical concepts both during and outside of the seminar by making certain kinds of writing, erasing, gesturing, and interacting possible and meaningful.

audience members in their choice of seats and occasionally demanding that the room be reconfigured to improve the board's visibility.[21] They presage the seminar's rhythm, its steady alternation of marking, talking, moving, and erasing. They are perpetually at hand: even in conference talks, whose frenetic pace tends to preclude blackboard exposition, they are occasionally mobilized to expand on a point missing from a speaker's prepared slides; in the tea room, conceptual discussions sometimes find their way to the room's otherwise rarely used boards; in offices, boards serve as notepads for nonmathematical ephemera (such as telephone numbers) in addition to mathematical jottings.

More features appear when blackboards are in use. They are big and available: large expanses of board are visible and markable at each point in a presentation, and even the comparatively small boards in researchers' offices are valued for their relative girth. Blackboards are common and copresent—multiple users see blackboard marks in largely the same way at the same time. They are slow and loud: the deliberate tapping and sliding of blackboard writing forces the sequential coordination of depiction and explanation at the board, pacing and focusing speaker and audience alike. They are robust and reliable. And, as noted above, they are ostentatious—so much so that colleagues in shared offices expressed shyness about doing board work when office-mates are present.

As a semiotic technology, the blackboard is as much a stage as a writing surface. That is, boards constitute spaces for mathematical performances that are not reducible to the speaker's chalk writing. Speakers frequently dramatized particular mathematical phenomena, using the board as a prop, setting, or backdrop.[22] Most seminar gestures, however, index rather than indicate mathematical phenomena, exploiting the spatial configuration of the blackboard to organize concepts and settings. That is, rather than *depict* particular phenomena such as taking limits, tracing paths, or comparing magnitudes, the vast majority of observed gestures *pointed to* those phenomena's past or present physical location in the blackboard record of the foregoing exposition—indexing place rather than indicating properties or procedures.[23] Proofs are explained with reference to their initial assumptions by pointing at or tapping boards filled with lists of conditions, which are typically placed at the tops of boards even when space remains at the bottom of the board at which the speaker had been writing.

When an argument is invoked for the second time in a lecture, the speaker's hand can trace its earlier manifestation from top to bottom as a substitute for rereading or rewriting it. A question from the audience frequently prompts the speaker to return a previously worked sliding board to its position at the time of its working in order to answer queries about the writing thereon, even if no additional marks are then made. It is not uncommon to see the speaker's eyes casting about the board for an earlier statement before deciding how to proceed with the next. On multiple occasions, a speaker gestured at a particular statement's former place on the board even after it had been erased, rather than reproduce the statement in another part of the board for the purpose of referring to it.[24]

Specific board locations can carry mathematical significance. Parts of an expression can be separated visually, and corresponding terms are often aligned or written over each other, even when this requires the writer to sacrifice some of the marks' legibility. For instance, when a new bound was introduced for an analytic expression, many speakers simply erased the bounded expression and contorted their writing so that the new bound would fit in its place. Similarly, when a proof hinged on the proper grouping or regrouping of terms in an expression, speakers exaggerated the physical spacing between the relevant terms when writing them. Thus spatialized, statements can be mobilized or demobilized by emphatic or obfuscatory gestures. Multiple speakers, for example, mimed erasing an expression or simply blocked it with their hands in order temporarily to exclude it from consideration or to show that an explanation strategically ignores it.

And what of the marks themselves? One rarely thinks of what *cannot* be written with chalk, a tool that promises the ability to add and remove marks from a board almost at will. The chalk's shape, its lack of a sharp point, and the angle and force with which it must be applied to make an impression, all conspire to make certain kinds of writing impossible or impractical. Small characters and minute details prove difficult, and it is hard to differentiate fonts in chalk text. Board users thus resort to large (sometimes abbreviated) marks, borrow typewriter conventions such as underlining or overlining, or employ board-specific notations such as "blackboard bold" characters (e.g., $\mathbb{Z}$, $\mathbb{R}$, and $\mathbb{C}$) to denote specific classes of mathematical objects.

Not every trouble has a workaround. Similar to a ballpoint pen or pencil on paper, chalk must be dragged along the board's surface to leave a trace. Entrenched mathematical conventions from the era of fountain pens, such as "dotting" a letter to indicate a function's derivative, stymie even experienced lecturers by forcing them to choose between a recognizable dotting gesture and the comparatively cumbersome strokes necessary to leave a visible dot on the board.

The consequences of chalk for mathematics are not just practical but ontological and epistemological. As Livingston (1986, 171) observes, mathematical proofs are not reducible to their stable records. Arguments are enacted and validated through their performative unfolding—an unfolding as absent from circulable mathematical texts as it is essential to the production and intelligibility of their arguments. Like the proofs it conveys, blackboard writing travels only through rewriting. Unlike the marks in books, papers, or slides, blackboard inscriptions can only ever unfold at the pace of chalk sliding against slate. The intrinsic necessity of bit-by-bit unfolding in mathematical exposition is thus built into chalk as its means of writing.

This unfolding matches the "following" mode discussed above, and extends to the audience's listening practices. Few audience members took notes during the Analysis

Group's seminar. Most who did made only an occasional jotting of a theorem or reference to pursue afterward. But those who did take extensive notes endeavored to make a near-exact transcription of what the speaker wrote on the board, reproducing a routine practice from their early mathematics coursework and training. The expectation of transcription obliges the speaker to make the board's text self-contained and accountable, leading to a striking duplication of effort between writing and speech whose epitome is the stereotypical speaker who reads his talk off the board as he writes it. The practice of "following" thus impinges both on the global narrative of the talk and on the textual subnarrative confined to the speaker's marks on the board.

The mutability of blackboard writing, moreover, enacts a specifically Platonist ontology of mathematics. In this view, mathematical objects and systems have an independent existence that is separate from their descriptions, and the same entity can be described in a variety of ways. On a blackboard, lecturers frequently amend statements and definitions about mathematical entities as their specific properties and constraints are made relevant by the exposition or by audience interrogation. In such a medium, the fact that the once-written text does not tell the final story about a mathematical concept allows a potentially infinite variety of descriptions simultaneously to apply to an object or situation under consideration. Where Suchman (1990, 315) and Suchman and Trigg (1993, 160) depict the board as the medium for making objects concrete, we would stress the board's corresponding ability to make those concepts mutable without threatening their persistence as Platonic entities. Thus, when a speaker returns later to add a necessary condition to a definition or theorem statement, it can be seen as an omission rather than an error in the speaker's argument—the condition can be made to have been there all along at any point where that anteriorized conceptual vestment is required for the lecture to go forward.

The logic of blackboard writing governs misstatements as well as omissions. When a speaker reconsidered a statement and deemed it false, the offending marks could be rubbed out without incident, preserving the veracity of the blackboard record. The dusty traces of the statement's removal cue those few in the audience taking notes by pen or pencil as to which items have been removed so they can appropriately modify their own transcripts. In other situations, a statement was not necessarily judged false but, usually after an audience enquiry, was judged to be either misleading or beyond the scope of the presentation. In these cases, the speaker could cross out the statement, removing it from the accountable portion of the talk but preserving it among the lecture's mathematical residues.

The availability of different modes of erasure also has narrative consequences. Minor corrections can be made using the side of one's hand to erase small areas of the board while producing an audible thud that preserves the ongoing sequence of words and board sounds in the speaker's story. Larger erasures, however, must be made with a separate instrument whose use requires the interruption of such discursive sequences—a desirable effect at the end of a planned segment of a talk and an appropriate one where the speaker must "reset" an argument after a significant lapse. The narrative break of clearing a board establishes a board sequence division that holds even when a new board is available. Before embarking on a new part of an argument, speakers sometimes clear multiple boards to avoid having to erase one in mid-sequence. Conversely, if a narrative sequence overruns its allotted board space, the speaker sometimes squeezes the remaining text in blanks on the current board rather than moving to a new one, often at the cost of legibility.

A final point interweaves the ontological, epistemological, and practical significance of blackboards. In seminars and offices alike, blackboards are used and experienced as places for translating complex, symbol-intensive ideas into a manipulable, surveyable form. Figure 6.2 shows an office board that had been used to work out a complicated expression from a published paper. The board shows evidence of insertion, annotation, and erasure. At the top, the researcher started to frame his ensuing writing by singling out the expression from the paper that he aimed to comprehend, labeling it with "To show." The expression of interest, the researcher realized in the midst of copying it out, was not so far removed from the chain of reasoning used to demonstrate it, so

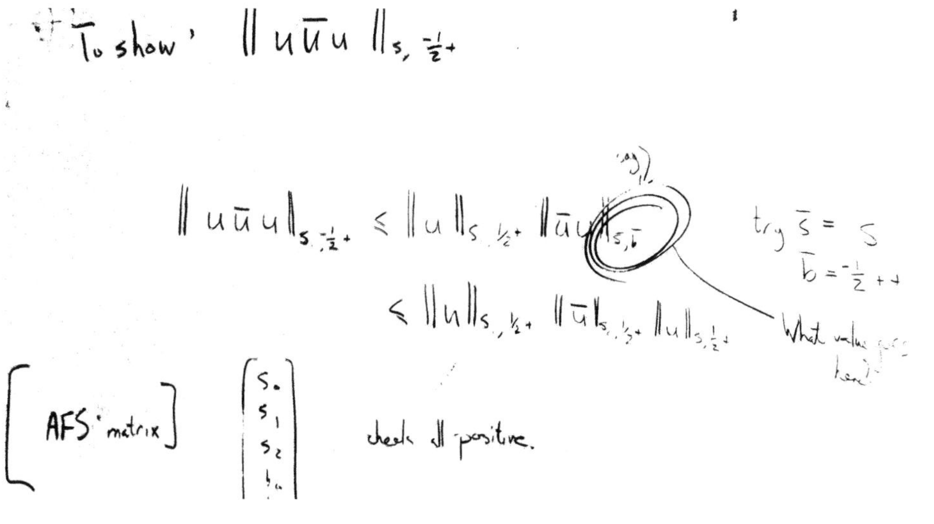

Figure 6.2
In a mathematician's office, the blackboard becomes a site for rendering mathematical expressions in a manipulable and surveyable form distinct from the tidy formalisms of published mathematics. The blackboard markings here (digitally reduced, for the purposes of legibility, to black and white and inverted so that white chalk marks on the board appear as black marks on a white background) show insertion, annotation, and erasure from the researcher's attempts to comprehend and make use of a published result.

he moved to the center of the board and wrote (in appropriate shorthand) the entire chain of reasoning. Here, as he described it, the challenge was not to grasp a particularly complex series of manipulations, but rather to understand a complicated array of indices as a whole.

After copying what he identified as the relevant expressions from the paper, the researcher proceeded to annotate it in terms of questions that would need to be satisfied for the chain of reasoning to be valid and in terms of techniques that could answer those questions. On the right of the expression, the researcher attempts to aid his understanding by specifying more features of the calculation than are present in the more general form in the paper, qualifying this specification with the note "say." Through this blackboard work, a supposedly abstract datum of certified knowledge becomes a self-identical yet pliable chalk instantiation. We were told that only in this latter form could the researcher comprehend and hope to use that expression, and yet that very form and all its advantages were stuck, for all practical purposes, on the board.

5   Proofs and Reformulations

A dominant theme in sociological accounts of laboratory sciences is the remarkable amount of labor and machinery—in Lynch's (1990, 182) formulation, taken-for-granted "preparatory practices"—devoted to producing texts which can materialize and stabilize unruly natural phenomena in the form of data, plots, and other representations—what Latour (1990) called "immutable mobiles." Mathematicians face, in a sense, the opposite problem: the phenomena they study are not unruly enough. Mathematicians thus spend remarkable amounts of labor to materialize their objects of study, but with the goal of coaxing those objects to behave in some new way, rather than disciplining them to hold some stable and circulable form.

There are thus two fundamentally different kinds of mathematical texts. There are papers and reports akin to journal articles in the natural sciences, but there are also tentative, transitory marks that try to produce new orders out of old ones (with a crucial stage of disorder in between). Blackboards, we have suggested, are the iconic site of this second sort of text making. Like the natural phenomena scientists try to tame, blackboard writing does not move well from one place to another. This spatial fixity contrasts with the flexibility afforded by blackboard writing's seemingly vast openness to annotation, adaptation, and reconfiguration. Symbols and images can be erased, redrawn, layered, counterposed, and "worked out" on the board's surface. Such "immobilized mutables" form a constitutive matrix for mathematical creativity.

This "blackboard" way of working with texts is not, therefore, limited specifically to board writing. Asked, while away from his office, to describe his work space, one interviewee began with the piles and piles of paper covering his desk (figure 6.3).

Figure 6.3
Creative research, in mathematics, relies on the researcher's ability to transform mobile and stable formalisms into transient and context-bound representations, like those on a blackboard, from which new results can be formulated. A mathematician's desk thus reflects the crucial and often-disordered mingling of articles, books, notes, and scrap paper, the last two of which are used in remarkably blackboard-like ways.

Populating those piles are reprints of articles, teaching notes, and, most importantly, page after page of scrap paper. The inscriptions of mathematical research, while implicating blackboards, whiteboards, computers, and other media, seem mostly to subsist in the sort of notes that suffuse the spare sheets of paper from our respondent's desk.

Scrap paper writing shares many characteristics with chalk writing. Both rely on augmentations, annotations, and elisions as concepts are developed through iterated inscriptions designed to disrupt the formal stability of mathematical objects. Such iterated efforts at proving, most of which are seen as unsuccessful, produce a long paper trail.[25] One would expect this scrap paper trail, at least, to be somewhat more mobile than blackboards. Not so: for the purposes of research, the process of writing appears to matter more than the record it produces. Scrap paper is almost never mobilized beyond its initial use. One respondent explained that "I don't tend to look back very much." Another has a policy of saving notes until he no longer recognizes the calculations, but confesses that he too rarely looks back at them. "I do a lot of stuff in my head," a third researcher recounted, and his research notes reflected this self-conception by rarely traveling beyond the sites in which they were produced.

Merz and Knorr Cetina (1997, 87, 93) describe mathematical work as a process of "deconstruction," where equations from problems are successively transformed through a variety of techniques until they yield a new theoretical insight. One of our

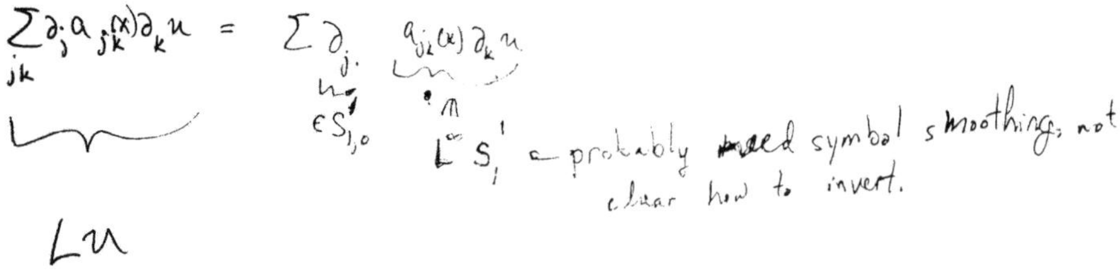

$$\sum_{jk} \partial_j a_{jk}(x) \partial_k u = \sum_j \partial_j \quad q_{ji}(x) \partial_k u$$

$$\in S_{1,0}^1 \qquad L^\infty S_1^1 \quad \text{a probably used symbol smoothing, not clear how to invert.}$$

$$L u$$

Figure 6.4
Researchers in mathematics continually materialize and reformulate promising or troublesome expressions with the hope of making sense of the "ocean of terms" that arise in their research process. Here, spacing and annotation in a researcher's notes help him to marshal a range of methods, identifications, and speculations.

subjects described the process perfectly: "I'm going to keep doing the calculations again, only now trying to look for terms of this form. . . . I have an ocean of terms like this, and the problem in some sense is how do you put them together so that they make some sense." Consider how terms are put together in the research notes excerpted in figure 6.4. This researcher's deconstruction begins with the operator L, whose effect on a function u is first written compactly on the left-hand side of the equation in his notes. (The brackets identifying this expression as Lu were added during the course of an interview as the researcher explained his inscriptions.) On the right there appears a nearly identical expression, with a space opened up between the ∂_j and the rest of the expression's summand (that is, $a_{jk}(x)\partial_k u$). Brackets beneath the two sets of symbols identify them as members of specific families of mathematical objects, respectively $S_{1,0}^1$ and $L^\infty S_1^1$, and the latter identification merits a written-out speculation about a technique ("symbol smoothing") and a desired outcome (inversion). All the while, these textual tokens are experienced and described as ideas. In this example: "We have some variable coefficient operator [L] that looks like the Laplacian, and so . . . [we] split it up into a sum of pieces, I guess a product of two things. In my case, the first product . . . is just a derivative, and . . . the second factor, less is known about."

In addition to being regrouped, symbols can be transformed according to mathematical principles and with the help of auxiliary equations and images. Notations and framings are often adapted to particular approaches. "There's a lot of notation, and it does help to go back and forth between them," offered one researcher. Moving between different variables and expressions can coax a troublesome formulation to resemble a familiar one or allow researchers to break a problem into smaller parts. Annotations can also declare aspects of a problem to be difficult, promising, or solved. In one particularly dramatic example of this, an interviewee recounted how "I put that in a red box because I was very excited when I realized that. . . . In my mind it moved us closer to completion of the project." As concepts are continually rematerialized, salient details are expanded or omitted, much as they would be on the blackboard. One researcher's notes had the word "factor" in place of a positive constant whose particular value was not relevant at that stage of the investigation. He expected that he might ultimately "see sort of which ones [factors] are ones that are helping you prove your result and which ones are the obstacles," and could then manage the obstacles separately. A process described by multiple respondents involved successive attempts to develop and refine a proof, with each attempt aimed at managing a new set of constraints after one is convinced of the proof's "main idea."

One should not get the impression, however, that the only papers of significance in a mathematician's office are scrap papers. A large amount of space is devoted to storing books and articles that contain mathematics in its most stable and circulable form. These are achieved through the "writing up" process, which (in our subjects' consensus) takes place strictly after the genuinely creative part of mathematical research—though all admitted that the form and often the substance of a result were liable to change substantially during or even after writing up. The work of writing up deserves a separate study—parts of it are addressed lucidly by Rosental (2008) and Merz and Knorr Cetina (1997). For our purposes, we would like to expand upon the inverse phenomenon: the less-recognized reading practices that convert "written-up" prose into a form usable in mathematical research—practices that might be called "reading down."

There is a crucial difference between mathematical papers and reports of scientific experiments. Where the latter are understood to depend on the credible reporting of experimental outcomes, the former are seen in principle to contain all the apparatus required for their verification. That is, where scientists must describe experiments and plot data, mathematicians are expected to reproduce in meticulous detail each of the novel rational steps behind their conclusions. The time and thought required to understand and verify each such detail makes mathematical papers subject to similar issues of trust, credibility, and reproducibility as have been described for the natural sciences, but the presentation of mathematical texts as (in principle) self-contained means that their circulation and deployment can have a decidedly different character.

In particular, mathematical texts present readers with two kinds of usable information. They establish lemmas and theorems that can be invoked as settled relations between specific mathematical phenomena, and they present methods and manipulations that can be used by others to establish different results. Researchers access others' papers through preprint and citation databases, and in smaller specialisms researchers will simply send preprints to a regular list of colleagues. They approach their stream of available papers using successive filters to identify where the two foregoing types of information most relevant to their research will be found. The process of perusing a database, for instance, might start with reading the titles of articles in relevant subject areas, the abstracts of articles with relevant titles, and so forth. The mathematicians with whom we spoke almost never read papers in their entirety—and certainly not with the goal of total comprehension.

When information from a paper is deemed immediately relevant to an ongoing project, it is finally read for its technical detail. Rather than attempt to digest every claim, however, readers try to identify concepts, formulations, and conclusions that are recognizable in the context of their own work. These identifications begin a process of rerendering papers' key passages in terms readers hope may ultimately advance their instrumental research goals.

This process of reformulating official papers into research instruments can span several media. A single page of one researcher's notepad, shared during an interview, visibly manifested a series of translations from an article to penned equations, to an email, to spatial gestures, and then to further writings. Interviewees reported experiencing mathematical concepts in terms of formalisms, properties, or operations. One described an equation by placing invisible terms in the air, one by one, in front of him. "I've written it down so many times," he explained, that he instinctively saw "the first-order terms appear here and the second-order terms there." Another used a box of tea on his desk as an impromptu prop for explaining a source of theoretical consternation from a recent effort. Different modes of mathematical cognition must necessarily interact to produce the transformations that bring about original proofs and theorems—transformations that would not generally be possible within a single framework of representation. Moreover, they must interact in a way that enables the coordination of mathematical understanding between different researchers in a variety of settings.

This leaves mathematical ideas in a strange position. Particular and idiosyncratic inscriptions and realizations are utterly central to the practice of mathematics. Paradoxically, mathematical inscriptions (especially on blackboards) work in ways that specifically (and, as we have argued, misleadingly) assert the opposite—that ideas somehow do not depend on the ways in which they are mobilized. The flexibility of mathematical representations obscures the sociomaterial coordination necessary to move concepts so freely from one form to another. Mathematical work rests on self-effacing technologies of representation that seem to succeed in removing themselves entirely from the picture at the decisive junctures of mathematical understanding. It is only by virtue of these disappearing media that one can be said to understand a concept itself rather than its particular manifestations.

Except when one cannot. Like scientific instruments, mathematical representations are subject to "troubles," flaws, and shortcomings (see Lynch 1985). The vast majority of attempts to use material proxies in one form or another to elucidate a concept are not counted as successes within a program of research. Seminars are among the rare displays of mathematical semiosis in a research setting where it is understood and expected that the signs will work. Mathematical research is marked by the constant struggle to create viable signs. As one of our subjects put it: "It's largely having a model and trying to get the new thing to fit into the old model, and at certain points that simply fails, and at that point you sort of mess around and think about . . . the old one a slightly different way, sometimes just calculating [and] seeing what comes out."

6   Representation in Mathematical Practice

In most people's experience, mathematics is a static body of knowledge consisting of concepts and techniques that are the same now as they were when they were developed hundreds or thousands of years ago, and are the same everywhere for their users and nonusers alike. Little would suggest that there are corners of mathematics that are changing all the time, where as-yet-unthinkable entities interact in a primordial soup of practices that constantly struggle to assert their intelligibility. Such is the realm and such are the objects of mathematical research.

The relationship between mathematicians and their objects of study is anything but straightforward. There is no mathematical concept whose formal immediacy or self-evidence stands beyond media and mediation. As a science of ideals, mathematics rests on the capacity of mathematicians to legitimize and manipulate particular representations of mathematical phenomena in order to elucidate rigorous mathematical knowledge.

In contrast to well-worn accounts of representation in the natural sciences, the story of mathematics is less about the hidden work of taming a natural phenomenon according to ideals than about the very public work of crafting those putatively independent ideals from their always-already-dispensable material manifestations. We have proposed chalk as both a literal and a figurative embodiment of that work. As a physical means of representation, chalk and blackboards entail a potent but highly circumscribed means of publicly materializing mathematical concepts. Their mode of representation, moreover, defines and influences mathematical practices far beyond those relatively limited circumstances where the mathematician actually has chalk in hand.

Mathematical writing and the mathematical thinking that goes with it are markedly dependent on the media available to the mathematician. Mathematical work traces the contours of its surfaces—there is little that is thinkable in mathematics that need not also be writable, particularly in the mathematics that is shared between mathematicians.[26] Blackboards, paper, and other media make certain forms of writing, and hence certain kinds of arguments and approaches, more feasible than others. Without having to assert that the limits on mathematical inscription definitively foreclose many potential truths from *ever* being described and accepted mathematically, it is manifestly clear that those limits *can* and *do* imply corresponding constraints on the lived and daily course of mathematical research. As De Millo, Lipton, and Perlis (1979,

274–275) put it, "propositions that require five blackboards or a roll of paper towels to sketch—these are unlikely ever to be assimilated into the body of mathematics."

Even when a viable constellation of representations is found, the mathematician's work is not done. These multifarious semiotic entities must then be made accountable to the equations, syllogisms, and arguments found in the published literature that compose the official corpus of mathematical knowledge—a project for which they are poorly adapted. A staggering portion of mathematicians' work goes into decoding published papers to create functional intuitions and understandings and, conversely, into encoding those intuitions in the accountable forms in which they will be credited as genuine. This is why chalk and seminars are so important. They give researchers shared partial access to what is so obviously missing from official accounts of completed work: namely, the experienced material performance of mathematics in action. The tension between circulation and application in mathematics is a real one. Mathematical ideas are not pregiven as the universal entities they typically appear to be. The most important features of mathematics can be as ephemeral as dust on a blackboard.

Acknowledgments

We would like to thank Alex Preda, Mike Lynch, and our three anonymous referees for their insights and suggestions, as well as our interviewees from the Analysis Group for their willingness to share their time and energy for this project.

Notes

1. Valson (1868), 1:253; our translation from the French original.

2. For a detailed account of the study's methods and findings, see Barany (2010). Barany observed the weekly seminar and conducted a series of interviews exploring the everyday research practices and diachronic research developments of a group of early- and mid-career mathematicians studying partial differential equations and related topics at a major British research university.

3. Of particular note are Ochs et al. (1994, 1996, 1997), Galison (1997), Merz and Knorr Cetina (1997), and Kaiser (2005).

4. Suchman (1990) and Suchman and Trigg (1993).

5. See Rosental (2004, 2008) and Greiffenhagen (2008).

6. Woolgar (1982) offers an early assessment of the literature; Lynch (1985) and Latour and Woolgar (1986) are two influential examples, treating "rendering" and "inscription" respectively; see also Lynch (1990) on the mathematical ordering of nature and Woolgar (1990) on documents in scientific practice.

7. For example, Lakatos (1979), Bloor (1973, 1976, 1978), Mehrtens (1990), Pickering (1995), Netz (1999), Jesseph (1999), and Warwick (2003).

8. Lave (1988), Kirshner and Whitson (1997), Greiffenhagen and Sharrock (2005).

9. Livingston (1999, 2006), Bloor (1976), Rotman (1988, 1993, 1997).

10. Livingston (1986), MacKenzie (2001).

11. E.g., Netz (1999), Lakoff and Núñez (2000). See also Hutchins (1995), Mialet (1999).

12. Prominent ones include De Millo, Lipton, and Perlis (1979), Davis and Hersh (1981), and Thurston (1994); see also Heintz (2000) and Aschbacher (2005).

13. Though women occasionally were present at the Analysis Seminar, all of the speakers during the period of Barany's study were men.

14. Thurston (1994, 165 and passim) notes something similar, and Merz and Knorr Cetina (1997, 74) identify a comparable phenomenon in theoretical physics.

15. In this sense, the mathematical seminar offers an alternative to the classical typology of lecturing proposed by Goffman (1981), featuring a form of "fresh talk" that is neither presented nor understood as spontaneous but is simultaneously quite distinct from memorized or strictly rehearsed lecture talk.

16. The project orientation of labor and narrative seems quite natural for neuroscientific research, with its vast assemblages of researchers, technicians, and apparatus. Given the stereotype of the lone mathematician and the importance of breakthrough stories in post-facto accounts of mathematical innovation, however, the predominance of project work in mathematics is considerably more surprising.

17. The observations in this section should be compared to Ochs, Jacoby, and Gonzales's (1994) discourse analysis of physicists' use of "graphic space" to narrate their work and to Suchman and Trigg's (1993) analysis of whiteboard work among artificial intelligence researchers.

18. We can only note here that blackboards' iconicity is vast. They are ubiquitous props in portraits of theoretical researchers in mathematics and physics—on which, see Barthes (1957, 104–105)—and a widely traded symbol of pedagogic authority and intellectual inspiration, from *Good Will Hunting* to Glenn Beck. On the nineteenth-century pedagogic history of the blackboard, see Kidwell et al. (2008) and Wylie (2011); on the blackboard's earlier history in music composition and instruction, see Owens (1998, 74–107).

19. We do not have the space for a systematic discussion of competing technologies to chalk and blackboards, which include alternative writing surfaces as well as tools for projecting text and images. See, however, Barany's (2010, 43–44 and passim) discussion of these technologies, with reference to adaptations that reinforce the disciplinary centrality of the blackboard even when it is not in use.

20. It should be said that blackboards have been made predominantly out of materials other than slate for most of their history. The paradigmatic relationship between blackboard and slate has, however, fundamentally shaped blackboards' social meaning and material development. Nor, for that matter, are blackboards always black. The seminar boards at the heart of this study were a dark shade of green.

21. Suchman (1990, 315) notes a related phenomenon of whiteboards orienting researchers in a shared interactional space in the more intimate settings of research discussions—a phenomenon we also noted among the mathematicians in our study.

22. These are gestures of the "iconic" class identified by Schegloff (1984, 275). Greiffenhagen (2008, par. 29–66) and Greiffenhagen and Sharrock (2005) make comparable observations for logic instruction. Núñez (2008) offers a contrasting approach to gestures in mathematical performance, seeking fundamental cognitive mechanisms underlying gestures and metaphors used in mathematics. We thank an anonymous referee for pointing out that these gestures, which are either audience-facing or board-facing, take place in a lecture context where nearly all writing is done while the speaker faces away from the audience. Thus, the physical constraints of the board provide a stage that markedly limits the timing and orientation of the gestures available to the speaker at any given point of the lecture.

23. In particular, this observation contrasts with the emphasis of Greiffenhagen and Sharrock (2005) on indicating gestures that enact or lend intuition to mathematical or logical phenomena. Greiffenhagen (2008) notes both indexing and indicating gestures in the sense we describe but does not indicate their comparative prevalence. Even in research settings, we found board positioning to be a significant but easily overlooked instrumental feature of board inscriptions, an observation consonant with Suchman (1990, 315–316).

24. These associations between particular gestures, inscriptions, and places in the speaker's "projection space" extend Schegloff's (1984, 270, 281–282, and passim) analysis of the gestural spatialization of speech and narration in the absence of media such as blackboards.

25. Latour (1990, 52) identifies the production and legitimation of such cascades of inscriptions as a decisive puzzle for the anthropology of mathematics.

26. Rotman (1993, x) likewise asserts an interweaving of thought and inscription, though his focus is on the semiotics of mathematical abstractions rather than the particularities of mathematical research and communication.

References

Aschbacher, Michael. 2005. Highly complex proofs and implications of such proofs. *Philosophical Transactions of the Royal Society A* 363:2401–2406.

Barany, Michael J. 2010. Mathematical research in context. MSc diss., University of Edinburgh Graduate School of Social and Political Science. Available at http://www.princeton.edu/~mbarany/EdinburghDissertation.pdf

Barthes, Roland. 1957. *Mythologies*. Paris: Éditions du Seuil.

Bloor, David. 1973. Wittgenstein and Mannheim on the sociology of mathematics. *Studies in History and Philosophy of Science* 4 (2):173–191.

Bloor, David. 1976. *Knowledge and Social Imagery*. London: Routledge.

Bloor, David. 1978. Polyhedra and the abominations of Leviticus. *British Journal for the History of Science* 11 (3):245–272.

Davis, P. J., and R. Hersh. 1981. *The Mathematical Experience*. Boston: Birkhäuser.

De Millo, Richard A., Richard J. Lipton, and Alan J. Perlis. 1979. Social processes and proofs of theorems and programs. *Communications of the ACM* 22 (5):271–280.

Galison, Peter. 1997. *Image and Logic: A Material Culture of Microphysics*. Chicago: University of Chicago Press.

Goffman, Erving. 1981. The lecture. In *Forms of Talk*, 160–196. Philadelphia: University of Pennsylvania Press.

Greiffenhagen, Christian. 2008. Video analysis of mathematical practice? Different attempts to "open up" mathematics for sociological investigation. *Forum Qualitative Sozialforschung* 9 (3):32.

Greiffenhagen, Christian, and Wes Sharrock. (2005) Gestures in the blackboard work of mathematics instruction. Paper presented at Interacting Bodies, 2nd Conference of the International Society for Gesture Studies (Lyon, France).

Heintz, Bettina. 2000. *Die Innenwelt der Mathematik: Zur Kultur und Praxis einer beweisenden Disziplin*. Vienna: Springer.

Hutchins, Edwin. 1995. *Cognition in the Wild*. Cambridge, MA: MIT Press.

Jesseph, Douglas M. 1999. *Squaring the Circle: The War between Hobbes and Wallis*. Chicago: University of Chicago Press.

Kaiser, David. 2005. *Drawing Theories Apart: The Dispersion of Feynman Diagrams in Postwar Physics*. Chicago: University of Chicago Press.

Kidwell, Peggy Aldrich, Amy Ackerberg-Hastings, and David Lindsay Roberts. 2008. The blackboard: An indispensable necessity. In *Tools of American Mathematics Teaching, 1800–2000*, 21–34. Washington, DC: Smithsonian Institution; Baltimore: Johns Hopkins University Press.

Kirshner, David, and James A. Whitson, eds. 1997. *Situated Cognition: Social, Semiotic, and Psychological Perspectives*. Mahwah, NJ: Lawrence Erlbaum Associates.

Lakatos, Imre. 1979. *Proofs and Refutations: The Logic of Mathematical Discovery*. Ed. John Worrall and Elie Zahar. Cambridge: Cambridge University Press.

Lakoff, G., and R. Núñez. 2000. *Where Mathematics Comes From: How the Embodied Mind Brings Mathematics into Being*. New York: Basic Books.

Latour, Bruno. 1990. Drawing things together. In *Representation in Scientific Practice*, ed. Michael Lynch and Steve Woolgar, 19–68. Cambridge, MA: MIT Press.

Latour, Bruno, and Steve Woolgar. 1986. *Laboratory Life: The Construction of Scientific Facts*. 2nd ed. Princeton: Princeton University Press.

Lave, Jean. 1988. *Cognition in Practice: Mind, Mathematics and Culture in Everyday Life*. Cambridge: Cambridge University Press.

Livingston, Eric. 1986. *The Ethnomethodological Foundations of Mathematics*. London: Routledge.

Livingston, Eric. 1999. Cultures of proving. *Social Studies of Science* 29 (6):867–888.

Livingston, Eric. 2006. The context of proving. *Social Studies of Science* 36 (1):39–68.

Lynch, Michael. 1985. *Art and Artifact in Laboratory Science: A Study of Shop Work and Shop Talk in a Research Laboratory*. London: Routledge and Kegan Paul.

Lynch, Michael. 1990. The externalized retina: Selection and mathematization in the visual documentation of objects in the life sciences. In *Representation in Scientific Practice*, ed. Michael Lynch and Steve Woolgar, 153–186. Cambridge, MA: MIT Press.

MacKenzie, Donald. 2001. *Mechanizing Proof: Computing, Risk, and Trust*. Cambridge, MA: MIT Press.

Mehrtens, Herbert. 1990. *Moderne Sprache Mathematik: Eine Geschichte des Streits um die Grundlagen der Disziplin und des Subjekts formaler Systeme*. Frankfurt am Main: Suhrkamp.

Merz, Martina, and Karin Knorr Cetina. 1997. Deconstruction in a "thinking" science: Theoretical physicists at work. *Social Studies of Science* 27 (1):73–111.

Mialet, Hélène. 1999. Do angels have bodies? Two stories about subjectivity in science: The cases of William X and Mister H. *Social Studies of Science* 29 (4):551–581.

Netz, Reviel. 1999. *The Shaping of Deduction in Greek Mathematics: A Study in Cognitive History*. Cambridge: Cambridge University Press.

Núñez, R. 2008. A fresh look at the foundations of mathematics: Gesture and the psychological reality of conceptual metaphor. In *Gesture and Metaphor*, ed. A. Cienki and C. Müller, 93–114. Amsterdam: John Benjamins.

Ochs, Elinor, Patrick Gonzales, and Sally Jacoby. 1996. "When I come down I'm in the domain state": Grammar and graphic representation in the interpretive activity of physicists. In *Interaction and Grammar*, ed. Elinor Ochs, Emanuel A. Schegloff, and Sandra A. Thompson, 328–369. Cambridge: Cambridge University Press.

Ochs, Elinor, and Sally Jacoby. 1997. Down to the wire: The cultural clock of physicists and the discourse of consensus. *Language in Society* 26 (4):479–505.

Ochs, Elinor, Sally Jacoby, and Patrick Gonzales. 1994. Interpretive journeys: How physicists talk and travel through graphic space. *Configurations* 2 (1):151–171.

Owens, Jessie Ann. 1998. *Composers at Work: The Craft of Musical Composition 1450–1600*. Oxford: Oxford University Press.

Pickering, Andrew. 1995. *The Mangle of Practice: Time, Agency, and Science*. Chicago: University of Chicago Press.

Rosental, Claude. 2004. Fuzzyfying the world: Social practices of showing the properties of fuzzy logic. In *Growing Explanations: Historical Perspectives on Recent Science*, ed. M. Norton Wise, 159–178. Durham: Duke University Press.

Rosental, Claude. 2008. *Weaving Self-Evidence: A Sociology of Logic*. Porter, Catherine, (trans.). Princeton: Princeton University Press.

Rotman, Brian. 1988. Toward a semiotics of mathematics. *Semiotica* 72 (1/2):1–35.

Rotman, Brian. 1993. *Ad Infinitum . . . The Ghost in Turing's Machine: Taking God out of Mathematics and Putting the Body Back In, an Essay in Corporeal Semiotics*. Stanford: Stanford University Press.

Rotman, Brian. 1997. Thinking dia-grams: Mathematics, writing, and virtual reality. In *Mathematics, Science, and Postclassical Theory*, ed. Barbara Herrnstein Smith and Arkady Plotnitsky, 17–39. Durham: Duke University Press.

Schegloff, Emanuel A. 1984. On some gestures' relation to talk. In *Structures of Social Action: Studies in Conversation Analysis*, ed. J. Maxwell Atkinson and John Heritage, 266–296. Cambridge: Cambridge University Press.

Suchman, Lucy A. 1990. Representing practice in cognitive science. In *Representation in Scientific Practice*, ed. Michael Lynch and Steve Woolgar, 301–321. Cambridge, MA: MIT Press.

Suchman, Lucy A., and Randall H. Trigg. 1993. Artificial intelligence as craftwork. In *Understanding Practice: Perspectives on Activity and Context*, ed. Seth Chaiklin and Jean Lave, 144–178. Cambridge: Cambridge University Press.

Thurston, William P. 1994. On proof and progress in mathematics. *Bulletin of the American Mathematical Society* 30 (2):161–177.

Valson, Claude-Alphonse. 1868. *La vie et les travaux du Baron Cauchy, membre de l'Académie des Sciences*. Paris: Gauthier-Villars.

Warwick, Andrew. 2003. *Masters of Theory: Cambridge and the Rise of Mathematical Physics*. Chicago: University of Chicago Press.

Woolgar, Steve. 1982. Laboratory studies: A comment on the state of the art. *Social Studies of Science* 12 (4):481–498.

Woolgar, Steve. 1990. Time and documents in researcher interaction: Some ways of making out what is happening in experimental science. In *Representation in Scientific Practice*, ed. Michael Lynch and Steve Woolgar, 123–152. Cambridge, MA: MIT Press.

Wylie, Caitlin Donahue. 2011. Teaching manuals and the blackboard: Accessing historical classroom practices. *History of Education*. doi:10.1080/0046760X.2011.584573 accessed August 2011.

7 Networked Neuroscience: Brain Scans and Visual Knowing at the Intersection of Atlases and Databases

Sarah de Rijcke and Anne Beaulieu

1 Introduction: Seeing the Brain

Brain scans have been in heavy circulation these past two decades as some of the most fascinating and ubiquitous digital images in scientific and cultural spheres. Such images contribute to the constitution of the brain as an object of knowledge (De Rijcke 2008a, 2008b), and of the mind-in-the-brain (Beaulieu 2000). Part of their influence stems from how brain scans tie in to the biologization and digitalization of behavioral and psychological processes. The discrete, mapped-out bright bits seem to provide visual proof for the existence of material substrates of behavioral mechanisms, and for the claim that the basis of the mind is biological. The scans reify the brain as a locus of control and as a site of neurological, neuropharmacological, and neurosurgical intervention, or even of self-improvement (Brenninkmeijer 2010). Their importance should also be understood in relation to what has been hailed as "neurosociety" (Schleim 2010) or a "neuroturn" requiring scrutiny (Littlefield and Johnson 2012). However, these important critical reflections on the proliferation of neurodiscourses in the humanities and social sciences do not fully address the particular roles played by the visuality of brain scans. This visuality intensifies the focus on the neurosciences by other scholarly fields, as part of a sociotechnological turn toward visualization (Joyce 2006) and, in particular, in relation to possibilities for circulation, transformation, and manipulation of digital media. Brain scans purportedly make up ideal boundary objects between disciplines, and between specialists and a lay public (De Rijcke and Beaulieu 2007). They enable the multiplication of witnesses of neurological conditions and states of mind, while at the same time grounding such conditions in the empirical, the observable. As icons of neuroscientific progress in an extremely networked and visually oriented culture, they proclaim new developments for ever-wider audiences.

Given the wide and potentially weighty circulation of brain scans, how are we to understand them in relation to scientific and broader visual culture? In this chapter, we address this question by analyzing the development of authoritative collections of brain scans known as "brain atlases" since the beginning of the "Decade of the Brain" (the 1990s). We set out to investigate the conditions that make brain scans authoritative visual objects and analyze three important dimensions of scans. We show how scans are increasingly parts of *suites* of networked technologies, rather than stand-alone outputs. We then trace the increasing presence of databases of scans in the constitution of atlases, and outline the consequences of what we call *database logic* for visualizations of the brain. The third development we discuss is the role of scans as *interfaces* that serve to open up a range of possibilities, rather than to stand in as fixed representations. Together, these dimensions help characterize the visual in digital and networked settings of contemporary science, and they enable us to trace how the very concept of the authoritative image has been transformed.

Brain scans and atlases as visual evidence

The point has often been made that brain scans are not snapshots, and cannot and should not be understood in terms of a photographic register. Several analyses have shown that a number of assumptions of mechanical objectivity (Daston and Galison 2007) associated with photographic realism do not hold for brain scans. For example, the possibility of relying on the physical truth chain established by light particles touching an object and moving to a photographic plate has been challenged, in the context of digital imaging and data reconstruction needed for scanning (Beaulieu 2002a; De Rijcke and Beaulieu 2007). The realism associated with photographic representation has also has been critically scrutinized (De Rijcke 2010) as well as the faith in a standardized mechanical process that is free of intervention (Dumit 1994; Joyce 2008). Last of all, the overinvestment in the "observer's" ability to detect meaningful differences between scans has also been exposed when it comes to complex digital scans (Dumit 2004). In addition to deconstructing brain scans, these studies emphasize the complexity of brain scan images, the importance of context in constituting their meanings, and the work that goes into presenting them as autonomous proofs of intrinsic conditions in the brain and of the localization of particular functions.

While it is important to deconstruct the representational idioms associated with brain scans—and we have contributed to such work ourselves—we are also eager to develop critical work on brain scans that pays attention to the media and mediation (Bolter and Grusin 2000) and the spatialities (Lynch 1991; Hine 2006) of their production and use. Scans are both digital and networked images that depend on "suites of technologies" (Shove et al. 2007) for their constitution and meaning, rather than on single devices. These suites of technologies include digital images, data models, databases, screen-based interfaces, and electronic networks. The entwinement of these tools signals the limitations of a conception of scientific research that emphasizes fixed, authentic, mechanically obtained objective representations, to be consulted and evaluated by an observer.

The development and use of atlases is an important scientific activity that involves the networking of brain scans in and through databases and interactive interfaces. Atlases are authoritative kinds of images, and much is at stake in their creation. They make public the aspirations of the scientists who produce them, and reveal preferred epistemic and ontological stances of a scientific field (Daston and Galison 2007). Given the common understanding of digital images as mutable, and given that they do not partake in many of the assumptions of photorealism, an important issue concerns how these images can still be authoritative.

Furthermore, by focusing on the atlas, a form that embodies and shapes scientific authority, we are able not only to discuss the normative potential of brain scans, but also to link that discussion to the history of atlases and of objectivity. Not only are some images more or less objective, according to specific standards, but these standards also vary from discipline to discipline, between time periods, and according to context—whether they are used for scientific or for clinical purposes, for example. We therefore propose an approach that examines how the authoritative status of these images is actively constituted, rather than simply revealed through what they represent (also see De Rijcke 2010). We use the concept of "authoritativeness" to evoke an emergent, distributed and interactive mode of visual knowing.[1] We prefer that term to "objectivity," especially as articulated in the work of Daston and Galison, where objectivity is illustrated by visual forms and treated as an immanent aspect of images produced according to certain norm-laden practices.

In order to trace how authoritative brain images are produced through suites of technologies, database logic, and interactive interfaces, we draw on ethnographic fieldwork and archival research on atlases of the brain.[2] The empirical material we discuss concerns the International Consortium for Brain Mapping (ICBM), a North American-based research program that became internationally important through the dissemination of its standards, protocols, and reference works, including databases and atlases. The early years and policy context of this program are detailed elsewhere (Beaulieu 2001), but it is relevant to note that neuroscientific, clinical, and bioinformatics components were all represented within the consortium, with each component bringing its own visual culture to the project. In particular, both clinical-radiological traditions (Joyce 2006) and modeling and computer visualization approaches coexisted as frames of reference among the stakeholders of the program (Beaulieu 2002a).

By covering nearly two decades of work, we are able to analyze the incorporation of digital and networked images of the brain both in lab practices and in formal and informal scientific communications. In the late 1990s, the presence of the human observer was treated as somewhat residual, a glitch in the smooth pipeline that would soon be removed through the improvement of algorithms and processing abilities. The emphasis on automation in the development of new atlases was very much in line with more general moves to quantify and rationalize medicine (Berg 1997; also see Hine 2008 on computerization movements). Compared to previous brain-mapping practices, digital atlases introduced new elements of control and restraint in achieving what we then labeled "digital objectivity" (Beaulieu 2001). A decade or so later, user involvement had shifted. Far from being residual, it is healthily persistent. Interfaces to enable and support such human-observer involvement seem to be proliferating rather than disappearing. As technological developments have affected the way data is processed, integrated, and visualized, the role of the observer has not been left unchanged. At the same time, digital atlases continued to incorporate older conventions of imaging the brain: they "remediated" tradition (Bolter and Grusin 2000) as well as supported new practices. We now discuss these developments in more detail.

2   Scans in Suites

The notion of suites of technologies (Shove et al. 2007) has been fruitfully applied to design and interaction in general, and to radiological imaging in particular (Saunders 2009). It is also a powerful way to think about digital imaging and neuroinformatics, since it foregrounds the mutual dependence of images, screens, software, and interactions for creating and using new views of the brain. Observation and visualization around digital brain atlases increasingly take place behind a computer screen. This implies a shift in relations between observers, the brains they study, the technologies used for this purpose, and the institutional arrangements within which this happens (Beaulieu et al. 2012; Hand 2008). In the process, brain atlases are reconfigured as interfaces between different spatial realms: the brain as a newly constituted digital object, the material space occupied by embodied observers at the screen in a lab or hospital, and the expansive networked infrastructures through which the images circulate. Furthermore, the atlases function through interfaces that enable and demand interaction. By raising specific expectations and providing particular opportunities, these interfaces shape how the observer comes to know through images.

Starting in the mid-1990s, a number of US-based funding initiatives supported efforts to implement informatics approaches to neuroimaging. For researchers and funders involved in the ICBM, the prospect of accessing the human brain in vivo and mapping out various cognitive functions contributed to a renewed interest in producing atlases (Beaulieu 2001; De Rijcke 2010). The brain atlases that they built relied on a standardized space in which to measure brain function and structure (Beaulieu 2002b).

To compare this process to a more familiar geographical example, this move is similar to the development of a standardized projection of the earth and to the development of common coordinate systems. These ensure standardized georeferencing with GIS technologies, enabling them to work more or less seamlessly across platforms (and making possible the "switch" from "map" to "satellite" view on Google Maps).

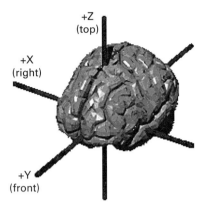

+Z
(top)

+X
(right)

+Y
(front)

Figure 7.1
An illustration of the implementation of Talairach space. This space was described using Carte-sian coordinates (along x, y, and z axes), which enabled researchers to transform and merge differ-ent kinds of information about brains, making different imaging modalities and different brains comparable. http://www.talairach.org/daemon.html.

The elaboration and adoption of a standardized "brain space" coevolved with the development of new digital brain atlases. Because of the possibilities for data integra-tion and computation offered by a quantified, grid-oriented brain space, what came to count as a "reference" brain changed. Rather than using one well-described specimen as a reference, many atlases developed during this period were made up of "averaged brains." The researchers considered these to be better reference points, that is, to offer a better baseline from which to pursue investigations. As the constitution of atlases changed, so did the kinds of uses to which they were put:

An atlas of the brain allows us to define its spatial characteristics. Where is a given structure, relative to what other features, what are its shape and characteristics and how do we refer to it? Where is this region of functional activation? How different is this brain compared with a nor-mal database? An atlas allows us to answer these and related questions quantitatively. (Toga and Thompson 2000, 635)

The emphasis in these new atlases came to lie in the possibilities for manipulating, generating, and displaying information spatially. A spatial organization of information in digital atlases was the meeting point of two important developments. First, images— or more precisely, voxels—were coupled to coordinates in digital media. Second, scans that were described in terms of spatial coordinates could be linked to database and computational possibilities. The use of informatics to constitute and manipulate digi-tal objects therefore changed the structure and content of atlases. The kinds of instru-ments that were needed to understand brain scans changed as these practices were adopted, since computational processes and workflows became important elements in the constitution of the brain and mind as objects of study. Similarly, new kinds of experts were recruited by neuroimaging labs in order to run experiments, since computer science expertise was increasingly needed to develop and manage databases and processing.

The atlas image is therefore not a normative scan to be compared visually to another scan. Instead, the atlas takes on features of both tool and representation, of epistemic and technical object (Rheinberger 1997), since it is both a tool that can be used to interrogate new instances and an object that improves by including new instances. The difference is not simply one between print on paper and pixel on screen. The scan to be interpreted is constituted through processing in relation to the "average brain" of the atlas, so that only certain kinds of differences will become apparent. Whereas a cascade of representations gained authoritativeness because of the growing distillation and augmentation of data in each subsequent step (Latour 1990), the ICBM atlas is authoritative because it enables further manipulations and evaluations.

Setting up "pipelines" to produce images

In the development of digital atlases, a standard space made it possible to manipulate different scans produced through different techniques (for example, different scanners) and from different subjects. By transforming them to a common "space" via a series of operations, they could become comparable and could be integrated into a single visualization. While there is some variation in the emphasis and techniques used, two challenges are especially prominent for those developing these new atlases. First, the right set of transformations must be identified and implemented through computational processes. For example, differences due to different scanners or even to different scanning sessions must be removed, since they are considered irrelevant to understanding the brain. Since brains vary in their anatomy, size, or even orientation in the scanner, all these elements, which are supposedly irrelevant, can be removed without loss of relevant information about the brain. Furthermore, the differences that *do* matter must be maintained. This means that the relative size or position of different parts of the brains must be preserved through and in spite of other transformations. Second, in the constitution of most atlases, other features of subjects assumed to be relevant to the brain's anatomy or organization are also recorded and taken into account. A whole list of characteristics, such as handedness, history of mental or neurological health, bilingualism, sex, and so on, come to function as metadata in the database that underlies these atlases. In more recent atlases, DNA samples are also collected for future correlation between scans, data about subjects, and genetic information. While no researcher or user of these atlases would claim that the characteristics are definitive and contain all explanatory information, this systematic correlation does enact the modernist ideal of the individual as the sum of its characteristics (Beaulieu 2000).

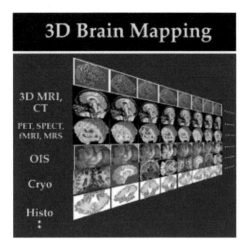

Figure 7.2
Each row represents data from one modality (kind of scanner or procedure). Each column represents data from one subject. Note the well-ordered presentation of data, which symbolizes the "pipelines," and the possibility for endless extension (open-endedness) of modalities and of subjects. http://www.loni.ucla.edu/~thompson/NEW/brain_mapping.gif.

This approach was developed in relation to "normal" brains, but also to variations considered relevant for understanding disease and cognition—for example, there is an atlas of the "Alzheimer's brain" (Thompson et al. 2003) and of children's brain development (Almli et al. 2007).

The set-up of pipelines as sets of standardized tasks and explicit operations to be performed by computers guarantees that the emerging object, made up of multiple kinds of data, will reveal the brain. These pipelines feed the atlas, providing the "raw" data that will constitute authoritative images. The pipeline ensures that the scans are purified of unnecessary information, that all metadata considered relevant are included, and that all voxels are made equivalent. Like the trust based on the mechanical objectivity of photography, the pipeline exudes standard handling and processing and constraining of subjective intervention. In some ways, these computational technologies push the values of control and restraint to new heights in the constitution of authoritative images. Yet these pipelines are also built with specific "manholes" that enable the user to remain involved in the process. In fact, they even demand that the user remain involved, in order to ensure the quality of the processing in the pipeline. Particularly interesting is the way a human observer can detect "garbage" in a way that a computer can't. Some of the visual inspections required in pipelines aim to identify when scans of phantoms (dummy objects used to calibrate scanners) rather than of human brains have wrongly been uploaded to data pipelines. (This is the neuroimaging equivalent of chucking out the photos taken with a thumb on the lens—a kind of image that a computer will happily process.) The latest developments in the production of these atlases offer an increase in the modularity and portability of pipelines, allowing users to tailor the pipeline to their own needs using "processing building blocks" (Van Horn and Toga 2009) and to interact with the infrastructure that runs the pipelines (such as the Grid) via graphical user interfaces. We return to the importance of interfaces later in this chapter. For now, we want to stress that, through the development of informatics, brain scans have become embedded in and reliant on a digital and networked context. The scan not only relies on processing via scanning and software technologies in order to then be compared to an atlas of the brain, but it can only be meaningfully generated and viewed in relation to the atlas and as part of a suite of technologies.

3 Deploying Database Logic and Probabilistic Thinking to Make Images

A second important development is the growing importance of database logic, which comes to the forefront of authoritative images. This development is related to digital media and the networked configuration of technologies, but it is distinct in the sense that it represents a particular investment in the organization and structure of knowledge produced through these images. Recall that when the scans processed through pipelines are aggregated to constitute a particular version of the brain, such aggregation can only be accomplished if the brain is represented as a set of voxels in a standard space, with voxels having n attributes (kind of tissue, site of activation, etc.). As noted above, each voxel in the resulting image is calculated based on the corresponding voxel values across the database of scans. This leads to a "naturalization" of the voxel as a component of the brain, turning neuroimages into thoroughly digital and informational objects.

The atlas relies on database logic for the constitution of authoritative images. This means that, rather than relying on "capturing" a good image, researchers invest in gathering and labeling data and relating them. Furthermore, the searchability of the metadata in the database contributes to the adaptability of the atlas. Ideally, for neuroscientists, atlases would be created based on a particular feature—for example, age—for which large archives of shared neuroimaging data are tagged (Van Horn and Toga 2009). Because the atlas is mutable, it can be shaped to best suit the analysis of a particular scan. The possibility of selecting the right subset of scans in order to adapt the norm makes the atlas more authoritative, because more relevant to a specific case.

Besides producing average atlases, another way researchers use the database is to develop probabilistic atlases. In these images, the voxels do not represent absolute values such as averages, but rather probabilistic outcomes of calculations across scans in the database. Such atlases will indicate, for example, that the probability of a particular coordinate being located in area X of the brain is 89%. It is also possible to draw on the "features" that have been recorded in relation to the scans. By articulating a set of variables, a query can be formulated that leads to the constitution of a specific atlas: for example, of the brain of a left-handed female in her thirties. However, such tailoring of the atlas can only be done by means of parameters that are already codified as differences that make a difference (Beaulieu 2001). In that sense, atlases function just like classification systems that enhance and obscure particular aspects of knowledge. The atlases also further shape other atlases, as they serve as baselines for identifying other features that may be relevant. The generation of further knowledge, whether through large-scale comparison or analysis, is always done according to the parameters for which data have been coded. Potentially, each new scan added to the database improves the atlas, giving the database a "generative" dynamic (Waterton 2010).[3]

This constitution and use of the atlas foregrounds the database. In contrast to the averaged brain, the image in the probabilistic atlas is not derived or distilled and then removed from the scans contained in the database. Probabilistic atlases highlight the range contained in the database. They depart radically from traditional atlases because of the way they foreground objects as both variable and relational, rather than as discrete "specimens" or autonomous instances that can be understood on their own (for example, Gong et al. 2009). While the figures thrown up by a probabilistic atlas are clearly visually codified, with colors referring to precise quantitative values, the observer is invited to consider them not as absolutes but within a register of

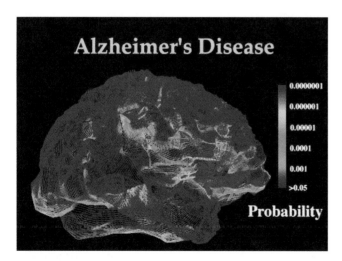

Figure 7.3
A probabilistic atlas from Thompson et al., *Journal of Computer Assisted Tomography* 21(4) (1997): 567–581. The recognizable shape of the brain is made up of colors that relate to a probabilistic key. It invites the viewer to consider the image as the outcome of calculations across a database, and in terms of a possible outcome—in this case, variation indicative of Alzheimer's disease.

possibilities. Accordingly, the register of possibilities is what lends authority to these atlases. By focusing on this aspect of images, we are drawn to consider the changing role of scans in relation to empirical, communicative, or analytic work in science. Furthermore, given the intensified relationship between image and database to which we draw attention, it seems urgent to consider how different kinds of databases will give rise to different kinds of visualizations.

4   The Atlas as Interface

The atlases constituted through databases embrace the range of possibilities contained in images of, and metadata relating to, the brain. These possibilities are not simply immanent. The interface, where these possibilities are presented *and* can be acted upon, becomes a crucial new aspect of visual knowledge production. It makes it possible to weigh different variables, or to alter the processing to explore how this affects the probabilities. This is more than a technical issue; interfaces link together investments in empirical investigations, mathematical and cognitive modeling, and the skills of the observer.

The increasingly networked settings in which brain atlases take shape facilitate the intensification of collaboration between different neuroimaging laboratories, but they also deepen compatibility issues on the level of the databases to be pooled, the image modalities to be merged, the scanning protocols to be followed, and the tools used to process and analyze the data. Although standardization and automation are still offered as complexity reduction strategies, at the same time the atlases are increasingly equipped with integrative, comparative, and interactive features. These atlases are therefore also interfaces, where an observer is to consider, judge, and interact with the image as database.

To demonstrate this, we turn to the analysis of the production of a networked probabilistic atlas of human white matter (the whitish connecting nerve tissue underneath the gray surface of the cortex) by the ICBM. In 1999, one of the ICBM's partner institutes at Johns Hopkins University developed a method for translating magnetic resonance imaging (MRI) data into three-dimensional reconstructions of neuronal tracts (Mori et al. 1999). The method, which became known as diffusion tensor imaging (DTI), was said to open up a whole new territory for digital, "in vivo" scrutiny. A couple of years later, its most highly developed application was that of fiber tracking in the brain (Le Bihan and van Zijl 2002). A new kind of epistemic object rapidly emerged on the basis of DTI, generated in the context of large-scale projects that gather and process data across centers, scanners, and subjects.

The appropriation of DTI as a new "window" on white matter anatomy and brain connectivity was largely instigated by the circulation of the first *MRI Atlas of Human White Matter* in 2005 (Mori et al. 2005). It consisted of high-resolution two- and three-dimensional visualizations of the major white matter tracts. Interestingly, the atlas was made available both electronically and in print. Using an electronic format was very much in concordance with ICBM's neuroinformatics goals. The networked context increasingly reconfigures the atlas to different spatial realms, ranging from the local "brain spaces" on the screens of observers to the larger, federated databases of which the brain data was meant to be(come) part. This made these and other digital atlases essentially emergent or—in the words of the ICBM—"evolving."[4] The fact that Mori and colleagues also saw fit to publish a printed edition of the atlas points to the contemporaneous clinical and laboratory practices in which the digital images were being integrated. The continued relevance of printed atlases as "stable" reference points in increasingly networked settings is an example of the ways in which these atlases "remediate" tradition even as they support new practices (cf. Bolter and Grusin 2000).[5] This remediation is visible not only in the publication strategy, but also in the digital atlas images themselves.

Atlases both reflect and shape a discipline's research objects. They also delimit what constitutes "proper" observation and visualization for practitioners, and new representational conventions can bring about a situation in which "everyone in the field addressed by the atlas must begin to learn to 'see' anew" (Daston and Galison 2007, 22). The forms and presentation of the white matter atlas show that new visual forms require a reconfiguration of the observer (cf. Crary 1991). This atlas was mainly aimed at a readership of radiologists and surgeons who regularly dealt with connectivity impairment. At the time, most of them were new to DTI. The atlas makers therefore carefully set the stage for DTI by means of short verbal descriptions and comprehensible visual juxtapositions. Among other things, they argued that the technique was capable of doing something "conventional MRI" could not do: provide good contrast at the level of individual voxels, so that variations in white matter structure are properly displayed: "From an MRI point of view, the white matter generally appears as if it were a fluid-like homogeneous structure, which, of course, is not the case" (Mori et al. 2005, 1). Borrowing a phrase from media theorists Bolter and Grusin, it is as though the authors justify the use of DTI "because it fulfills the unkept promise of an older medium" (Bolter and Grusin 2000, 60). To fully advertise the impact of the new modality, the authors presented a visual comparison (see figure 7.4). This triptych defines the parameters for understanding the two- and three-dimensional images in the rest of the atlas.

From left to right, we see a T1-weighted magnetic resonance image (which represents differences in "T1 relaxation time," i.e., the time it takes various tissues to return to an equilibrium after magnetization), followed by a representation of variations in the direction of the diffusion of water molecules, and a color-coded version of the same information. (Note how the colors correspond to arrows in the top right corner.) The oval shapes drawn above the middle and right image are connected to a specific area in

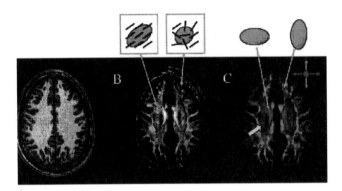

Figure 7.4
T1-weighted MRI, and two representations of anisotropy levels in the same slice. Water molecules preferentially diffuse anisotropically—toward and away from the cell body, parallel to the axons' length. The orientation of the restricted diffusion is therefore a factor in tissue characterization, next to its magnitude and shape (Le Bihan et al. 1991). The color-coded image on the right displays the direction of the measured anisotropy. For each of the slices (moving from left to right), a different level of detail, graphical explanation, and color are provided, emphasizing the increasing complexity and revelatory power of imaging modalities.

the brain through yellow arrows. They are meant to facilitate the visual appreciation of brain areas in which the diffusion is either anisotropic (unidirectional, elongated oval, brighter) or isotropic (random diffusion, round, darker).

The image on the left, which the authors designate as "conventional MRI," is the only one that does not have explanatory features. A large portion of the field of view is occupied by a flat, rather homogeneous mass. This "fluid-like structure" represents a segment of the brain's white matter. The authors deliberately chose an axial (horizontal) brain slice at a level where white matter dominates gray matter in terms of percentage. At no other slice level is this so obviously the case. The image is meant to provide a strong contrast with the diffusion-weighted anisotropy image in the middle, and to show how this "unique new MRI modality" provides a very different view on white matter structure, providing much more information about it.

The juxtaposition has yet more complex effects. Interestingly, all three images represent a constructed or simulated "digital space" (Lynch 1991, 63), since they are visualizations produced with a rendering algorithm. Despite major differences in data processing, the first and the second image are visually similar because both are represented in black and white, in contrast to the third image which is color-coded. What we see here is a *mise-en-scène* of a break between two imaging traditions. The black-and-white images evoke the immediacy and mechanical objectivity of photographic realism, while the color-coded image explicitly draws attention to its digital mediation.

The left and middle images remediate a long-standing radiological tradition, in which X-ray and computed-tomography scanning technologies have created familiar cultural objects (Joyce 2008). The "conventional MRI" is rooted in the opticist pictorial tradition that makes reference to "mechanical reproduction" (Lynch 1991, 72) and to "the position of the observer in a 'real,' optically perceived world" (Crary 1991, 1–2). These connotations are transferred to the diffusion-weighted image in the middle.

In the image on the right, the use of colors breaks with this tradition of invoking an "optical" truth. The colors are meant to be read as codifications of the direction of the largest principal vector in each particular voxel. The color-coded image is what Bolter and Grusin (2000) refer to as "hypermediated": it makes the viewer aware of the medium and the acts of representation that created it. This framing of DTI as part of a new epistemology is achieved through constant reference to "conventional" MRI, as a benchmark of anatomical "reality." The white matter atlas therefore emphasizes its digitality: its images are an outcome of novel forms of processing and of powerful calculation. The atlas's appeal is based on the standardization and pipeline calculations discussed above, but here we see how the observer needs to be taught to see anew, to appreciate that the visual features of the atlas rest on a new approach to visualization. This signals both the scale of changes in atlases and the enduring importance of the observer, who remains indispensable.

In the studio[6]

One of the purposes of the publication of the *MRI Atlas of Human White Matter* was to persuade researchers and radiologists to integrate DTI with their own practices and laboratories. As such, the atlas also played a role in coordinating professional activities and investments, becoming part of the pool of narratives that frame how images are to be interpreted (Cohn 2007; Roepstorff 2007; Saunders 2009). As we have seen, scans in digital atlases can act as intersections between existing pictorial traditions and new image parameters, depending on the visual form given to the image data.

Diffusion tensor imaging is not a straightforward task: image production with it requires "involved post-processing" behind a desk, after the "raw" data have been gathered by the scanner (Jiang et al. 2006, 106). This is one of the ways in which the configuration of the network in which DTI operates differs fundamentally from "conventional" MRI, which also requires processing but in which end users have less influence on image production. To make DTI available as a new tool for use with ICBM-networked brain databases, technological changes were required in the pipeline set-up, in the trajectory from image data acquisition by a scanner to the visualization of white matter tracts on a computer screen, and in the data exchange between the local image database and the ICBM-federated "reference system." Several software packages were specifically designed for these purposes. One of these packages, called DtiStudio, which had been used to produce the white matter atlas, became the ICBM standard processing program for tensor calculation, color mapping, fiber tracking, and 3D visualization.

DtiStudio prescribes how the "raw" diffusion-weighted image data should be handled and organized for further scrutiny. The software manual takes the reader through all the steps implicated in the "involved post-processing" procedure, but not before emphasizing that its "user-friendly interfaces" make DtiStudio the requisite image-processing tool for practitioners and researchers interested in brain connectivity (and disorders of it). A "user-friendly interface" in this case means a Microsoft Windows interface. The program mimics other Microsoft applications, and claims to combine smoothly with them. The manual addresses its intended audience in a straightforward manner, exuding fun and effortlessness. The authors stress that the software takes up only a few hundred kilobytes to install, and that after installation users are immediately "ready to play with the program." Provided that there is enough memory for the (much larger) datasets, the software can be used on virtually any computer, at home or in the laboratory. In addition, "most operations can be done with only a few clicks" (Jiang et al. 2006, 1). Users can cut and paste images from DtiStudio to the Word file of an article in progress, upload images to their web pages, or email interesting findings to colleagues after data processing. These and other elements of the software imply an interactive, dynamic process of image production and circulation.

One of the software's salient characteristics is that visual and manual processing is neither marginalized nor fully standardized: the interface provides possibilities for interacting with the data through the image, but does not dictate what should be done (cf. Coopmans, this volume). Users are invited to engage with the images displayed on their computer screen by adding or extracting colors, defining regions of interest, or subjecting parts to several algorithmic filters. The shift from a situation in which mediation was erased as much as possible (Bolter and Grusin 2000) to one of "hyper-mediacy" deeply affects both image production and the status of the images themselves. In DtiStudio, "real" or "objective" images are not static visualizations of brains, but flexible tools (cf. Rheinberger 1997) that enable further processing. The images are amalgamations of standardized calculations and physical interventions (such as mouse movements). Although the process is partly standardized, users maintain an interactive relationship with the images on the screen. Researchers or clinicians who use the software can generate and isolate various white matter tracts, have them appear and disappear on the screen, change their color, or play with the three-dimensional renditions of white matter structure. They can choose certain modifications on the basis of their experience with brain anatomy, or create visualizations of white matter that have never been created before. By presenting an indeterminate range of imaging modalities and options for further processing, the software allows users to tailor their experience to the purpose of their research or clinical study.

The scans are thus placed in a complex infrastructure that enables visual knowing in a manner distinctly different from the consultation and evaluation of fixed, mechanically obtained objective representations by an observer. However, despite the emphasis on hypermediated, interactive visualization, the "immediacy" of mechanical objectivity continues to have some influence on conventions for proper observation and visualization. We showed how the apparent transparency of "conventional" MRI served to position DTI in the existing visualization traditions and knowledge structures of radiology. In addition, DTI makes use of familiar anatomical conventions for displaying and analyzing preserved slices of the brain and for composing anatomical atlases. What is more, by using the "conventional" MRI as a reference, the pictorial and observational conventions of clinical radiology act as an important filter to the digitally constituted and hypermediated images. Furthermore, these conventions are inscribed in the new DTI database logic by software packages such as DtiStudio, by enabling the weighting of different features and the selection of subsets for visual inspection, among other things. The authoritativeness of brain scans and atlases is increasingly produced through such database logic as well as through the interactive interfaces and suites of technologies in which brain imaging data are embedded and which shape their use.

5 Conclusion

By examining present-day brain imaging atlases, we were able to focus on changes in existing configurations of users, visualization technologies, and atlas images, and to relate these to the transformation of the concept of the authoritative image. The atlases we analyzed are increasingly taking shape at the intersection of digital, networked, and computational technologies. As we have seen, huge investments are made in the development of standards and protocols to make scans aggregable and comparable. These are then leveraged to impart authority to the atlases. The authoritative status of these new atlases can be achieved through averaging, or through probabilistic approaches that link an "end" or "target" image with the underlying data. The integrity of this underlying data is in turn warranted by the implementation of the "pipeline" that organizes and purifies these scans, and by sophisticated transformation algorithms. At the same time, a number of interfaces are specifically built into this process, interfaces where an encounter takes place with a human observer who orients to these scans as visual evidence. Consequently, these atlases draw on both the radiological tradition and the computational, data-driven approach. While these various epistemic regimes coexist, and can even depend on each other, the kinds of brains produced are changing along with the kinds of instruments, kinds of work, and kinds of experts needed.

An important characteristic of these brain atlases is that they act as bridges between images in databases, types of data, technological platforms, image modalities, scales, and dimensions. They remain in continuous dialogue with the databases that under-pin them. Adjustment always remains a possibility, and can be spurred by temporal, computational, or visual considerations, depending on the practices of clinicians and researchers. The images themselves also act as interfaces (cf. De Rijcke and Beaulieu 2011). They do not only reveal epistemic objects; they also constitute relations and opportunities—ranging from local interfaced practices of handling the data to aggre-gate-level large-scale federated databases. The interactive dynamics between local and federated databases increasingly complicate clear-cut distinctions between these levels. Together with the scanning technologies, software, screens, databases, and the role of the observer in relation to them, the images partake of the dynamic of "generative exploration" that Waterton (2010, 654) identifies with archives. Yet the flexibility and creativity this suggests are not boundless. Atlases become entwined with networked infrastructures and become increasingly obdurate as their implementation spreads.

To reveal the epistemic and ontological assumptions around images of the brain, it is important to recognize their embeddedness in suites of technologies (Shove et al. 2007), as well as the remediation (Bolter and Grusin 2000) at work in rendering them meaningful. The iterative and relational aspects of neuroimaging are central to the spe-cific kind of authoritativeness that comes into play when these images are used in mak-ing claims about the brain, the mind, and the self. With brain scans in pipelines, the brain becomes a spatially configured set of voxels, whose tissue and other features can be computed, highlighted, compared, filtered, labeled, and drawn upon interactively. It becomes possible to switch between kinds of data, to go back and forth, to retrieve a brain as specimen in the average, to bring out the specific object in relation to a "population," and even to shape what counts as the "population."[7] Given the growing scope of the terrain of the brain, it is crucial to understand these dynamics in the cre-ation of these images. Not only do they purport to show the structural characteristics of the brain, but they also claim to convey function, cognition, and sociality. Genetics are also increasingly coupled to neurobehavioral assessments and become metadata to the voxels in the space inside the skull, so that regarding the brain in these atlases is an increasingly layered exercise. It requires a specific sensibility to the particularities of networked databases of the brain—their size, the quality of images, the way in which digitization was implemented, and the practices of looking enacted through them—to see how these factors shape the constitution of the brain itself. The analytical tools we have put forth in this chapter contribute to developing such a sensibility. They clarify the role of the visual and provide an additional angle of critical examination of the neuroscientific turn, besides those of biologization of the self or bio-governmentality.

The approach we present also shows how to understand the entwinement of digi-tization, pipelines, database logics, and interfaces. We have traced how they come together to support an emerging kind of looking[8] and a new way to create images that matter. Interfaces such as the DTI and other probabilistic atlases demand "relational looking," through which the image is seen as a dataset in relation to the parameters of the atlas/database. In these atlases, observer involvement is distributed and iterative.

The database logic and the networked context of these atlases also add to a tendency toward iteration, malleability, and aggregation.[9] Crucially, these features are the result of important differences between images constituted through cascades of inscriptions (Latour 1990) and those produced through the alignment of computation and digital infrastructure and the development of images as interfaces. The possibilities of digital media contrast with the tendency to build unidirectional cascades in the use of print-on-paper inscriptions, which can be cascaded but not so easily reconstituted, nor re-formed along changing parameters.[10]

We should note that the role of interfaces and suites of technologies in producing authoritative images is not confined to brain imaging but can also be found in other forms of visual knowing (around databases of images) that could be labeled "mundane," such as getting to know a museum's collection (De Rijcke and Beaulieu 2011). Similarly, an emphasis on the fluidity and openness to further scrutiny of digital images can be found in everyday snapshot photography (Rubinstein and Sluis 2008). Such links between scientific and cinematic practices, between professional and amateur spheres, between science and art (Cartwright 1995; Kember 1991; van Dijck 2005), are important because they situate scientific images within culture.

As images become interfaces to networked databases, the dynamics of knowledge production change. Several of the dynamics we identify in this chapter reach beyond the neurosciences, and beyond scientific visualization at large: investments in standardization, in metadata and curation, and in the personalization of data, norms, and even infrastructures can also be found in other networked contexts for digital knowledge. The way images become connected leads to interactions that exceed the limits of single databases, kinds of data, technological platforms, image modalities, scales, and dimensions. These transformations are not simply a question of databases providing information more effectively through digital media. Rather, we are witnessing changes in the interaction with visual information, in the evaluation of what constitutes information, and in the production of visual knowledge.

Acknowledgments

The work presented in this chapter spans many years, and benefited from the insights, critique, and support of participants in our fieldwork, of our colleagues at Science and Technology Dynamics, University of Amsterdam, and at History and Theory of Psychology, University of Groningen, and of the editors. The writing of this chapter was informed by our work in the project Network Realism, for which the Virtual Knowledge Studio for the Humanities and Social Sciences provided financial support and intellectual stimulation. Interactions with Paul Wouters, Susan Legêne, Annamaria Carusi, Sissel Hoel, Hannah Landecker, Lisa Cartwright, and Morana Alač in the course of this project were precious and much appreciated.

Notes

1. This mode of visual knowing is called network realism (http://networkrealism.wordpress .com/). The focus in this chapter is on one particular aspect of this mode of knowing, which is the production of authoritative images.

2. The fieldwork and archival research were conducted by Beaulieu and de Rijcke, respectively, for their PhD research in the late 1990s and late 2000s.

3. However, in order to enable clinicians to make diagnostic evaluations, or researchers to publish, the database is "frozen" in time to provide stability for a given period, and the atlas image recalculated only periodically rather than on an ongoing basis. The mutability of these atlases is therefore limited in practice.

4. http://www.loni.ucla.edu/ICBM/About/, accessed 28 January 2011.

5. Remediation is the dynamic by which digital (or other media) define themselves by borrowing from and refashioning other media forms such as radiology, photography, cinema, animation, television, etc. Remediation is not a one-way process, and we have already seen many examples of printed media remaking themselves in the image of websites or social media.

6. This section is named after a software program we discuss below, called DtiStudio. We aim to bring out the artistic connotation of the word "studio," which typically refers to a place to try things out and where room is provided for creative acts.

7. Recall the possibilities of having an average left-handed, 35-year-old female brain to use as baseline; personalization as a media option can also be found in scientific digital environments.

8. The framing of the observer's tasks also shapes visual practices: for example, in the case of the CT suites discussed by Saunders (2009) he observed "sacral looking," a search for revelation that will enable resolution of the intrigue posed in clinical diagnostics. In the cases we have analyzed, looking is configured either as detection of garbage, of blatant (for a human) noise in a given visualization, or as "a dynamic interaction between trying to find or to generate an image" (Cohn 2007, 99). In the former cases, the observer serves to bound or correct the exaggerations and overinclusive processing of the digital suites.

9. One of the most recent areas of development in these atlases is the aggregation and use of metadata as a way of further disciplining atlas data, or to link the data to what emerges as really important in the course of research, knowing that it cannot be articulated ahead of time. Tagging and using metadata to document provenance are common strategies, not only in digital brain atlases (Van Horn and Toga 2009) but also in data-sharing platforms across life and social sciences (Dormans and Kok 2010) and in social and cultural production (Beaulieu et al. 2012). These processes call attention to the increased importance of understanding how collections are

curated. They also point to the need for the management of databases across disciplines, institutions, and countries—to understand not only atlases but also broader (scientific) digital visual culture.

10. Of course we should not forget that responsibilities come with these possibilities, since observers/users are expected to make appropriate decisions regarding adjusting, selecting, rejecting, or evaluating them. This responsibility highlights the need to understand the skills required to deal with these images (De Rijcke and Beaulieu 2011). It also calls for further studies that trace how observers and users are assigned particular responsibility in the very course of gaining expertise (Goodwin 1994; Alač 2008) and how agency and digital techniques intertwine.

References

Alač, Morana. 2008. Working with brain scans: Digital images and gestural interaction in fMRI laboratory. *Social Studies of Science* 38 (4):483–508.

Almli, C. Robert, Michael J. Rivkin, and Robert C. McKinstry. 2007. The NIH MRI study of normal brain development (Objective-2): Newborns, infants, toddlers, and preschoolers. *NeuroImage* 35 (1):308–325.

Beaulieu, Anne. 2000. The space inside the skull: Digital representations, brain mapping and cognitive neuroscience in the decade of the brain. PhD diss., University of Amsterdam.

Beaulieu, Anne. 2001. Voxels in the brain: Neuroscience, informatics and changing notions of objectivity. *Social Studies of Science* 31 (5):635–680.

Beaulieu, Anne. 2002a. Images are not the (only) truth: Brain mapping, visual knowledge, and iconoclasm. *Science, Technology and Human Values* 27 (1):53–86.

Beaulieu, Anne. 2002b. A space for measuring mind and brain: Interdisciplinarity and digital tools in the development of brain mapping and functional imaging, 1980–1990. *Brain and Cognition* 49 (1):13–33.

Beaulieu, Anne, Sarah de Rijcke, and Bas van Heur. 2012. Authority and expertise in new sites of knowledge production. In *Virtual Knowledge*, ed. Paul Wouters, Anne Beaulieu, Andrea Scharnhorst, and Sally Wyatt, 25–56. Cambridge, MA: MIT Press.

Berg, Marc. 1997. *Rationalizing Medical Work: Decision-Support Techniques and Medical Practices.* Cambridge, MA: MIT Press.

Bolter, Jay David, and Richard Grusin. 2000. *Remediation: Understanding New Media.* Cambridge, MA: MIT Press.

Brenninkmeijer, Jonna. 2010. Taking care of one's brain: How manipulating the brain changes people's selves. *History of the Human Sciences* 23 (1):107–126.

Cartwright, Lisa. 1995. *Screening the Body: Tracing Medicine's Visual Culture.* Minneapolis: University of Minnesota Press.

Cohn, Simon. 2007. Seeing and drawing: The role of play in medical imaging. In *Skilled Visions: Between Apprenticeship and Standards*, ed. Cristina Grisseni, 91–105. New York: Berghahn Books.

Crary, Jonathan. 1991. *Techniques of the Observer: On Vision and Modernity in the Nineteenth Century*. Cambridge, MA: MIT Press.

Daston, Lorraine, and Peter Galison. 2007. *Objectivity*. New York: Zone Books.

De Rijcke, Sarah. 2008a. Light tries the expert eye: The introduction of photography in nineteenth-century macroscopic neuroanatomy. *Journal of the History of the Neurosciences* 17 (July):349–366.

De Rijcke, Sarah. 2008b. Drawing into abstraction: Practices of observation and visualization in the work of Santiago Ramón y Cajal. *Interdisciplinary Science Reviews* 33 (4):287–311.

De Rijcke, Sarah. 2010. Regarding the brain: Practices of objectivity in cerebral imaging, seventeenth century–present. PhD diss., University of Groningen.

De Rijcke, Sarah, and Anne Beaulieu. 2007. Taking a good look at why images don't speak for themselves. *Theory and Psychology* 17 (5):733–742.

De Rijcke, Sarah, and Anne Beaulieu. 2011. Image as interface: consequences for users of museum knowledge. *Library Trends* 59 (4):663–685.

Dormans, Stefan, and Jan Kok. 2010. An alternative approach to large historical databases: Exploring best practices with collaboratories. *Historical Methods* 43 (3):97–107.

Dumit, Joseph. 1994. Desiring a beautiful image of the brain: A cultural semiotic inquiry. Paper presented at the Conference on Visual Representation in Scientific Practice, 28 April–1 May. Institute for Medical Humanities, Galveston, Texas.

Dumit, Joseph. 2004. *Picturing Personhood: Brain Scans and Biomedical Identity*. Princeton: Princeton University Press.

Gong, Gaolang, Yong He, Luis Concha, Catherine Lebel, Donald W. Gross, Alan C. Evans, and Christian Beaulieu. 2009. Mapping anatomical connectivity patterns of human cerebral cortex using in vivo diffusion tensor imaging tractography. *Cerebral Cortex* 19 (3):524–536.

Goodwin, Charles. 1994. Professional vision. *American Anthropologist* 96 (3):606–633.

Hand, Martin. 2008. *Making Digital Cultures: Access, Interactivity and Authenticity*. Aldershot, UK: Ashgate.

Hine, Christine. 2006. Databases as scientific instruments and their role in the ordering of scientific work. *Social Studies of Science* 36 (2):269–298.

Hine, Christine. 2008. *Systematics as Cyberscience: Computers, Change, and Continuity in Science*. Cambridge, MA: MIT Press.

Jiang, Hangyi, Peter C. M. van Zijl, Jinsuh Kim, Godfrey D. Pearlson, and Susumu Mori. 2006. DtiStudio: Resource program for diffusion tensor computation and fiber bundle tracking. *Computer Methods and Programs in Biomedicine* 81 (2):106–116.

Joyce, Kelly A. 2006. From numbers to pictures: The development of magnetic resonance imaging and the visual turn in medicine. *Science as Culture* 15 (1):1–22.

Joyce, Kelly A. 2008. *Magnetic Appeal: MRI and the Myth of Transparency*. Ithaca: Cornell University Press.

Kember, Sarah. 1991. Medical imaging: The geometry of chaos. *New Formations* 1 (5):55–66.

Latour, Bruno. 1990. Drawing things together. In *Representation in Scientific Practice*, ed. Michael Lynch and Steve Woolgar, 19–68. Cambridge, Mass.: MIT Press.

Le Bihan, Denis, Robert Turner, Chrit Moonen, and James Pekar. 1991. Imaging of diffusion and microcirculation with gradient sensitization: Design, strategy, and significance. *Journal of Magnetic Resonance Imaging* 1 (1):7–28.

Le Bihan, Denis, and Peter van Zijl. 2002. From the diffusion coefficient to the diffusion tensor. *NMR in Biomedicine* 15 (7):431–434.

Littlefield, Melissa, and Jenell Johnson. 2012. *The Neuroscientific Turn in the Humanities and Social Sciences*. Ann Arbor: University of Michigan Press.

Lynch, Michael. 1991. Laboratory space and the technological complex: An investigation of topical contextures. *Science in Context* 4:51–78.

Mori, Susumu, B. J. Crain, V. P. Chacko, and Peter van Zijl. 1999. Three-dimensional tracking of axonal projections in the brain by magnetic resonance imaging. *Annals of Neurology* 45 (2):265–269.

Mori, Susumu, Setsu Wakana, Peter van Zijl, and Lidia Nagae-Poetscher. 2005. *MRI Atlas of Human White Matter*. Elsevier Science.

Rheinberger, Hans-Jörg. 1997. *Toward a History of Epistemic Things: Synthesizing Proteins in the Test Tube*. Stanford: Stanford University Press.

Roepstorff, Andreas. 2007. Navigating the brainscape: When knowing becomes seeing. In *Skilled Visions: Between Apprenticeship and Standards*, ed. Cristina Grisseni, 191–206. New York: Berghahn Books.

Rubinstein, Daniel, and Katrina Sluis. 2008. A life more photographic: Mapping the networked image. *Photographies* 1 (1):9–28.

Saunders, Barry. 2009. *CT Suite: The Work of Diagnosis in the Age of Noninvasive Cutting*. Durham: Duke University Press.

Schleim, Stephan. 2010. *Die Neurogesellschaft: Wie die Hirnforschung Recht und Moral herausfordert*. Hannover: Heise.

Shove, Elizabeth, Matthew Watson, Martin Hand, and Jack Ingram. 2007. *The Design of Everyday Life*. Oxford: Berg Publishers.

Thompson, Paul, Kiralee Hayashi, Greig de Zubicaray, Andrew Janke, Stephen Rose, James Semple, David Herman, et al. 2003. Dynamics of gray matter loss in Alzheimer's disease. *Journal of Neuroscience* 23 (3):994–1005.

Toga, Arthur W., and Paul M. Thompson. 2000. Image registration and the construction of mul-tidensional atlases. In *Handbook of Medical Imaging: Processing and Analysis*, ed. Isaac Bankman, 635–658. San Diego: Academic Press.

Van Dijck, José. 2005. *The Transparent Body: A Cultural Analysis of Medical Imaging*. Seattle: University of Washington Press.

Van Horn, John Darrell, and Arthur W. Toga. 2009. Is it time to re-prioritize neuroimaging data-bases and digital repositories? *NeuroImage* 47 (4):1720–1734.

Waterton, Claire. 2010. Experimenting with the archive: STS-ers as analysts and co-constructors of databases and other archival forms. *Science, Technology and Human Values* 35 (5):645–676.

8   Rendering Machinic Life

Natasha Myers

1   Introduction

"Who here has taken a biology course before?" Dan,[1] a professor of biological engineering, looked up at the eighty or so students who had crowded into a too-small lecture hall on the first day of spring semester classes at this private university on the east coast of the United States. They had arrived for a freshman seminar aimed at recruiting a new cohort of students into the school's brand-new biological engineering major. Save one or two, all the students put up their hands. "Good," he responded. "But this will be a little different from what you learned in your other courses." Dan was the coordinator for this half-semester course that featured lectures by biological engineers drawn from departments across the institution. He turned to introduce the director of the program, Stan, who offered the students a taste of what this new major would offer. "Biology has changed," Stan told the class. "When I was your age biology was *just starting* to be on the verge of being quantitative and designable." According to him, the molecular and genomics revolutions transformed biology by making biological "parts" and "components" available for manipulation at the molecular scale. "Biology today is at the point where getting the parts and manipulating them is relatively easy. Now, the hard part is: How do they *work*? Now that you know what the components are, how do they *work*? Well," he announced to the class, "they work as machines."

Stan turned to the projection screen that displayed a black-and-white, time-lapse movie of a macrophage cell migrating across a slide. He and the students watched its magnified, animal-like body undulate as it pulled itself across the screen.

If you look at a picture of a cell here migrating across a surface, you want to know how to make that cell migrate faster, to colonize a biological material, or slower to prevent a tumor from metastasizing. You have to look *inside the machine* for how the molecular components work together as a machine to transmit forces to the environment; to pull on the environment, pull the rest of the cell along. There's the actin cytoskeleton, and all sorts of proteins that link the actin cytoskeleton to receptors across the cell membrane. These all work as an exquisite, *many, many, many, many-*molecule machine.

This moving image was juxtaposed with other projected images, including colorful cartoons of the "molecular machinery" of the cell, and engineering-styled electrical circuit diagrams that traced the intracellular "regulatory circuits" that "govern" the cell's large machine "assemblages." "Now that we have the components," Stan explained, "biology needs to be studied the way engineers look at things."

These freshmen, interpellated as would-be biological engineers ("*you* have to look inside the machine"), were instructed to see this cell as engineers engage their objects. Over the duration of the course, the classroom became a training ground for new students to learn to see through the obscuring density of the seething cellular masses that constitute living bodies. The instructors aimed to instill in them the desire to get at the underlying components and devices that "do work" in the cell to "drive" cellular life.

"These are very appealing metaphors and this is engineering language," Stan explained. Indeed, molecular machine analogies are alluring to many life scientists. They have become pervasive in the conventional forms of writing that appear in scientific texts, as well as in pedagogical contexts and in popularized accounts of the contemporary life sciences. In these contexts, proteins are ubiquitously rendered as "molecular machines," "the machinery of life" (Goodsell 1993), and even as "nature's robots" (Tanford and Reynolds 2001). Biological molecules assemble into the mechanical levers, hinges, switches, motors, gears, pumps, locks, clamps, and springs that "transduce" forces, energy, and information inside living cells (e.g., Hill and Rich 1983; Bourne 1986; Hoffman 1991; Kreisberg et al. 2002; Harrison 2004; Chiu et al. 2005). These components form complex interlocking devices that act to build and maintain the cell as a higher-order machine. In the hands of some biological engineers and structural biologists, cells have become the factory floors of nanoscale industrial plants (see also Calvert 2008; Fujimura, 2005; Roosth 2010).

This chapter draws attention to the conjoined material and semiotic labor involved in the work of *rendering life molecular,* and more specifically it takes a close look at the machinic renderings that propagate so widely through the life sciences today. I begin from the premise that the manufacture of machinic renderings is an exquisite material-semiotic achievement (see Stengers 2008; Haraway 1997). Rather than delivering a polemic against the reductive logics of machine analogies and their contribution to the instrumentalization of life, this chapter offers an ethnographic account of protein modelers' conceptual and haptic creativity with both words and things as they learn to *put machines to work* in living organisms. I argue that rendering molecules as machines is a craft practice, one that makes it possible for practitioners to visualize and intervene in molecular worlds in particularly effective ways. This chapter examines the many layers of this achievement by taking a close look at specific renderings of machinic life at distinct moments in the history of protein science and examining the expertise required to make these renderings do work in research and teaching contexts.

In a more cautionary mode, the chapter draws critical attention to moments when machine analogies collapse in on their referents and *literalize molecules as machines*. Take the example of the cell displayed in the introductory class for biological engineering students. To the uninitiated, including those who showed up for the first lecture, the gooey substances that churned inside the writhing cell looked nothing like machines. Yet for those trained in the practical arts of machinic modeling, the seething cell isn't merely *like* a machine; it has become one in their hands. In such situations, practitioners' semiotic labor and their technical dexterity with machine metaphors are at risk of being rendered invisible. In what contexts does the creative work of rendering molecule-as-machine get obscured? When do analogies become so conventional that they get "frozen into literal expressions" and become "dead metaphors" (Lakoff and Johnson 1999, 124)? In addition to the achievements protein modelers gain by modeling molecules as machines, this chapter also tracks contexts where machine analogies cease to be recognized as the crafty work of their makers and moments when researchers disavow their own ingenuity and contributions in the visualization process. Here, I take heed of Emily Martin's (1991, 501) astute insight that perhaps the analogies that animate life science practice are not so much dead metaphors as "sleeping metaphors," and I follow through on her invitation to "wake up" these slumbering tropes to see what kind of work they are doing for these scientists. The overarching aim of this chapter is to innovate an approach to scientific representation that can keep the creativity of practitioners' rendering practices in the foreground as remarkable achievements.

2 Rending Representation

Protein models are undeniably representations of otherwise imperceptible phenomena. Yet standard accounts frame scientific representation as a practice of describing objects that exist "out there" in the world. In this view, to represent the world well scientists must work hard to reduce the epistemic uncertainties that plague their experiments: they must design better tools and employ more effective language to enable them to deliver a clearer picture of those ready-made objects just waiting to be discovered. Such an approach, however, assumes the world has a fixed ontology that preexists its encounter with the scientist (for further discussion see the introduction to this volume). This chapter offers a counterpose to such "representationalist" analyses by approaching the problem of model building with an attention to how phenomena, like the subvisible substances of life, are *rendered*. The aim is to shift how we think about practices of scientific visualization. To engage model making as a rendering practice acknowledges the ways in which protein models do more than just re-present molecular phenomena. Indeed, these models *rend* the world in particular ways: they pull, tear, and torque the world in some ways (if not others). In the process they shape how and what we come know.

Renderings, in this sense, are *performative*; that is, they recursively transform the ways we see and intervene in the world. This approach to rendering draws on a long genealogy of feminist and queer theories of performativity (e.g., Haraway 1991; Butler 1993; Barad 1996; Herzig 2004). According to feminist science studies scholar Karen Barad (2003, 802), "The move toward performative alternatives to representationalism shifts the focus from questions of correspondence between descriptions and reality (e.g., do they mirror nature or culture?) to matters of practices/doings/actions."[2] Take a close look at the term rendering. As my dictionary reminds me, a rendering can indeed be a representation of something, as in a translation, a work of art, or a detailed architectural drawing. But a rendering is not just an object that can stand in for something else: "rendering" as a verb is also the activity of producing these representations. In this active sense of the term, a rendering is a performance, as in the rendering of a play or musical score. Renderings thus carry the mark of the artist, such that, as the performer enacts it, a musical score is inflected with unique tones, textures, and affects. Another use of the term is in the field of computer modeling, where a rendering is "the processing of an outline image using color and shading to make it appear solid and three-dimensional" (*Oxford American Dictionary*). In this sense, a rendering is the modeler's elaboration, addition, or augmentation of a simpler thing. To render is also to provide, hand over, or submit (as in "to render up" a verdict or a document); these are all performative gestures that pass an object or communication from one person to another. Heard in a different register, to render is also to tear or rip things apart. What holds all of these uses of the term together is that each refers not just to the object that is rendered, but also to the subject, the one who renders, and the activity of rendering (see also Myers 2007, 2012).[3]

A protein model, then, is not just an object at the end stage of model building; the modeler articulates her knowledge of molecules through the rendering process, and in so doing inflects her models with kinesthetic and affectively charged knowledge (Myers 2008a). A model is a rendering in the sense that it embodies, performs, and sediments a modeler's form of knowing. Moreover, when a protein is rendered as a machine, something more than just a likeness to machines is produced. Renderings are simultaneously material, semiotic, and performative; to render is to *world* new phenomena and forms of life. This means that it matters how molecules are rendered: different renderings can activate a scientist's imagination and shape her perceptions; and at the same time, the models and meanings that she mobilizes can act recursively to sediment particular ways of seeing and storying life. Thus, when practitioners figure molecules through machine tropes and stories, they are not just artfully describing the molecular phenomena that appear at the end point of their experiments. Their renderings transform how the substances of life are made into visible, tangible, and workable objects.

Which renderings come to matter? This is a crucial question as renderings can contour the conditions of possibility for *what can be seen, said, imagined, and felt as fact* at particular moments in the history of a science. Machinic renderings have played a central role in shaping how molecules have come to matter as experimental objects in the history of the life sciences, but they are not the only renderings that animate the life sciences. Molecular life has been and continues to be figured in registers that oscillate rapidly between machinic and animistic forms.[4] There are, however, particular contexts in which practitioners explicitly put the metaphor of molecular machines to work and denounce more wily figurations. This chapter documents sites where conventions of scientific writing, pedagogies, and popularized accounts encourage practitioners to clamp down on machinic renderings, and examines the effects of this literalization and naturalization of machines in cells. One crucial insight generated in this study is that machine analogies have special salience for some practitioners. Once deployed as recruitment devices, machine analogies can lure practitioners and students with engineering expertise into the study of protein biology. The effect is that machinic renderings are contouring "thought collectives" in the life sciences today (Fleck 1979).

3   The Invention of Molecular Machines

In the late nineteenth century, Thomas Henry Huxley drew on a long history of mechanistic reasoning when he argued that life must be analyzed according to its chemical and physical properties. In the 1870s he developed a mechanical theory of the cell that he peddled as the "protoplasmic theory of life" (see Gieson 1969; Huxley 1878). In 1880, the *Encyclopedia Britannica* published Huxley's definitive entry on "Biology," in which he displayed for a wide audience his new way of thinking about the stuff of life. He explained:

A mass of living protoplasm is simply a molecular machine of great complexity, the total results of the working of which, or its vital phenomena, depend, on the one hand upon its construction, and, on the other, upon the energy supplied to it; and to speak of vitality as anything but the name of a series of operations, is as if one should talk of the "horology" of a clock. (cited in Beale 1881, 297; see also Huxley 1878, 15)

Huxley was influenced by his Cartesian predecessors who conjured pliers, springs, pumps, bellows, cords, retorts, and hydraulic systems inside the bodies of living organisms to help them interpret the mechanical functions of organs and tissues (see Canguilhem 2008, 78–79).[5] The difference Huxley made was to effect a scale shift that rendered machines inside the subvisible recesses of the cell.

Huxley's parsing of the protoplasm as a working machine whose aggregated parts must be supplied with energy would probably not make today's audiences swerve. Huxley's contemporaries, however, took great exception to this analogy and its implications for vital phenomena. Lionel Beale, a vitalist who argued against the mechanization of life, was the president of the Royal Microscopical Society and one of Huxley's most prominent opponents (see Gieson 1969). In his 1881 annual address to the Society he voiced this strong objection to Huxley:

It is not most wonderful that Professor Huxley can persuade himself that a single reader of intelligence will fail to see the absurdity of the comparison he institutes between the *invisible, undemonstrable, undiscovered* "machinery" of his suppositious "molecular machine" and the *actual visible works of the actual clock, which any one can see and handle, and stop and cause to go on again.* (Beale 1881, 297, emphasis added)

Beale's primary complaint was that, given the limits of microscopic vision at the time, "molecular machines" could be no more than an elaborate fantasy. For molecular machines to exist they had to have, like a clock, "actual visible" workings into which one could intervene. According to Beale, "[m]agnify living matter as we may, nothing can be demonstrated but an extremely delicate, transparent, apparently semi-fluid substance" (ibid., 279). Beale deplored subjective interpretations and rejected the use of what he saw as figurative language. Refusing a machinic analogy, he placed his faith firmly in a "mechanical objectivity" that entrusted his vision to the limited power of his microscope (see Daston and Galison 2007). Beale's objectivity hinged on a practice of detachment and neutrality, and he vociferously disavowed the figurative nature of his own rendering practices. Better representation would have to wait until he could augment the magnification of his microscope; only then would the true nature of living substance be confirmed.

What shifts when scientific representation is construed as a performative practice that relies as heavily on semiotic invention as on optical devices like microscopes? From this perspective it is possible to see the irony of Beale's denunciation. He apparently did not recognize the genius of what might be called Huxley's "working conceptual hallucination" (Gilbert and Mulkay 1984, ch. 7, quoted in Lynch 1991, 209). In contrast to Beale's intractable vision of the delicate, transparent substance of cells, Huxley's molecular machine could become an alluring object of analysis for the exact scientist. By conjuring the cell as a complex molecular machine, Huxley conceived of a biological object whose properties could, theoretically at least, be quantified. The metaphor of machinery offered Huxley a bridge he could traverse in his imagination between the visible, tangible, and manipulable world in which he lived and the invisible, intractable world of biological molecules.[6] The prominence of the molecular machine analogy today suggests that this indeed was the enduring metaphor that could lure would-be engineers into the sciences of life.

Huxley's theory of molecular machines was not vindicated in the late nineteenth century. In recent years, however, structural biologists have been augmenting the resolution of their optical systems to visualize molecules at atomic scale, and increasingly they are modeling large assemblages of proteins. Oddly enough, the closer they look, the more machines they seem to be discovering "at work" in cells. Some might read this as a triumphalist tale in which molecular visualization technologies have finally vindicated Huxley's daring and provocative *premonition* of the underlying nature of molecular life. Indeed, it seems as if the mechanical "works" of the cell have finally been made as "actual" and tangible as Beale's clock.

Yet, as Donna Haraway reminds us, there can be no unmediated access to the molecular realm: visual systems are "active" and "partial" ways of "organizing worlds" (1991, 190). To insist that all molecular visions are *renderings* is to acknowledge that those who practice molecular visualization have never relied exclusively on visual evidence in order to construct models of subvisible worlds. That is, visualizing molecules has always involved invention: modelers must continually conjure new figural vocabularies; and these analogies can grant them at best a tenuous, tentative link between what is visible, imaginable, and speakable at that particular moment in the history of their science (see Foucault 1973).

From this perspective it becomes possible to see that the visualization technologies researchers deploy do not merely "unveil" fully functional molecular machines within the body of the cell. Machinic analogies are also not merely aesthetic flourishes of language or attractive figures of speech. Rather they can be seen as powerful devices for rendering new views of life. Moreover, these renderings can also be seen as "lures" that "vectorize" practitioners' imaginations and experimental inquiry (Stengers 2008). What, then, is the allure of the machine for these practitioners in the sciences of life?

4   Twenty-first-Century Molecular Machines

The 1980s and 1990s saw living phenomena rendered as a problem of coding to be solved in the language of cybernetics and informatics (see Kay 2000). In her reflections on the sequencing craze spurred on by the rise of corporate biology and genetic engineering, Donna Haraway suggested that "the living world" had become a "command, control, communication, intelligence system," or "C³I in military terms" (1987; 1991, 150; 1997, 97). In 1987, in an early publication of her "Manifesto for Cyborgs," Haraway remarked: "The new machines are so clean and light"; "cyborgs are ether, quintessence" (1987, 7). In the era of C³I, materiality was disavowed, and the "depths" of the organism were elided by the glinting "surfaces" of silicon chips and bodyless codes.

In the twenty-first century, our machines, sciences, and economies are transforming. Practitioners in this postgenomic era of computer-intensive, atomic-scale, 3D modeling and simulation refuse to flatten molecular phenomena into thin threads of information read as genetic scripts. Structural biologists are involved in the work of reconstituting the materiality of cells and molecules by modeling them as machines with functional architectures. In the process, a distinct set of machines have come to stand in as figures for contemporary molecular phenomena (see Fujimura 2005). Stephen Harrison, a prominent protein crystallographer whose Harvard-based laboratory builds atomic-resolution models of protein molecules, has seen "hints" in recent years that

specific kinds of control logic are embodied in specific kinds of molecular architecture. . . . Thus, structural biology must seek to understand information transfer in terms of its underlying molecular agents by analyzing the molecular hardware that executes the information-transfer software. . . . The architectural principles of the cell's control systems and the dynamics of their operation are no less proper studies of structural biology than are the organizational and dynamical properties of the molecular machines that execute the regulated commands. (Harrison 2004, 15)

Harrison remains invested in a view of the cell as a hub of information transfer modeled on a militarized computing command, control, and communication system. Yet for Harrison, life is denser than code. In his iteration, analyzing the "execution" of a cell's "regulated commands" requires an approach that goes beyond cracking codes scripted by genetic messages. Such analyses must be buttressed by structural studies of the "architecture" and "hardware" of living cells. He keeps his eye on the organization and dynamics of the "molecular agents" that underlie the transfer of information and the transduction of forces and energy in the cell. These "molecular machines" are the cell's hardware, and thorough investigation of their physical and chemical properties is required in order to give full form to this model of cellular life.

As Harrison and his contemporaries invent novel analogies, new machines begin to accumulate inside cells. Like cars, cells are now constituted by "biological parts," "components," and "devices." Molecules have become mechanical objects that exert force and perform "work" in order to "drive" cellular life. These machines are built from a range of familiar devices including computer hardware, electronic circuits, and the springs, locks, clamps, pumps, and motors of modern-day mechanical devices. Machinic renderings also gear into older imaginaries, including cogs and wheels of early industrial capitalism. One such rendering appears on the cover of a 2006 issue of the journal *Cell,* which features a research paper describing a "cogwheel for signal transduction across membranes" (see figure 8.1).

5   Renderings as Materialized Refigurations

Nature is . . . about figures, stories, and images. This nature, as *trópos,* is jerry-built with tropes; it makes me swerve. A tangle of materialized figurations, nature draws my attention. (Haraway 1994, 60)

Haraway's remarkable insights into machines and metaphors in the life sciences make her especially good to think with on the topic of rendering life as machine. She invites her readers to take a closer look at the tropes and stories that organize our optics and shape how and what we know about the living world. She asks: "How do we learn *inside the laboratory and all of its extended networks* that there is no category independent

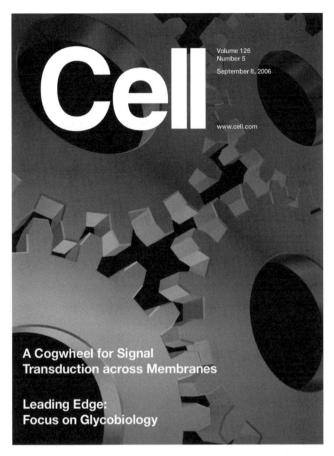

Figure 8.1
The cover of *Cell* 126 (5) (8 September 2006): "A Cogwheel for Signal Transduction across Membranes." Used with permission.

of narrative, trope, and technique" (1997, 161, emphasis in the original)? She explains that in Greek, "*trópos* is a turn or a swerve"; tropes spin meanings off in new directions; in so doing they "mark the nonliteral quality of being and language" (ibid., 135). To bring attention to the figures that shape how and what we see and know is not to reduce the world to text or language. Tropes, metaphors, and analogies are not immaterial utterances or "mere textual dalliances" for her: they are "material-semiotic actors" (ibid., 64; 1991, 200; see also Latour 1990).

Technoscience is a site where things and words can be made to "implode" (1997, 97). The machinic renderings so central to the life sciences today are the products of what Haraway would call *materialized refiguration*. This a concept that makes palpable the ways that tropes are sedimented or "corporealized" in objects and forms of life (ibid., 141). Materialized refiguration is about more than "metaphor and representation" for Haraway:

Not only does metaphor become a research program, but also, more fundamentally, the organism for us is an information system and an economic system of a particular kind. For us, that is, those interpellated into this materialized story, the biological world *is* an accumulation strategy in the fruitful collapse of metaphor and materiality that animates technoscience. (1997, 97)

Renderings are like the materialized refigurations that corporealize life in the form of information systems; they also embody this "fruitful collapse of metaphor and materiality." As we shall see, machinic renderings not only constitute molecular visions; they also reconfigure entire research programs, and, crucially, they inform *who is recruited to do the work* of modeling life-as-machine.

It turns out that the work of rendering cells as factories and proteins as machines hinges on the productive meeting of biologists and engineers. During an interview with Joanna, a postdoctoral fellow trained in the sciences of protein folding, I learned about one such productive meeting. Jim, the principal investigator in Joanna's lab, had invited Geoff, a mechanical engineer, to join their weekly lab meetings. Jim had sought out his expertise for help in working on the lattice structure of viral coat proteins. According to Jim, "No wet biochemist could deal with a lattice. Who knows about lattices? Engineers know." On his first day at the group's lab meeting, Geoff completely refigured how Joanna thought about a protein with which she was already quite familiar:

Just as we were all sitting around the table describing [the protein], and talking about [it], all Geoff did was take a paper clip that was sitting at the table . . . and he took the paper clip and he's like "Okay. I need to understand what you guys are talking about." And he just folded the paperclip into a three-dimensional representation of what we were talking about. And we were all sitting there going, "Wow." It was just kind of so bizarre that after twenty some odd years of working on [this protein], [we] had never thought about it in this way.

And Geoff, as soon as we started talking about the structure, he wanted to see a model of it. It was like, immediately, "Let's make a model." He came back the next day with more elaborate wire that he had taken and molded at his house. . . . And he came in and said, "Oh! There's a clamp! This is a lock. I mean this is lockin' that molecule right in place." And now there are all these papers they've published on the molecular clamp. It's a lock! It's a clamp! And it's so exciting. And it's funny, his whole approach was entirely different . . . I'd looked at that structure a million times. You know. And I was like, "Oh yeah. It's a lock! I mean that's a clamp!" . . . I wasn't working on that project, but it opened a whole door of experiments that would have not happened otherwise. An entire postdoc was hired to work on this. And it hadn't been called a clamp until Geoff came to the meeting.

A modeler by training, this engineer needed to hold the structure in his hands. Simple modeling tools would suffice: a paper clip or wire that he could turn and twist.

But he also had other tools at his fingertips that were integral elements of his visualization apparatus: he had a facility and a familiarity with clamps as mechanisms, and rendering the protein as a molecular clamp came easily to him. As Haraway (1997, 97) suggests, a metaphor can even drive an entire research program. In this sense, the livelihood of a postdoc, the scientific significance of a protein, and the productivity of a research laboratory all *turned* around the figures of the clamp, the engineer, and his jerry-built model.

6   Cultivating a Feeling for Machines

It takes work to build machines into the bodies of organisms. Haraway's call to attend to the materiality of language is thus also a call to make visible how technoscientific tropes are put to work, for whom, and at what cost. Today, structural biologists use a vast range of interactive computer graphics media and multiple conventions for depicting molecular structures (as wire frame models, ribbon structures, space-filling models, etc.). These distinct media produce numerous opportunities for modelers to express their own molecular aesthetic. Modelers can render proteins in a range of aesthetic forms: in ways that make them appear to have glinting, metallic architectures; or as globular, gooey bodies that wriggle when they are animated on screen (see Myers 2006, 2012). Thus, while the rhetoric of the molecular machine is pervasive, it is by no means the only way proteins are figured. When, how, and for whom do proteins cease to oscillate between lively body and machine? When do life scientists' discourses and practices *clamp down* on the molecule as a machine? What other figurations of life cease to exist?

Fernando is a fifth-year PhD student working in a protein crystallography laboratory. In one of our several interviews I asked him if he ever used metaphors other than machines to talk about his proteins. I mentioned that I heard structural biologists talk about proteins as wily, lively bodies. This suggestion put him on edge a little, and his response was firm: "A protein by itself is not a living thing," he tells me. "It is . . . it is a machine. And it will break down, just like machines do. Okay? And if something is not there to repair it, another machine, another piece of machinery" (the whole system) will "break down." At the suggestion that proteins had lively qualities, Fernando clamped down firmly on the metaphor of molecular machines.

Fernando is fluent in the rhetoric of molecular machines. Yet machine metaphors are not just pedagogical devices for him. He likes to use the metaphor in part because he has a particularly nuanced feel for machines and their parts. He is a latecomer to science, and at forty he is significantly older than most of the graduate students in his cohort. He grew up in a working-class Hispanic family and spent his twenties working as a plumber, manual laborer, and pizza delivery boy, and took much pleasure in building cars. He later went back to school, and started teaching CAD drawing to architecture and engineering students at a community college. Machines are familiar to Fernando: they are, like Beale's clock, "actual visible works" into which he can see and intervene. He understands how they work, how their parts fit together, and what keeps them ticking. Our conversation produced dizzying Alice-in-Wonderland effects of scale as we zoomed along what seemed to be a continuum of visibility, from human-scale machines down to the scale of molecular machines and back again.

As a protein crystallographer he builds models of proteins to figure out what the "machinery" of the cell looks like and how it works. For him, X-ray crystallography is a visualization tool that he uses to get a "snapshot of the machine." He describes his job as a protein modeler through an allegorical tale that took us to the factory floor of a robotics-mediated automotive assembly line:

So you know, you are talking about the machine that screws in the fender at the Ford car plant. We're studying that machine because we are trying to find out what it does. And without [the X-ray crystal] structure we are just *feeling* it, just tentatively, sometimes with big thermal gloves. So we can't really get to *feel* the intricacies or the nuances of the drill bits. And all of a sudden crystallography is a snapshot of the machine. Okay. It [the machine] can even be in multiple states. Standing still turned off. In a state when there is a screw being drilled into the fender. You know, it can be somewhere in between. Alright? But because we've seen a similar machine in another company, we kind of have an idea of what the machine does. We've seen the individual parts and stuff like that. I'm not going to mistake the machine for drilling for the machine for welding. Okay. What crystallography allows you to do is to say, "Hey, that is a drilling machine, not a welding machine." Okay. And by looking at certain parts of the machine you can tell whether the drill bit is six inches long or two inches long or whether it has a neck that moves up and down or whether the neck is static. That's the sort of stuff you get in a crystal structure that you don't have before.

Intense in his delivery, Fernando successfully sustained the analogy of the cell as the factory floor of the Ford car plant throughout his story. He had such a strong grip on the analogy that there was eventually a slippage from the machine as a metaphor for the molecule to the molecule that had actually become a machine, in this case a (robotics-mediated) machine that could do highly specialized kinds of work in a (capital-intensive) cell.

Fernando's image of a human worker whose tactile and visual acuity is dampened by wearing "big thermal gloves" is an effective analogy for how hard it is for a structural biologist to make sense of molecules without both the resolving power of X-ray crystallography and an elaborate figural vocabulary. Crystallographic modeling gives Fernando both a three-dimensional visualization of the spatiality of the molecule and a "nuanced" "feeling" for its "intricate" structure. As he made clear during another interview, crystallographic modeling with interactive computer graphic interfaces is, for him, a craft practice through which he has been able to develop what he calls a kind of "touchy-touchy-feel" for the molecular model as he builds it on screen: "I don't want to say touchy-touchy-feely like that, but that sort of holding on to something and getting a feel of it." Taking off his thermal gloves, so to speak, he uses the interactive computer graphics interface to bring molecules into haptic sensation, as well as into view. And yet he goes a step further: he draws on his feeling for machines to complete his mechanistic model of the molecular structure.

Though he is quite taken by X-ray crystallography, Fernando is ambivalent about his future in the field. "You can get so fascinated by the intricate gear work of a particular piece," he told me, "that you never learn how to operate the whole machinery." He finds that he's attracted to developments in biological engineering that promise the possibility of "operating" the "whole machinery" of the cell. These are the same promises that have enticed a new generation of students to sign up for the biological engineering major at his university, the one directed by Stan. The "molecular machine" is thus a powerful lure for recruiting would-be engineers into life science practice. Rendering molecules as machines gives engineers something they can get their hands on, something they can grasp. And it is through this metaphor that biology has become quantifiable, manipulable, and redesignable, in ways that have enabled engineers to *rework biology*; literally and figuratively they have *put life to work* at the molecular scale.

7   Engineering Biological Engineers

A biophilic analogy might figure metaphors as enzymatic catalysts: that is, as dense, fleshy substances that can activate, congeal, and precipitate new kinds of bodies and new kinds of meanings. Like enzymes, metaphors left to their own devices don't automatically crystallize into forms that can produce new meanings and material effects. Crystallographers expend great effort to coax proteins to form "living, breathing" crystals; similarly, metaphors must be nourished and sustained within the context of a practice and a culture that can keep them alive. This raises a problem for the emerging discipline of biological engineering. While machines are prominent companions in daily life, as technical objects they tend to fall outside the expertise of the classically trained biologist and, more to the point, outside the technical competence of many freshmen and sophomores taking a seat in introductory biological engineering courses. Few would likely have had as much experience with machinery as Fernando.

For life scientists to effectively *work the machine into the cell*, their expertise also must be reconfigured. In order to properly build and deploy molecular machines, biological engineers must enlist and train a new generation of scientists to think and work like engineers. They do this by luring new recruits with the promise of technological precision in manipulating biology at the molecular scale. In order to make machine analogies do work, however, they must *engineer biologists* who not only have a "feeling for the organism" but also have a feeling for the machine.[7] Biological engineers' kinesthetic and affective dexterities must be calibrated to machines if they are going to be successful in their efforts to render living substances as workable machines.

Ethnographic observations in a semester-long undergraduate biological engineering teaching laboratory taught me that cultivating a feeling for machines is not a straightforward task. This was a required laboratory course for sophomores who wanted to major in biological engineering. The third of this four-module course focused on what the instructors called "systems engineering." Over the course of six labs, students helped fine-tune a "bacterial photography" system that used the simple principles of pinhole camera technology to generate images on the surface of bacterial colonies. The goal was to optimize the conditions under which the engineered light-sensitive bacteria would change color when exposed to light. In order to analyze this system they needed to understand the cellular "circuitry" that controlled the engineered bacteria's light sensitivity. Their laboratory manual offered this insight:

Biology is particularly well suited to model building since many natural responses appear digital. . . . The digital responses of cells to perturbations, combined with lab techniques for moving DNA parts around, allow logic functions and circuits to be constructed in living cells.

But what is a circuit? Electrical engineers are of course quite familiar with circuits; the concept of a cellular circuit is drawn from electrical engineering, and uses the same iconography and nomenclature as electronic circuit diagrams (see for example Gilman and Arkin 2002; Dumit 2010). In the lab, the students were introduced to an analog— what could be considered a materialized figuration—of their photosensitive bacteria. This analog was itself a circuit board. The students were expected to apply electrical engineering concepts and complete a light-sensitive electronic circuit in order to demonstrate their understanding of the circuitry they were actively building into their bacterial cells. Each pair of students had at their desk a partly assembled electronic "solderless breadboard" that included a photodiode and an LED, which, if the circuit was completed correctly, would turn on and off in response to changes in light intensity sensed by a light-sensitive photodiode (figure 8.2).

Students were asked to complete the circuit by connecting resistors of varying strength to appropriate sites on the breadboard. Some of these students, however, struggled with basic electrical engineering concepts. Meera, who was trained in computer science before coming into biological engineering, was the TA for this module of the lab. She ran a remedial tutorial in electrical engineering several times over for small groups of students. Looking rather confused, they gathered around her at the white board. The laboratory director, herself not trained as an engineer but as a molecular biologist, also joined the lesson. Current flow, resistors, converters, photodiodes, signal matching, and ground all had to be explained. Meera, who had assembled all the circuit boards herself, seemed a little surprised by how hard it was for the students to get the concepts: "Inverters . . . you all know what that is? . . . Okay? . . . Does it make sense when I say current flows through a wire? . . . Does that make sense?"

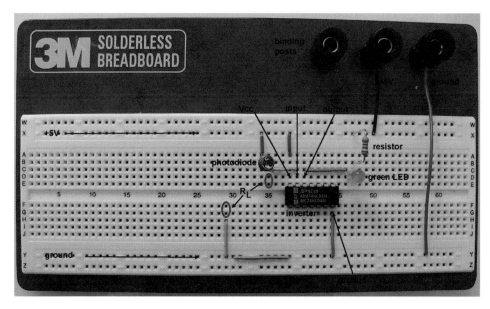

Figure 8.2
A "breadboard" wired up as an analog of the bacterial photography system. Used with permission.

The students' blank stares and repeated questions gave the lie to the excited statement in their lab manual: "Notice how much easier it is to assemble electrical circuits as compared to biological circuits. It takes seconds to swap in a new resistor into your circuit but a few days to assemble" a couple of "biological parts." Apparently it was not that easy: what they were being asked to do was to make sense of what was a rather dense material-semiotic tangle. Modeled on a circuit diagram, the bacteria had been engineered from "standardized biological parts"; that is, proteins and genetic sequences were being modeled as input, output, and signal-matching devices, resistors, inverters, terminators, and protein generators. The bacteria were in one sense already an analog, or rendering, of an electronic circuit. In another sense, the circuit the students fumbled with at their lab benches was an analog of a cell that had been rendered and built on the model of the circuit board. Their attempts to wrap their heads and hands around this veritably loopy set of renderings did indeed give them cause to swerve.

One might expect that analogies are most useful when they draw on knowledge of a familiar realm to illuminate another, less well-known realm. Cellular signaling and regulation were consistently rendered as circuits in classroom lectures, yet the students did not yet have an appreciation of its full import. The bacterial system they had been using throughout the module depended on an in-depth understanding of electrical circuits. However, the students were not yet fluent in the proper terminology and techniques. The circuit-building exercises were, in this regard, quite productive: they were diagnostic of where students' understanding of the analogy of cellular circuits came undone; and they also offered to remedy the situation by enabling students to cultivate a feeling for circuitry.

One of the lessons learned by the students, their instructors, and their ethnographer was that circuits are not self-evident figures for cells, and that the skills to render cells as machines must be cultivated. It takes work to build machines into organisms, and this work must be supported by a practical, conceptual, and material culture.

8   Deadpan Literalism

In an article in *Nature* in 1986, cell biologist Henry Bourne remarked astutely that in the life sciences "argument by analogy, like gambling, was once practiced behind closed doors" (Bourne 1986, 814). Bourne was claiming that by the mid-1980s, analogical reasoning in biology had finally been "elevated into respectability" with rich "payoffs" (ibid.). In this article Bourne could be said to have "outed" analogy as an integral practice in the work of science. This was a remarkable declaration in a context that doesn't normally reward efforts to acknowledge the figurations and fantasies that shape scientific knowledge.

What is curious, however, is that just as Bourne put the rich productivity of some analogies on display, he simultaneously obscured others. Notably he made no reference to the machinic figures that promiscuously populated his essay. One hundred years after Huxley first introduced the machine metaphor, it seems to have lost its punch. Bourne's molecular machines are no longer animating figures that enable the scientist to take a leap across the divide between the visible and the invisible; with atomic-resolution molecular vision, molecular machines have been forged as fact and become unremarkable things-in-themselves.

Haraway is wary of such moments in which the richly "tropic" and figurative nature of technoscientific vision is erased or denied (1997, 133–137). By naturalizing and literalizing machines in the bodies of organisms and asserting the neutrality of visualization technologies, practitioners risk giving the impression that they are merely unveiling the underlying machinery of life. In so doing they disavow the power of their own renderings. In this light, their creative labor is at risk of being made invisible and drawn back behind closed doors.

When are molecular machines naturalized as nature's tools, rather than recognized as the elaborately constructed figural machinery of the investigator? This phenomenon can be seen clearly in the context of the high-stakes debates unfolding in the US between advocates of evolutionary biology and those who argue for "intelligent design" in the origins of life. Remarkably, proponents of both intelligent design and neo-Darwinian evolutionary theories deploy the metaphor of molecular machines with serious deadpan literalism.[8] The question of "who" (God) or "what" (nature) made these molecular machines is anything but a trivial matter. What is ironic is that both sides continually defer the responsibility for engineering these machines to higher powers, either natural or divine. Neither the creationists nor the evolutionists want to take credit for this crafty work. Neither are in a position to laugh at the absurdity of their denial; their silence on the matter reveals the depth of their investments in the work of either naturalizing or deifying molecular machines. But the joke is on them: these are neither God's clever little devices nor evolution's sometimes-clumsy concoctions. Once they are recognized as renderings, molecular machines can be seen as none other than modelers' own inventions; indeed, machinic renderings are marvelously crafted devices that enable both evolutionists and creationists to materialize and manipulate the molecular world to their own ends.

In the end, it is the biological engineers who recognize the absurdity of this denial. For, though they might nervously muffle their laughter, biological engineers do get the joke: as they struggle to reassert the respectability of "design" and "designers" in the realm of life science, they do, after all, want recognition for their labors—those massive, micro-scale engineering projects that they have rigged up within living cells. They are keenly aware of how machinic renderings have been productive of new objects, meanings, lines of research, and forms of life. In the face of pressure to patent their designs and inventions, they understand well how these renderings sustain their very livelihoods.

9   Conclusion

If, as Haraway suggests, materialized refiguration is a practice of "worlding" that gives substance and significance, body and meaning, to emerging technoscientific objects, then how researchers render the stuff of life matters: this is a practice that materializes some kinds of bodies and meanings rather than others. So in what other ways might molecules be rendered? The account I have just offered is complicated by the fact that machines are not the only figures that populate molecular imaginaries. Ethnographic attention to the wide range of analogies in use in laboratory practice, and to the specific moments in which machinic renderings are performed, shows that the machine analogy does not resolve fully mechanical objects in the bodies of organisms. In spite of attempts to clamp down on the figural vocabularies of proteins, to render them as deterministic machines, life scientists' animated performances of the analogy produce biological machines that are undeniably *lively*.

Indeed, the biological engineers that designed the laboratory course recognized the limits of their electrical circuit analogies. After describing the "easy" features of the circuits they had engineered in bacterial cells, their laboratory manual provided this caveat:

In practice, spatial and temporal factors hamper even simple designs. The *cell is a messy circuit board* without the static physical separation you could find between electronic circuit elements. Proteins are made and roam the cell, invariably interacting with nucleic acids and with other proteins in unpredictable and unspecified ways [emphasis added].

Proteins in this "messy circuit board" can "roam the cell," and in their meanderings they escape full characterization and predictive analysis. In practice, biological engineers' machines are indeterminate and unpredictable. Modelers animate their machinic renderings as if they were simultaneously breathing, desiring, writhing bodies with chemical affinities and anthropomorphized affects. This performative practice makes the craftwork and creativity of their figurations even more palpable. Biological engineers and structural biologists thus articulate a far more "wily biology" (Dumit 2003) than their mechanistic discourse avows.

In the end, to literalize machines in the body of the cell is to refuse to recognize the creative labor involved in rendering machinic life. Taken as a profound achievement, in Stengers's (2008) sense, it is possible to track the "intense pleasure in skill" required to render molecules as machines. In Haraway's cyborg figuration: "The machine is not an *it* to be animated, worshiped, and dominated. The machine is us, our processes, an aspect of our embodiment" (Haraway 1991, 180). The machinic renderings documented here remind us that there is, after all, "no fundamental, ontological separation in our formal knowledge of machine and organism, of [the] technical and organic" (ibid., 178). And yet machinic renderings must always be seen as artfully engineered devices that have worked effectively to secure some kinds of matter, meanings, and forms of life, if not others. Perhaps it is by acknowledging these creative labors as an achievement that we might be able to keep open what it is possible to see, say, imagine, and feel as fact in the twenty-first-century life sciences.

Acknowledgments

I'd like to thank the scientists and engineers who graciously allowed me to spend time in their laboratories and classrooms. Thanks also to the editors of this volume, the anonymous reviewers, and Michelle Murphy and Astrid Schrader who offered crucial feedback on this chapter at various stages. An excerpt of an earlier version of this chapter was published in 2008 in the journal *Spontaneous Generations* (see Myers 2008b).

Notes

1. In this chapter, the names of all ethnographic informants have been changed.

2. "Discourse" in this context is not merely "what is said" (Barad 2003); it includes the conditions of possibility for *what can be seen, said, imagined, and felt*. Discursive practice is an "emi-

nently solid process" (Haraway 1997, 64) that materializes some kinds of scientific objects, some forms of life for practitioners, and some worlds, if not others.

3. While my use of the concept of rendering is developed out of a genealogy of feminist and queer theory, it is important to note that a conception of rendering and rendering practices has long been in use in STS and ethnomethodology. See especially Garfinkel (2002) and Lynch (1985, 1990).

4. Recall the time-lapse image of the cell that Stan rendered as a machine. That movie was generated in Dan's laboratory, and in other contexts Dan can be seen animating the cell as if it were a rock climber inching its way along a sheer surface (see Myers and Dumit 2011). Indeed, there has been a continuous oscillation between the figure of the protein as an animate body and as machine (see Myers 2006, 2012). At the turn of the twentieth century, however, mounting evidence of the chemical and physical basis of vital phenomena contoured a discursive field that favored mechanistic accounts (see for example Jacob 1973; Lenoir 1982; Kay 1993; Keller 1995). This was facilitated in part by efforts already well under way among chemists to figure molecules as the mechanical building blocks of life (Meinel 2004).

5. Mechanism has long held a prominent place in the history of inquiry into the nature of living organisms (see for example Gieson 1969; Hopwood 1999; Keller 1995, 2002; Pauly 1996). Georges Canguilhem (2008) documents the history of machinic figurations that have shaped concepts of the organism. These tropes and logics date as far back as Aristotle, who likened animal movements to the parts of "war machines," specifically the "arms of catapults" (Canguilhem 2008, 78–79).

6. This is a phenomenon that Alberto Cambrosio and his colleagues (1993) have examined in their historical study of the cartoon diagrams Paul Ehrlich drew in the process of his inquiry into the otherwise imperceptible chemical interactions among antibodies and antigens. In this case, Ehrlich's "inventive" diagrams conjured antibodies in ways that allowed him to hypothesize their interactions with other molecules. While these diagrams were not grounded in empirical evidence, they did help him to organize his experimental protocols and to establish a successful research program.

7. One may see a "feeling for the machine" as the masculine counterpart to Evelyn Fox Keller's (1983) study of Barbara McClintock's "feeling for the organism." However, it is crucial to avoid conflating these forms of "feeling as knowing" with gendered stereotypes. Gender is part of the story in all forms of knowing, but the scene is much more complex. Feminist science studies scholars have long shown that gender is itself in the making in the laboratory (e.g., Haraway 1997). In molecular biology, for example, a "feeling for the organism" becomes a "feeling for the molecule" (Myers 2008a). Ethnographic research shows that male scientists and students actually have more permission—more physical and discursive space—in which to articulate their "feeling for" molecular structure (Myers 2010, 2012). The "feeling for the machine" documented in this chapter must be understood in the context of larger fields of power: who does and who does not get recruited to reengineer biology? Indeed, more often than not, boys and men are encouraged to develop the capacity to design, make, and build things. And yet, where I conducted my research, women comprised nearly half of the faculty and more than half the students. Whether

they are given the same opportunities to cultivate their feeling for the machine is the crucial question at hand.

8. For insight into how molecular machines are used by proponents of intelligent design, see the Access Research Network website: http://www.arn.org/mm/mm.htm (accessed 30 April 2013).

References

Barad, Karen. 1996. Meeting the universe halfway: Realism and social constructivism without contradiction. In *Feminism, Science, and Philosophy of Science*, ed. Lynn Hankinson Nelson and Jack Nelson, 161–194. Boston: Kluwer.

Barad, Karen. 2003. Posthumanist performativity: Toward an understanding of how matter comes to matter. *Signs* 28 (3):801–831.

Beale, Lionel S. 1881. The address of the President of the Royal Microscopical Society. *Science* 2 (52):294–297.

Bourne, Henry. 1986. One molecular machine can transduce diverse signals. *Nature* 321:814–816.

Butler, Judith P. 1993. *Bodies that Matter: On the Discursive Limits of "Sex."* New York: Routledge.

Calvert, Jane. 2008. The commodification of emergence: Systems biology, synthetic biology and intellectual property. *Biosocieties* 3 (4):383–398.

Cambrosio, Alberto, Daniel Jacobi, and Peter Keating. 1993. Ehrlich's "beautiful pictures" and the controversial beginnings of immunological imagery. *Isis* 84:662–699.

Canguilhem, Georges. 2008. *Knowledge of Life*. 3rd ed. New York: Fordham University Press.

Chiu, Wah, Matthew Baker, Wen Jiang, Matthew Dougherty, and Michael Schmid. 2005. Electron cryomicroscopy of biological machines at subnanometer resolution. *Structure* (London, England) 13:363–372.

Daston, Lorraine, and Peter Galison. 2007. *Objectivity*. New York: Zone Books.

Dumit, Joseph. 2003. How to do things with science: Facts as forces in an uncertain world. Paper presented at the Patient Organization Movements Symposium in Gothenburg Sweden.

Dumit, Joseph. 2010. True demons of cognition: When computers were glad, sad, and mad yet logical: A brief exploration of experimental epistemology at the end of cognitive science. Colloquium presentation, Department of Anthropology, University of California, Davis (12 November).

Fleck, Ludwik. 1979. *Genesis and Development of a Scientific Fact*. Chicago: University of Chicago Press.

Foucault, Michel. 1973. *The Birth of the Clinic: An Archaeology of Medical Perception*. New York: Pantheon Books.

Fujimura, Joan H. 2005. Postgenomic futures: Translations across the machine-nature border in systems biology. *New Genetics and Society* 24 (2):195–225.

Garfinkel, Harold. 2002. *Ethnomethodology's Program: Working Out Durkheim's Aphorism*. Lanham, MD: Rowman and Littlefield.

Gieson, Gerald. 1969. The protoplasmic theory of life and the vitalist-mechanist debate. *Isis* 60 (3):272–292.

Gilbert, G. Nigel, and Michael Mulkay. 1984. *Opening Pandora's Box: A Sociological Analysis of Scientists' Discourse*. Cambridge: Cambridge University Press.

Gilman, Alex, and Adam Arkin. 2002. Genetic "code": Representations and dynamical models of genetic components and networks. *Annual Review of Genomics and Human Genetics* 3:341–369.

Goodsell, David. 1993. *The Machinery of Life*. New York: Springer-Verlag.

Haraway, Donna J. 1987. Manifesto for cyborgs: Science, technology, and socialist feminism in the 1980s. *Australian Feminist Studies* 4:1–42.

Haraway, Donna J. 1991. *Simians, Cyborgs, and Women: The Reinvention of Nature*. New York: Routledge.

Haraway, Donna J. 1994. A game of cat's cradle: Science studies, feminist theory, cultural studies. *Configurations* 2 (1):59–71.

Haraway, Donna J. 1997. *Modest_Witness@Second_Millennium.FemaleMan_Meets_ OncoMouse: Feminism and Technoscience*. New York: Routledge.

Harrison, Stephen. 2004. Whither structural biology? *Nature Structural Biology and Molecular Biology* 11:12–15.

Herzig, Rebecca. 2004. On performance, productivity, and vocabularies of motive in recent studies of science. *Feminist Theory* 5 (2):127–147.

Hill, Robert, and Peter Rich. 1983. A physical interpretation for the natural photosynthetic process. *Proceedings of the National Academy of Sciences of the United States of America* 80:978–982.

Hoffman, Michelle. 1991. An RNA first: It's part of the gene-copying machinery. *Science* 252 (5005):506–507.

Hopwood, Nick. 1999. "Giving body" to embryos: Modeling, mechanism, and the microtome in late nineteenth-century anatomy. *Isis* 90:462–496.

Huxley, Thomas Henry. 1878. *A Manual of the Anatomy of Invertebrated Animals*. New York: D. Appleton.

Jacob, François. 1973. *The Logic of Life: A History of Heredity*. Princeton: Princeton University Press.

Kay, Lily E. 1993. *The Molecular Vision of Life: Caltech, the Rockefeller Foundation, and the Rise of the New Biology*. New York: Oxford University Press.

Kay, Lily E. 2000. *Who Wrote the Book of Life? A History of the Genetic Code*. Stanford: Stanford University Press.

Keller, Evelyn Fox. 1983. *A Feeling for the Organism: The Life and Work of Barbara McClintock*. New York: W. H. Freeman.

Keller, Evelyn Fox. 1995. *Refiguring Life: Metaphors of Twentieth-Century Biology*. New York: Columbia University Press.

Keller, Evelyn Fox. 2002. *Making Sense of Life: Explaining Biological Development with Models, Metaphors, and Machines*. Cambridge, MA: Harvard University Press.

Kreisberg, Jason, Scott Betts, Cameron Hasse-Pettingell, and Jonathan King. 2002. The interdigitated Beta-helix domain of the P22 tailspike protein acts as a molecular clamp in trimer stabilization. *Protein Science* 11:820–830.

Lakoff, George, and Mark Johnson. 1999. *Philosophy in the Flesh: The Embodied Mind and Its Challenge to Western Thought*. New York: Basic Books.

Latour, Bruno. 1990. Drawing things together. In *Representation in Scientific Practice*, ed. Michael Lynch and Steve Woolgar, 19–68. Cambridge, MA: MIT Press.

Lenoir, Timothy. 1982. *The Strategy of Life: Teleology and Mechanics in Nineteenth Century German Biology*. Boston: D. Reidel.

Lynch, Michael. 1985. Discipline and the material form of images: An analysis of scientific visibility. *Social Studies of Science* 15 (1):37–66.

Lynch, Michael. 1990. The externalized retina: Selection and mathematization in the visual documentation of objects in the life sciences. In *Representation in Scientific Practice*, ed. Michael Lynch and Steve Woolgar, 153–186. Cambridge, MA: MIT Press.

Lynch, Michael. 1991. Science in the age of mechanical reproduction: Moral and epistemic relations between diagrams and photographs. *Biology and Philosophy* 6:205–226.

Martin, Emily. 1991. The egg and the sperm: How science has constructed a romance based on stereotypical male-female roles. *Signs (Chicago, Ill.)* 16 (3):485–501.

Meinel, Christoph. 2004. Molecules and croquet balls. In *Models: The Third Dimension of Science*, ed. Soraya de Chadarevian and Nick Hopwood, 242–275. Stanford: Stanford University Press.

Myers, Natasha. 2006. Animating mechanism: Animation and the propagation of affect in the lively arts of protein modeling. *Science Studies* 19 (2):6–30.

Myers, Natasha. 2007. Modeling Proteins, Making Scientists: An Ethnography of Pedagogy and Visual Cultures in Contemporary Structural Biology. PhD diss., Massachusetts Institute of Technology, Program in History, Anthropology, and Science, Technology, and Society.

Myers, Natasha. 2008a. Molecular embodiments and the body-work of modeling in protein crystallography. *Social Studies of Science* 38 (2):163–199.

Myers, Natasha. 2008b. Conjuring machinic life. *Spontaneous Generations: A Journal for the History and Philosophy of Science* 2 (1): 112–121.

Myers, Natasha. 2010. Pedagogy and performativity: Rendering lives in science in the documentary *Naturally Obsessed: The Making of a Scientist. Isis* 101 (4):817–828.

Myers, Natasha. 2012. Dance your PhD: Embodied animations, body experiments and the affective entanglements of life science research. *Body and Society* 18 (1):151–189.

Myers, Natasha, and Joseph Dumit. 2011. Haptic creativity and the mid-embodiments of experimental life. In *A Companion to the Anthropology of the Body and Embodiment*, ed. Frances E. Mascia-Lees. Chichester, UK: Wiley-Blackwell.

Pauly, Philip J. 1996. *Controlling Life: Jacques Loeb and the Engineering Ideal in Biology*. Berkeley: University of California Press.

Roosth, Hannah Sophia. 2010. Crafting life: A sensory ethnography of fabricated biologies. PhD diss., Program in History, Anthropology, and Science, Technology and Society, Massachusetts Institute of Technology.

Stengers, Isabelle. 2008. A constructivist reading of process and reality. *Theory, Culture and Society* 25:91–110.

Tanford, Charles, and Jacqueline Reynolds. 2001. *Nature's Robots: A History of Proteins*. Oxford: Oxford University Press.

9 Nanoimages as Hybrid Monsters

Martin Ruivenkamp and Arie Rip

1 Introduction

In nanoscience, images play an important role. Among these there are visualizations of nanoscale realities assumed to be "down there,"[1] based on data provided by instruments. At the same time, there are graphic designs that convey the message about the nanoscale more clearly, or indicate actual or potential nanoachievements. Nanoimages go further in expressing such design achievements[2] than do traditional representations in science that attempt to be faithful to nature or to experimental findings. In this chapter we explore whether a new genre of representation is emerging in which design and vision are integral elements. While we discuss this with particular reference to nanoscience, it is possible to speculate about nanoimaging as heralding a new mode of representation in science more generally. Lorraine Daston and Peter Galison (2007, ch. 7) have recently made a similar argument concerning an emerging new mode of representation; they refer to new possibilities afforded by information and communication technologies that enable interactive imaging. The other point they make is about a shift in nanoscience, with scanning probe microscopy enabling a move from "seeing" to "feeling" the world out there (Daston and Galison 2010, 383). They caution that their claim is speculative because the modes of representation they indicate are not yet regular practices of representation in science generally. In that respect, the genre we identify, which integrates design and vision as part of current work within nanoscience, would be a better candidate for a new mode of representation.

Nanoimages are hybrid, in the sense that they combine elements of traditional scientific representation (in which resemblance to a world out there is the ideal) along with elements that anticipate what an invisible world might look like. One could argue that such further elements are not actually representations, and so cannot form the basis for speculations about a new mode of representation. Such an argument, however, hinges on a specific notion of representation-as-resemblance, which is just one version of the more general notion of representation-as-standing-for.[3]

Resemblance as an ideal implies reference to things and properties "out there," which may be already visible, newly made visible, or invisible but inferred. However, actual representations are constructed in practice, and their acceptance as representative of what is "out there" is achieved as the outcome of a process rather than a simple assertion of resemblance between a purported original and what an image shows: "to analyze representation is to expose the conjurer's tricks through which chains and networks of similitude are laboriously built-up and then 'forgotten' in the presumptive adequacy of their reference to an 'original'" (Lynch and Woolgar 1990, 7).

In this chapter, we work with the broader perspective on representation as "standing for" to capture what images do in nanotechnology.[1] An example of what such a perspective affords is provided by Hans-Jörg Rheinberger's (2002, 519) discussion of "layers of representation" in the case of the presumably complete sequence of a human genome, which was announced in June 2000 and visualized as sequences of letters standing for the nucleic acid building blocks. The sequence of letters also represented the claim that genomes could be traced, and it signaled the achievements of the two competing groups working on the sequence. It stood not just for scientists' capacity to decipher the genome, but for the "legibility of the world."

Within this broader perspective, it need not be problematic that nanoimages move from scientific pictures to graphic designs and artist's impressions without any discrete categorical breaks, and that they can combine what is present with what may (or may not) lie in the future. Their hybridity may actually be a positive feature, similar to the way the "hybrid monsters" of human and nonhuman agency, which Bruno Latour (1991) discusses, sustain modernity without necessarily being recognized for what they are. This is not to say that the notion of images as hybrid monsters is generally accepted. In the next four sections (2–5), we will discuss why nanoimages may be considered hybrid and consider some responses to such hybridity. We then discuss, in section 6, the design orientation that characterizes visual representation in nanoscience disciplines, before returning to the question of whether this amounts to a new mode of representation.

2   Images of Nanotechnology

Images of nanotechnology that circulate in the world of nanotechnology can be characterized along two dimensions, used by practitioners in their discussions of such images: one related to whether or not an image is derived from data, and the other related to the nature and extent of its orientation toward the future. The first dimension ranges from visualizations of the nanoscale adapted from original instrument-based data to graphic designs (and artist's impressions) of what the nanoscale *could* look like, or how audiences may expect that it *should* look. In practice, nanoimages tend to fall somewhere on the continuum between these poles. This is apparent when

Figure 9.1
A continuum ranging from a visualization of the nanoscale to a graphic design. (A) A visualization of carbon nanotubes produced by an imaging device. (B) A color-enhanced picture of carbon nanotubes. (C) Graphic design of double-walled carbon nanotubes. Source: (A) http://www.msm .cam.ac.uk/polymer/members/iak21.html (Ian Kinloch); (B) http://www.techdigest.tv/corked -nanotubes.html (anonymous); *Nature Nanotechnology*, vol. 4, no. 1 (2009), http://www.nature .com/nnano/journal/v4/n1/covers/largecover.gif (Karen Moore).

we consider common images of carbon nanotubes (figure 9.1). The first image is created on the basis of data obtained from an imaging instrument (figure 9.1A). The production of such visualizations is guided by expectations about what can be "seen" at the nanoscale and rules how the nanoscale should be visualized. Such expectations are, by now, internalized (Ruivenkamp and Rip 2010). The data and the initial "piction"[5] are further manipulated to highlight the information they should convey, or to increase the image's aesthetic appeal.[6] The second visualization of carbon nanotubes (figure 9.1B) tells a story about the nanoscale that is easier to understand for broader audiences (and is aesthetically more pleasing), due to the highlighting and the colors added to the image. The argument of scientific practitioners is that manipulation of the initial "piction" is acceptable as long as the scientific content remains visible. This creates ever more distance from the original data. Going one step further, there are graphic designs that do not necessarily rely on data generated through the use of imaging tools, but still offer an impression of what the nanoscale *could* look like (figure 9.1C).

Of course, the composition of a given image depends on the audience to which it is presented. Visualizations of the nanoscale produced with imaging tools and graphic designs are published in scientific journals as part of the presentation of research results. In addition, the same journals may use "pretty" images such as figures 9.1B and 9.1C as cover illustrations to attract potential readers' attention. The graphic design type (figure 9.1C) also appears in texts that explain nanotechnology to lay audiences. Scientific journals publish requirements for authors about how to present scientific visualizations such as figure 9.1A. By doing so, they maintain their role as assurers of scientific integrity (see Frow, this volume), but in practice they accept a certain amount of hybridity.

There are further situations where visualizations of the nanoscale are used as intermediaries to create visibility for research outcomes, e.g., when promoting programs and conferences; or to demonstrate what imaging tools can do (Mody 2004), and "to sell [imaging] machines" (Daston and Galison 2007, 397; see also Ruivenkamp 2011 on image galleries of firms such as Veeco and FEI).

The second dimension along which images in and around nanoscience can be characterized is the nature and extent of how they are oriented toward the future. The use of images with a future orientation is a familiar feature of scientific work; it is visible in the uses of images as semiotic devices, indicating possible routes for tinkering in the lab (Pombo 2010), and in the continual movement between "data" and "images" in physics (Galison 2002).[7] In nanoscience, this dimension runs from graphic designs indicating, for example, how a rotaxane molecule might work as a molecular motor (figure 9.2) to artist's impressions envisioning future applications of nanorobots. One widely circulated image is commonly called the Nanolouse (which is not a label approved by the artist Coneyl Jay). Originally titled Nanoprobe, this computer graphic creation won the 2002 "Visions of Science Award" sponsored by the *Daily Telegraph* (London) and Novartis Corporation (see figure 9.4). Though originally created as art, it is now commonly featured on websites of scientists and in university press releases and strategy documents about nanomedicine (Ruivenkamp 2011). Such images import the future into the present, and may—as in the case of the Nanolouse—come to stand for the promise of nanotechnology, or, occasionally, for concerns about its impacts.

Based on the foregoing, images can be located on a two-dimensional grid (see table 9.1) in terms of whether or not they use instrument-based data and the extent to which they show a future orientation.[8] On such a grid, practitioners' characterizations of nanoimages mainly would be located in the upper left and lower right quadrants.

In actual practice, there is no such simple clustering on the two end points of the main diagonal, separating instrument-based visualizations that focus on the present from future orientations that are exclusive to artist's impressions. Images tell a variety of stories about what is "out there" and of how nanotechnology reaches out into the future. Accordingly, practitioners produce "hybrids" (Latour 1991), which are accepted in practice, even when the discourse about the images can revert to dichotomies between the present and future, and between scientific images and artist's impressions. Take the image in figure 9.2—a design for a molecular motor—which is understood as a scientific visualization but with a definite future orientation. Placed in table 9.1, it would go somewhere in the bottom-left quadrant. There is now also a movement to appreciate the aesthetic qualities of scientific images, sometimes as works of art; such cases would be located in between the two quadrants at the top level in table 9.1.

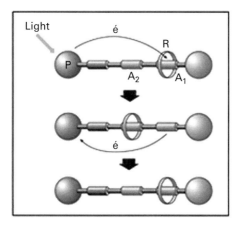

Figure 9.2
Image of a rotaxane: a molecular machine which consists of a dumbbell-shaped molecule surrounded by a macrocyclic compound that can move along the string of atoms (axle) between the two bulky groups at the string's termini (wheels). Source: *Nature Materials*, 5, 165.

Table 9.1
Mapping nanoimages on a two-dimensional grid, with X indicating the way practitioners tend to characterize the images. Note that both horizontal and vertical dimensions are continuums.

	Scientific presentation	Artist's impression
Present orientation	X	
Future orientation		X

While the hybrid character of images such as the one in figure 9.2 is accepted in practice, some images that lie outside the main diagonal in table 9.1 (top left–bottom right) are seen as "hybrid monsters." These evoke different types of reactions, which we discuss in the sections that follow.

3   Hybrid Monsters and Attempts at Domestication and Purification

There are good reasons to understand nanoimages as "hybrid monsters," in Bruno Latour's (1991) sense that they proliferate, that there are attempts to purify them, and that they are domesticated in scientific practices, including practices of visualization and clarification for wider audiences (Ruivenkamp and Rip 2010).

Hybridity arises when imaging instruments are used to transform invisible nanoscale phenomena into visible forms. Imaging instruments, such as scanning tunneling microscopy (STM) and atomic force microscopy (AFM), produce digital data, which are turned into pictures through software programs that go with those instruments. It is not clear what the "right" picture should look like; actually, there is no "right" picture, because there is no independently visible referent with which to compare an image (Ruivenkamp and Rip 2010). Practitioners bootstrap the phenomena into visibility: starting with an idea or experiment, and sharing interpretations, they find out what works when building on preliminary insights to achieve a degree of stabilization in the relevant communities of practice. In other words, there will be expectations and tacit rules shaping the visualizations (Ruivenkamp and Rip 2010). Practices involve coloring and highlighting of the images produced by instruments, such as with astronomical images (Lynch 1991), brain scans (Dumit, this volume), and satellite images (Phipps and Rowe 2010), and now also with graphic designs for visualizing what practitioners imagine is happening at the nanoscale.[9]

The term "hybridity" usually suggests a mixture of elements or a crossing of species that had been separate.[10] Latour's monsters are hybrid because they cross modernist dichotomies between subjects and objects, humans and nonhumans. However, as he points out, those dichotomies are a *consequence* of purification rather than an original condition. Latour still wants to use the word "hybrid" (as a *nom de gueux* as it were) to challenge received (modernist) views.

"Monster" is used in the sense Mary Douglas (1966) made famous, as something that threatens received boundaries and distinctions and thus needs to be "purified," backgrounded, or negated. The militant response "Purify!" is not the only response to monsters when they become visible—that is, when they are noticed and recognized for what they are. Domestication, turning them into something that can be accommodated within existing practices, is another response, and it becomes apparent when scientific representations, graphic designs, and artists' impressions are used without any need to draw a boundary between them. Still, rules can emerge defining what is acceptable as a scientifically adequate representation, just as there are rules for graphic design. The net effect would then be "soft" purification, as is visible in flexible standards in the lab or for a scientific journal. Besides *purifying* and *domesticating*, a third response to hybridity is to *embrace* it: to enjoy it, play with it. We don't see much of that response in the world of science, which remains strongly modernist, even while knowledge production is becoming hybrid (Rip 2002, 2011). Still, scientists have always enjoyed playing with the images they create. Thus, nanoscientists such as Chris Ewels who put a lot of effort into composing what are essentially graphic artist's creations, such as planet Earth depicted in a buckyball, are not exceptional. His creations can be taken up widely, because it is clear that no literal resemblance is intended.[11]

Of course, the three responses are not always clearly differentiated. We now offer some examples of debates about, and practices of, scientific visualization in which domestication and purification responses can be recognized.

4   Images of Science and the Tension with Fiction

Julio M. Ottino's (2003) attempt to purify visualizations, so that they would only have epistemologically adequate features, ran aground because scientists were not interested. Ottino stipulated that "images should not conflict with or violate known physics" (Ottino 2003, 475). This is not a simple requirement when unknown territory is being explored, as in some areas of nanoscience. One of his examples of an unacceptable image was a cover illustration for *Science* magazine, which referred to an article by Bachtold et al. (2001) (see figure 9.3). Ottino criticized this and other purportedly scientific images like it for blurring science and fantasy, and called for clear rules to address the tension between the scientific content and the aesthetic attractiveness of images. The scientists, however, were not overly concerned.[12] When asked about Ottino's critique, one of the authors of the article (Cees Dekker, a nanoscientist with an outstanding scientific reputation) offered a range of arguments. First, he emphasized that every image of nanoscale phenomena must be incomplete, since light wavelengths are larger than the relevant dimensions of structures at that level. He then argued that, just as with metaphors, images can convey an essential meaning and make it understandable. Aesthetic attractiveness is a means to present scientific content to a broader audience. "While pretty images as such do not get your paper published since it is peer-reviewed by referees, nice images that make it to the cover of journals function as good PR for your research."[13]

This is not an idiosyncratic response. Manuals now include advice about the best ways to make images visually appealing, to improve the quality of images for submission to peer-reviewed academic journals, and to create an image suitable as a cover illustration (Frankel 2002). Scientists accept, even embrace, the hybrid character of the images they use and produce, as long as it serves their interests. They are willing to engage graphic designers to increase the clarity and aesthetic appeal of images. At the same time, most of them also use modernist dichotomies, emphasizing that their

Figure 9.3
Graphic design of an array of nanotube FETs overlaid with gold source and drain electrodes. This image, which was used on the cover of *Science*, is criticized by Ottino, who argued that "if the carbon atoms are visible, then the much larger gold atoms in the structure should also be on view" (Ottino 2003, 476). In this image gold is depicted as if it is macroscale rather than a nanoscale structure. Courtesy of C. Dekker, TU Delft/Tremani. Source: Bachtold et al. (2001).

images stand for scientific facts, not speculation. For example, David S. Goodsell distinguishes the fantasy and speculation conveyed by images from prosaic "scientific facts" when he warns that

images are so compelling that they may compete with the scientific facts. Images may be created that cover the spectrum from fantasy to science, and trouble may occur when the distinction is not clear. . . . Without careful description of what is shown, it is often difficult to separate fact from fantasy, science from speculation. (Goodsell 2006, 47)

The question about the validity or "correctness" of visualizations of the nanoscale is thus broadened to questions about keeping science "scientific" and maintaining distance from visions and popular impressions.

5   Images Importing the Future into the Present

At the moment, nanoscience lives on promises. In contrast to physics and chemistry, "nanotechnology seems decidedly non-presentist" (Mody 2004). Images of buckyballs, molecular machines, and new uses of carbon nanotubes refer to anticipated future nanoachievements and make them visually present, as though saying "this is what the future looks like." Such images can be compared to designs (see section 6, below), because design, by definition, must anticipate future situations, and the work of designing is often supported by images (Henderson 1999). Images represent a future, often quite specific, which can be discussed as if it were present. Such images are about imagined possibilities, and can be visions rather than actual construction plans for future achievements.

There are tradeoffs involved in the creation of such future-oriented images. On the one hand, there is a high degree of freedom in articulating and presenting future possibilities, which can be exploited for spectacular effect, as in the notorious images of molecular machines on the website of the Foresight Institute, and the Nanogear image (which has become iconic for the Drexlerian vision of mechanical engineering transported to the nanoscale). On the other hand, there is the need to come across to intended audiences as plausible, depending on the genre of the images. Plausibility is always performed in concrete situations and for particular audiences, and it can be enhanced by drawing on accepted cultural repertoires. For example, the cover image of the brochure "Nanotechnology: Shaping the World Atom by Atom" (National Science and Technology Council, 1999) shows a nanoscale landscape depicted as a lunar surface, against the dark blue of outer space. Alfred Nordmann (2004, 48) interprets the image as nanotechnology's claim to open up a new space, which "draw[s] on traditional imagery that, through redeployment in a novel context, acquires new meaning."

Borrowing from science fiction, creators of nanoimages strive to reconcile freedom and plausibility through one of two "diegetic" strategies.[14] In the first, the narrative simply stipulates the existence of new scientific and technological possibilities, using them as props that enable and support the story. Video games in which nanotechnology assists the protagonists are an example (Milburn 2010; see also Milburn 2004a). In the other strategy, the narrative offers background for the new technology, and its regular use is filled in, making the extraordinary ordinary (Bleecker 2010).[15]

Another tradeoff involves the tension between exaggeration and modesty. Often there is a simplistic reference to a lack of realism: "this isn't science, it's just science fiction." But this is a move in a struggle, rather than an observable difference. Inflated promises are a regular feature when scientists and spokespersons for science mobilize support. This was very visible in the promises about recombinant DNA and biotechnology in the 1970s and 1980s, and recently appears to have reached new heights for nanotechnology.[16]

Nanoscientists and policy makers can (and often do) attempt to draw a boundary between "science" and "science fiction." This can be a tactic for promoting one vision at the expense of others, as Arie Rip and Marloes van Amerom (2010) show was the case with Eric Drexler's visions of molecular manufacturing. It can also be a defensive move to avoid being called to account for possible negative implications of investing in nanotechnology research. While media reporting about nanotechnology can focus on futuristic visions, it is interesting that, with the advent of nanotechnology-enabled products on the market, there has been some "defuturization" (Lösch 2010).

Brigitte Nerlich (2005) observes how the "most iconic and most recent representations of nanotechnology," depicting robotic submarines traveling through our blood vessels, fall back on historical imagery "surrounding the various 'submarines' that have travelled through popular imagination, [such as] Jules Verne's Nautilus, driven by Captain Nemo." Such reaching for the past has a hold on the present when the future is being imaged and narrated. Nerlich notes further that such visions have a profound effect on public imagination and serve to "sciencefictionalise science fact and blur the boundaries between cultural visions and scientific reality."[17] This can be read as a call for purification, but Nerlich actually sees some value in the blurring. Others like David Goodsell are less sanguine. He notes that lay audiences' interest in nanotechnology is spurred by fictional imagery emphasizing both hopes and fears about future nanotech developments, and he calls for "good images" to dispel myths:

Of course, there is always the danger of inhibiting progress with irrational fears of alien invasion or rampant disassemblers. Here, however, imagery again comes to our rescue and provides the best way to dispel myths and present the successes and dreams of practical nanotechnology and nanomedicine. (Goodsell 2009, 56–57)

Imagery now becomes part of the struggle to present nanotechnology to wider audiences. In depicting possible futures, there is no sharp boundary between images that articulate directions of further research or designs of molecular machines and images that present open-ended visions of new functionalities that might be achieved.

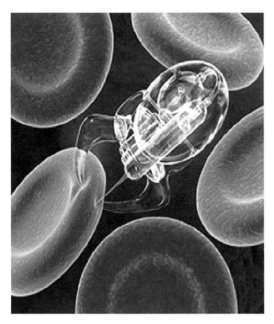

Figure 9.4
Nanolouse handling red blood cells—or attacking them? Derived from Micro-syringe, by Coneyl Jay. Science Photo Library T395/0126, http://www.sciencephoto.com/media/348229/view.

A case in point is the Nanolouse image by Coneyl Jay (see figure 9.4), an artist's impression that is widely reproduced, but for different purposes. The same nanoimage "piction" can convey different messages depending on how it is framed by the accompanying text. When the image was presented on the BBC website, it was linked to contrasting visions of nanotechnology. One framing of the image positioned it as winner of the Visions of Science award and described it as hype: "The more over-hyped applications see tiny machines roaming the body to cure diseases" (news.bbc .co.uk/2/hi/technology/2981480.stm). Another framing, however, emphasized realistic possibilities: the image of the Nanolouse "intends to show one of the possible applications of nanotechnology in medicine in the future—microscopic machines roaming the body, injecting or taking samples for tests" (http://www.bbc.co.uk/radio4/today/reports/archive/science_nature/nanotechnology.shtml). Whatever the textual framing, the Nanolouse image tells a story of its own: a semiotic reading points to the action by the nanobot as "Helper" directed at the blood cells as the "Subject" desiring to become well again (or the nanobot may be attacking the red blood cell; see Landau et al. 2009).[18]

The Nanolouse image is widely used (Nerlich 2008; Ruivenkamp 2011), and its imagery is an instance of a new visual language to present possibilities that may be emerging more widely within and beyond nanotechnology. In general, a fictional world functions as a background against which artifacts are positioned as diegetic prototypes (Kirby 2010) or design fictions (Bleecker 2010). The Nanolouse image features red blood cells that are depicted realistically. The message of the image is then that the nanobot is realistic (realizable) as well. Because of this message, actors will find it necessary to position themselves (as the BBC website did by adding text), indicating their view on whether the image presents something "realistic" or "fictive."

6   Design Orientation

It is clear that practitioners experience changes in the sorts of images that are produced and play a role in various practices, and they may just follow the trend without reflecting on these changes. This appears to be the case for the design orientation that increasingly characterizes visual representation in nanoscience disciplines. This orientation is apparent in the way nanotechnology deploys the visual languages of chemistry,[19] biology, and engineering. Though traditionally different, these are beginning to converge through the use of conventions for visualizing nanoscale phenomena as designable macroscale objects. This genre of representation is not new, but it is now used more widely than before, and the ideal of resemblance has increasingly given way to visualizations of functionality.

In figure 9.5, we reproduce examples of images of molecular machines that chemists and biologists construct, together with images of an engineering kind that chemists and biologists also use.[20] The traditional graphic formulas used by chemists to indicate actual or projected configurations of atoms are now adapted to the design of large molecular and supramolecular structures. A rotaxane molecule is depicted on the left in figure 9.5. Each section of the molecule is designated with a function, which can then be pictured as a graphic formula or as a mechanical schematic. Such sketches of possible rotaxanes can be used as blueprints for synthesis.[21] The engineering-type image of rotaxane at bottom left is a graphic design. As figure 9.2 shows, similar drawings of rotaxanes appear in scientific articles, where more detail is shown.

Biologists have a tradition of using visualizations of functions and mechanisms. At right in figure 9.5, myosin powered by ATP is depicted (with a text, "cargo," at the top) in a way that is reminiscent of engineering. At the bottom, one sees how the visual language of space-filling models is adapted to visualize the bioengineering that is going on within the cell.

As Joachim Schummer (2006) has emphasized, images of molecular structures based on molecular models are widely used in chemistry, and are now presented as ordinary objects abstracted from their chemical context. This suggests that molecular structures are able to perform functions much like macroscale objects do, and thus speak directly to a design orientation.[22]

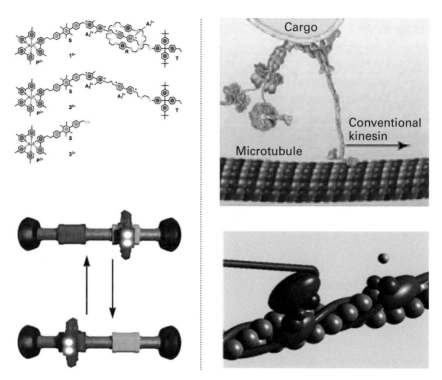

Figure 9.5
The visual languages of supramolecular chemists (left) and molecular biologists (right) become increasingly similar to engineering models and present nanoscale phenomena as designable like macroscale objects.

The visual language of design engenders visions of possible achievements.[23] The potential of molecular machines like rotaxanes can then be presented with only the barest indication how they could work (figure 9.6). There is a continuum from traditional chemical formulas (which can already be used as blueprints for synthesis) to space-filling models, further designations in terms of functions, engineering-type images, and visions of applications.

Our brief discussion of a design orientation in images and visual languages in nanoscience disciplines indicates the increasing use of the representation genre of design-and-vision. This is linked to actual design approaches. For nanotechnology in the 2000s, a distinction was often made between miniaturization (making things smaller) as a top-down approach and bottom-up approaches that rely on chemical or biological (induced) self-assembly (see NSTC 1999). In the former approaches, it is possible to specify what has to be developed, whereas the latter have relied more on "design fictions" (Bleecker 2010, cf. note 75). This is of course very clear in Drexler's vision of

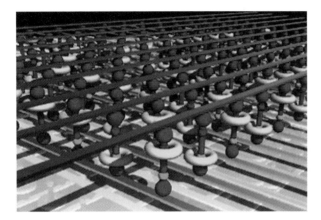

Figure 9.6
A visualization of the research goal to attach rotaxanes to two layers of molecules, to create moving surfaces enabling, according to Stoddart, "molecular memory." Source: http://masakilaboratory .blog129.fc2.com/blog-category-2.html.

building macroscale products from the bottom up, i.e., from the level of atoms (see Drexler 1986).

One also sees visionary images used to mobilize resources and public support more generally, without necessarily devoting much attention to design specifications. An interesting example is the US National Science Foundation's use of the image of the Vitruvian man, perfect in proportion (see figure 9.7A), to link their interest in NBIC (the convergence of nano-, bio-, info-, and cognitive sciences) to human enhancement.[24] Such images are not innocent. They inspire and guide research funding (see Bainbridge 2009). This is not unique to nanoscience, though it is very visible there.

More important for our overall question about representation is the possibility that the increasing use of the genre of design-and-vision reflects ongoing changes in how science is done. Alfred Nordmann (2008) discusses the tinkering and engineering/design orientation in ongoing nanotechnology research as a key element in a philosophy of nanotechnoscience.[25] In his recent work, in collaboration with Bernadette Bensaude-Vincent, he proposes that technoscientific objects are emerging which "are neither natural phenomena nor technical devices, and that enjoy considerable prominence in contemporary research" (Nordmann et al. 2011). Such objects include known objects that are of interest to multiple disciplines (stem cells, Arctic ice) and objects that do not exist yet but are to emerge in the long term from a convergence of research efforts (targeted drug delivery systems). Others are said to be just around the corner (lab-on-a-chip, BioBricks™ for synthetic bacteria). There are objects established in one research context that are of interest to others (mechanochemical molecular motors, marine

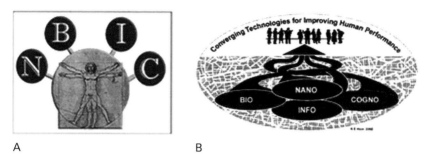

A   B

Figure 9.7
Textual and visual rhetorics tend to focus on human enhancement. Source: (A) ETC Group 2003; image originally used as the logo of the Los Angeles NBIC conference; (B) Roco and Bainbridge (2002) (image by R. E. Horn).

shells, microalgae), and objects that are engineered to provide a common referent for a variety of approaches (onco-mouse, carbon nanotubes, an artificial water catchment).

Even while one might hesitate to accept technoscientific objects as a new overarching category, it is clear that a new mode of doing science is emerging which integrates engineering and design, and one that will stimulate the use and further development of design-and-vision as genre of representation.

Our analysis of design-oriented nanoimages which integrate hybridities indicates that images may do more than follow the emergence and prominence of technoscientific objects. They can also *lead* such emergence, opening up possibilities that can be taken up in further exploration.[26] In other words, images and practices coevolve. Representation of a new mode of doing science may lead to a new mode of representation of science.

7   In Conclusion: A New Mode of Representation?

Images as hybrid monsters circulate in the nanoworld and become part of its regular practices and forms of representation. Artist's impressions are accepted as representations of nanoentities, and future possibilities are treated as actual. The traditional notion of representation, with its emphasis on resemblance to what is "out there," becomes less important.[27]

As the design-and-vision genre of representation in nanoscience practices (with its origins, as we have briefly alluded, mainly in chemistry) becomes more widely used, it is likely to be extended to other arenas, with concomitant extensions of modes of visualization that include graphic designs and artist's impressions to embody and disseminate promises.

The two other main genres of imaging—mapping and tracing—each exhibit their own hybridities. Mapping and other resemblance-induced representations have been important all along, but have become more sophisticated, and some of the mapping is now delegated to imaging devices. An example is the various medical imaging techniques for brain scanning discussed by Dumit and by de Rijcke and Beaulieu in this volume (also see Beaulieu 2002). In this genre of imaging, mapping what is "out there," encoding is an important aspect (Staley 2008), for while maps stand for the object or domain that is mapped, there is always a reduction in complexity, linked to intended uses and audiences.

The third genre of imaging is tracing, in the sense of following an entity with the help of the traces it leaves (see Ginzburg 1979). Françoise Bastide's (1992) analysis of how actors tried to make sense of the images sent by a space probe to Saturn is particularly illuminating. Hans-Jörg Rheinberger (2002) also discusses examples where samples (of proteins or nucleic acids) are made visible in their composition by turning them into traces with the help of chromatography.[28]

Peter Galison's analysis of ever more sophisticated image-making devices, in particular the elusive "electronic bubble chamber" to track elementary particles (Galison 1997, 807), shows that visualizing traces can become a mapping exercise. His main point, however, is the (productive) tension between "pictures and propositions" (Galison 2002, 316). In nanoscience, "propositions" (in Galison's sense of theoretical claims) are almost absent, because the interest is in exploring new phenomena that can better be captured in images. This point relates to the distinction between analytic and synthetic instruments (Böhme et al. 1978, 229–231): the former capture and measure what is out there (for example, a thermometer measuring temperature), while the latter produce phenomena (for example, a Leiden jar producing electrical sparks). Nanoscience is oriented toward the creation and study of novel effects, and thus to a specific kind of imaging that captures such effects.

The three genres, while distinct, overlap in practice, also because they all draw on an increasingly sophisticated arsenal of techniques to render and manipulate images and maps. Mary Tiles (2009), for example, discusses weather maps and satellite data as opportunities for simulation and prediction. This indicates that the long tradition of mapping, monitoring, and modeling (3M) sciences (Rip 2002) can include a future orientation, even when an engineering/design approach as found in nanoscience (while sometimes pushed) is not in order.

In this brief discussion of evolving genres of imaging, we have argued that a blurring has occurred between depicting what is "out there" and the practices of mapping and highlighting what is useful for relevant audiences, and showcasing what might, realistically or speculatively, be constructed. Clearly, more variety is now accepted in imaging practices. There is no simple answer, however, to the question of whether a new mode of representation is emerging. Strong claims like those of Daston and

Galison (2007, 383) about wrenching the image out of a long historical trajectory are speculative at best. There may be doubts about how "new" the mode of representation is (it definitely has a history), and as to whether there actually is a coherent new mode. What is clear, however, is that in nanotechnology and some other emerging fields resemblance, or better, being faithful to whatever is out there, whether known or unknown, is not the dominant requirement anymore. Images continue to be optimized, but now in terms of functionalities. And these functionalities range from resemblance, to design, to visions that mobilize.

These are the building blocks for an emerging new mode of representation, with the new genre of design and vision, where hybridity of images is an integral part, leading the way. In general, images are hybrid monsters because they include expectations, ranging from expectations about what entities in the world might look like, to how newly made technoscientific objects might evolve, to visions of possible entities and their functionalities. Whether a new mode of representation will become established will depend on how these hybrid monsters are handled. Even scientists are becoming more comfortable with the monsters, definitely so in the world of nanotechnology.

Acknowledgments

We are grateful to three anonymous reviewers for their constructive criticism, to Mike Lynch for his guidance when we were writing our text and for helpful comments, and to Catelijne Coopmans for important editing suggestions. The research on which this chapter is based was funded through the Dutch R&D Consortium NanoNed.

Notes

1. Cf. Feynman's 1959 dinner talk "There's Plenty of Room at the Bottom," which has become the origin myth of nanotechnology (McCray 2005; Toumey 2005).

2. Such combinations of design and vision are well known in architecture, and occur occasionally in the engineering world when a future plant or system is rendered in a drawing.

3. Brown (2009, 4–7) offers important insights about the genealogy of the concept of representation, starting from the Latin verb *repraesentare*: to make present or manifest or to present again. In medieval times, "it concerned the depiction or embodiment of a person or thing in language, art, theatre, and religion" (Brown 2009, 4). It started with the person or the thing that would be represented, so there was little doubt about what was represented. From the seventeenth century onward, scientific representations of nature and political representations of civil society were mooted and explored. Brown criticizes the "correspondence model" in science (in which representations should resemble nature out there) and in politics (in which the political will of civil society should be reflected), and proposes "a conception of political and scientific representation as practices of mediation that engage and transform what they represent" (ibid., 7).

4. We use the terms "nanoscience" and "nanotechnology" (and their combination) to refer to the fields that produce the images we study. Nanotechnology, often used to refer to both nanoscience and nanotechnologies, is an umbrella term covering research and development in which nanoscale phenomena play an important role, but without much further specification. It is also an umbrella term in the sense that its broad recognition derives as much from the way it is used in science policy documents and priority setting (where it often is promissory) as from its status as a newly emerging field of science and technology. One can argue that, as a generative umbrella term, nanotechnology was born in 2000, when US President Clinton established the National Nanotechnology Initiative (Toumey 2005; Ruivenkamp 2011). We will follow what actors are doing, and accept particular images as images of nanotechnology when actors present them as such. But it is not "anything goes"; what can be counted as nanotechnology, and thus as images of nanotechnology, has undergone articulation, convergence, and stabilization. We follow such articulations and stabilizations, without buying into them as the last word on nanotechnology.

5. We use the term "piction," in contrast to the commonly used term "depiction" which suggests that there is something out there that is being depicted. Speaking of "piction" is like drawing attention to the brush strokes and other material traces of the work of painting, and allows consideration of their effects. The "piction" is a material configuration that works, and is intended to work, but the work need not be a matter of resemblance.

6. This includes what Lynch (1991, 72) has called "epistemological nostalgia" (see also de Ridder-Vignone and Lynch 2012), the use of conventions of macroscale depiction (like shadow effects and clear boundaries) that are not applicable to the nanoscale, but which nevertheless are abundant in images of the nanoscale.

7. One can also refer to Latour's general claim: "An isolated scientific image is meaningless. . . . A table of figures will lead to a grid that will lead to a photograph that will lead to a diagram that will lead to a paragraph that will lead to a statement. The whole series has meaning, but none of its elements has any sense" (Latour 2002, 34). Actual as well as potential meaning of images are constituted in the context of such a series.

8. There have been other attempts to categorize nanotech images. For example, Chris Robinson identifies four overlapping categories of nanotech images "which should help us to understand how an image operates and what information it intends to convey." He distinguishes *schematics*, *documentations*, *fantasy*, and *fine art*. While the first category includes "scientific visualizations with little visual drama," the last category would include a type of image that "with respect to nanoscience seeks some form of meaningful and long-term effect on culture" (Robinson 2004, 168). These categories mix the nature of images with their functions and intended impacts.

9. See for example how Andre et al. (2010) offer AFM images of a peptide in a cell wall, and then use, in their figure 7, a graphic design to visualize what the architecture really looks like (it is labeled as a "schematic drawing" and also called a "cartoon"). The hybridity is not a problem for the authors, nor, presumably, for their readers.

10. There is the further term "monstrosity," a malformed newly born or grown organism that cannot survive on its own. Monstrosities can be created experimentally.

11. Chris Ewels's image gallery is available at: http://www.ewels.info/img/science/ (accessed 13 November 2012).

12. Some biology journals, however, have taken a stand against the increasing photoshopping of images. See Emma Frow's chapter in this volume.

13. This is our translation from an interview with Dekker in a Dutch newspaper (Persson 2003).

14. The diegesis of a narrative is the world created in the story (Kirby 2010). In cinema studies, the diegetic part is the story of the characters as shown on the screen, while voice-overs and background music are extradiegetic. They make the narrator present to the audience, though not to the characters. Movie makers play with the possibilities afforded by the combination of diegetic and extradiegetic elements.

15. Bleecker also introduces the notion of "design fiction": "[It] allows one to do the design work for things and ideas that are too speculative for reasonable, balanced people. They tell stories the same way as good science fiction does—immersive, imaginative and imminent" (Bleecker 2010, 3). It is a kind of prototype, through the stories it evokes and the conversations it starts.

16. See the intervention by Philip J. Bond, US Under-Secretary of Commerce, in the Swiss Re workshop of December 2004: "Nanotechnology offers the potential for improving people's standard of living, healthcare, and nutrition; reducing or even eliminating pollution through clean production technologies; repairing existing environmental damage; feeding the world's hungry; enabling the blind to see and the deaf to hear; eradicating diseases and offering protection against harmful bacteria and viruses; and even extending the length and the quality of life through the repair or replacement of failing organs. Given this fantastic potential, how can our attempt to harness nanotechnology's power at the earliest opportunity—to alleviate so many earthly ills—be anything other than ethical? Conversely, how can a choice to halt be anything other than unethical?" (Swiss Re 2004, 7).

17. She refers to Milburn (2004b).

18. This way of phrasing ("Helper," "Subject") draws on actantial analysis as introduced by Greimas (1983); compare also Greg Myers's translator's note included in Bastide (1992).

19. Chemists, and to an extent molecular biologists, are explicit about their visual languages, for example in the way they set out visualization exercises for students. See for example the tutorial by Kalju Kahn, Department of Chemistry and Biochemistry, UC Santa Barbara, Macromolecular Visualization Laboratory Exercise (2003–2006), available at: http://web.chem.ucsb.edu/~kalju/chem110L/public/tutorial/intro.html (accessed 13 November 2012). These languages draw upon the history of material models of molecules, such as space-filling or rod-and-ball models (Francoeur 1997; also see the famous picture of Watson and Crick with their double helix model). Interestingly, chemists play with their models (creating the ethanol dog), print models on T-shirts, and have jewelry made in the shape of (interesting) molecules. For them, molecules are their friends (see also Laszlo 2000).

20. As Natasha Myers (this volume) shows, the machinery metaphor for cellular processes is used with increasing frequency, and biology students are now trained to use analogies with circuit boards and to become "biological engineers."

21. Supramolecular chemists at an Italian university, who could not synthesize the projected supramolecular structures themselves, sent their images of rotaxanes as "blueprints" to their collaborators in the US. The synthesized structures were then shipped back to Italy so that they could study and use them. (Interview by Martin Ruivenkamp with Alberto Credi and Vincenzo Balzani, University of Bologna, 2008.)

22. Schummer (2006, 66) adds: "since the technomorph sign language is intuitively accessible by everybody, such images have become popular illustrations beyond academia, in popular science magazines and even newspapers." But such perception and interpretation can also play a role in "inspiring and guiding new research fields" (ibid., 69). Clearly, there is no strict separation between these images as used within and outside of science—a further example of hybridity.

23. In architecture, with its long tradition of presenting designs in images and mock-ups, the possible achievements are visualized at a reduced scale (see Yaneva 2005) but can be readily understood by most people. Readers of nanoscale design images, on the other hand, have to be tutored to understand the nature and purpose of the image. This point is actually a general one about images in science, and is not limited to lay readers (cf. Bastide 1990). Myers (this volume) adds an interesting example in which an engineer uses a paper clip, bending it to create a model for the folding of a protein molecule, and in doing so reconfigures the thinking of the biologist postdoc working on the protein.

24. Leonardo da Vinci's "Vitruvian man" shows the ideal proportions of man, in a geometry described by the ancient Roman architect Vitruvius.

25. This applies not only to nanotechnology but also to synthetic biology.

26. Compare this to Myers (this volume), who suggests that metaphors can drive an entire research program.

27. This is also the starting point for Daston and Galison (2007, 396): "along with this activity in the trading zone between the scientific and the engineered, a new role came into existence for the visual, one that is only awkwardly and irrelevantly reducible to faithful depiction—direct or indirect—of what can exist." They do not really discuss what the new role for the visual might be.

28. In these two genres, the epistemological issues of representation, in the sense of resemblance, are often foregrounded. They are slightly different, however, in that mapping raises questions about resemblance to the known (at least, known elsewhere), and tracing is about inference to the unknown (or as yet unknown).

References

Andre, G., S. Kulakauskas, M.-P. Chapot-Chartier, B. Navet, M. Deghorain, E. Bernard, P. Hols, and Y. F. Dufrêne. 2010. Imaging the nanoscale organization of peptidoglycan in living *Lactococcus lactis* cells. *Nature Communications* 1:27. doi:10.1038/ncomms1027.

Bachtold, A., P. Hadley, T. Nakanishi, and C. Dekker. 2001. Logic circuits with carbon nanotube transistors. *Science* 294:1317–1320.

Bainbridge, B. 2009. Personality enhancement and transfer. In *Unnatural Selection: The Challenges of Engineering Tomorrow's People*, ed. P. Healey and S. Rayner, 32–39. London: Earthscan.

Bastide, F. 1990. The iconography of scientific texts: Principles of analysis. In *Representation in Scientific Practice*, ed. M. Lynch and S. Woolgar, 187–229. Cambridge, MA: MIT Press.

Bastide, F. 1992. A night with Saturn. Trans. G. Myers. *Science, Technology and Human Values* 17 (3):259–281.

Beaulieu, A. 2002. Images are not the (only) truth: Brain mapping, visual knowledge, and iconoclasm. *Science, Technology and Human Values* 27:53–86.

Bleecker, J. 2010. Design fiction: From props to prototypes. Unpublished paper, Near Future Laboratory, Venice Beach, CA.

Böhme, G., W. van den Daele, and W. Krohn. 1978. The "scientification" of technology. In *The Dynamics of Science and Technology*, ed. W. Krohn, E. T. Layton, Jr., and P. Weingart, 219–250. Dordrecht: D. Reidel.

Brown, M. B. 2009. *Science and Democracy: Expertise, Institutions, and Representation*. Cambridge, MA: MIT Press.

Daston, L., and P. Galison. 2007. *Objectivity*. New York: Zone Books.

De Ridder-Vignone, K., and M. Lynch. 2012. Images and imaginations: An exploration of nano image galleries. *Leonardo* 45 (5):447–454.

Douglas, M. 1966. *Purity and Danger*. Harmondsworth: Penguin.

Drexler, K. E. 1986. *Engines of Creation: The Coming Era of Nanotechnology*. New York: Anchor Press/Doubleday.

ETC Group. 2003. The strategy for converging technologies: The little bang theory: A mix of bits, atoms, neurons, and genes (B.A.N.G.) make the world come 'round—for the USA! *ETC Group Communiqué*, issue 78.

Francoeur, E. 1997. The forgotten tool: The design and use of molecular models. *Social Studies of Science* 27:7–40.

Frankel, F. 2002. *Envisioning Science: The Design and Craft of Science Images*. Cambridge, MA: MIT Press.

Galison, P. 1997. *Image and Logic: A Material Culture of Microphysics*. Chicago: University of Chicago Press.

Galison, P. 2002. Images scatter into data, data gather into images. In *Iconoclash: Beyond the Image Wars in Science, Religion, and Art*, ed. B. Latour and P. Weibel, 300–323. Karlsruhe: ZKM; Cambridge, MA: MIT Press.

Ginzburg, C. 1979. Clues: Roots of a scientific paradigm. *Theory and Society* 7:273–288.

Goodsell, D. S. 2006. Seeing the nanoscale. *Nano Today* 1 (3):44–49.

Goodsell, D. S. 2009. Fact and fantasy in nanotech imagery. *Leonardo* 42 (1):52–57.

Greimas, A. J. 1983 [1966]. *Structural Semantics: An Attempt at a Method.* Trans. D. McDowell, R. Schleifer, and A. Velie. Lincoln: University of Nebraska Press.

Henderson, K. 1999. *On Line and on Paper: Visual Representations, Visual Culture, and Computer Graphics in Design Engineering.* Cambridge, MA: MIT Press.

Kirby, D. 2010. The future is now: Diegetic prototypes and the role of popular films in generating real-world technological development. *Social Studies of Science* 40 (1):41–70.

Landau, J., C. R. Groscurth, L. Wright, and C. M. Condit. 2009. Visualizing nanotechnology: The impact of visual images on lay American audience associations with nanotechnology. *Public Understanding of Science* 18 (3):325–337.

Laszlo, P. 2000. Playing with molecular models. *Hyle* 6 (1):85–97.

Latour, B. 1991. *Nous n'avons jamais été modernes. Essai d'anthropologie symétrique.* Paris: Editions La Découverte. (English translation, *We Have Never Been Modern.* Cambridge, MA: Harvard University Press, 1993.)

Latour, B. 2002. What is iconoclash? Or is there a world beyond the image wars? In *Iconoclash: Beyond the Image Wars in Science, Religion, and Art*, ed. B. Latour and P. Weibel, 14–37. Karlsruhe: ZKM; Cambridge, MA: MIT Press.

Lösch, A. 2010. Visual dynamics: The defuturization of the popular "nano-discourse" as an effect of increasing economization. In *Governing Future Technologies: Nanotechnology and the Rise of an Assessment Regime*, ed. M. Kaiser, M. Kurath, S. Maasen, and C. Rehmann-Sutter, 89–108. Dordrecht: Springer.

Lynch, M. 1991. Laboratory space and the technological complex: An investigation of topical contextures. *Science in Context* 4:51–78.

Lynch, M., and S. Woolgar. 1990. Introduction. In *Representation in Scientific Practice*, ed. M. Lynch and S. Woolgar, 1–18. Cambridge, MA: MIT Press.

McCray, W. P. 2005. Will small be beautiful? Making policies for our nanotech future. *History and Technology* 21 (2):177–203.

Milburn, C. 2004a. *Nanovision: Engineering the Future.* Durham: Duke University Press.

Milburn, C. 2004b. Nanotechnology in the age of posthuman engineering: Science fiction as science. In *NanoCulture: Implications of the New Technoscience*, ed. K. N. Hayles, 109–129. London: Intellect.

Milburn, C. 2010. Digital matters: Video games and the cultural transcoding of nanotechnology. In *Governing Future Technologies: Nanotechnology and the Rise of an Assessment Regime*, ed. M. Kaiser, M. Kurath, S. Maasen, and C. Rehmann-Sutter, 131–155. Dordrecht: Springer.

Mody, C. C. M. 2004. Small, but determined: Technological determinism in nanoscience. *Hyle* 10:99–128.

Nerlich, B. 2005. From Nautilus to nanobo(a)ts: The visual construction of nanoscience. *AZoNano—Online Journal of Nanotechnology* (December). Available at: http://www.azonano.com/article.aspx?ArticleID=1466 (accessed 14 November 2012).

Nerlich, B. 2008. Powered by imagination: Nanobots at the Science Photo Library. *Science as Culture* 17 (3):269–292.

Nordmann, A. 2004. New space for old cosmologies. *IEEE Technology and Society Magazine* 23 (4):48–54.

Nordmann, A. 2008. Philosophy of nanotechnoscience. In *Nanotechnology*, vol. 1, *Principles and Fundamentals*, ed. G. Schmid, 217–243. Weinheim: Wiley-VCH.

Nordmann, A., B. Bensuade-Vincent, and A. Schwartz. 2011. The genesis and ontology of technoscientific objects. Project description, Technische Universität Darmstadt. Available at: http://www.philosophie.tu-darmstadt.de/goto/goto/home/home.en.jsp (accessed 18 December 2011).

NSTC. 1999. *Nanotechnology: Shaping the World Atom by Atom*. Washington, DC: National Science and Technology Council.

Ottino, J. M. 2003. Is a picture worth 1,000 words? Exciting new illustration technologies should be used with care. *Nature* 421:474–476.

Persson, M. 2003. Fraaie platen in saaie bladen. *De Volkskrant* (20 December).

Phipps, M., and S. Rowe. 2010. Seeing satellite data. *Public Understanding of Science* (Bristol, England) 19:311–321.

Pombo, O. 2010. Epistemological features of nanoimages. Paper presented at the Second Annual Conference of the Society for the Study of Nanoscience and Emerging Technologies, Darmstadt, Germany (29 September–2 October).

Rheinberger, H. J. 2002. Auto-radio-graphics. In *Iconoclash: Beyond the Image Wars in Science, Religion, and Art*, ed. B. Latour and P. Weibel, 516–519. Karlsruhe: ZKM; Cambridge, MA: MIT Press.

Rip, A. 2002. Science for the 21st century. In *The Future of the Sciences and Humanities: Four Analytical Essays and a Critical Debate on the Future of Scholastic Endeavour*, ed. P. Tindemans, A. Verrijn-Stuart, and R. Visser, 99–148. Amsterdam: Amsterdam University Press.

Rip, A. 2011. Protected spaces of science: Their emergence and further evolution in a changing world. In *Science in the Context of Application: Methodological Change, Conceptual Transformation, Cultural Reorientation*, ed. M. Carrier and A. Nordmann, 197–220. Dordrecht: Springer.

Rip, A., and M. van Amerom. 2010. Emerging *de facto* agendas around nanotechnology: Two cases full of contingencies, lock-outs, and lock-ins. In *Governing Future Technologies. Nanotechnology and the Rise of an Assessment Regime*, ed. M. Kaiser, M. Kurath, S. Maasen, and C. Rehmann-Sutter, 131–155. Dordrecht: Springer.

Robinson, C. 2004. Images in nanoscience/technology. In *Discovering the Nanoscale*, ed. D. Baird, A. Nordmann, and J. Schummer, 165–169. Amsterdam: IOS Press.

Roco, M. C., and W. S. Bainbridge, eds. 2002. Converging technologies for improving human performance: Nanotechnology, biotechnology, information technology and cognitive science. NSF/DOC-sponsored report, National Science Foundation. Arlington, VA, June. Available at: http://wtec.org/ConvergingTechnologies/1/NBIC_report.pdf

Ruivenkamp, M. 2011. Circulating images of nanotechnology. PhD thesis, University of Twente, 2011.

Ruivenkamp, M., and A. Rip. 2010. Visualizing the invisible nanoscale: Visualization practices in nanotechnology community of practice. *Science Studies* 23 (1):3–36.

Schummer, J. 2006. Gestalt switch in molecular image perception: The aesthetic origin of molecular nanotechnology in supramolecular chemistry. *Foundations of Chemistry* 8:53–72.

Staley, T. W. 2008. The coding of technical images of nanospace: Analogy, disanalogy, and the asymmetry of worlds. *Techne* 12 (1):1–22.

Swiss Re. 2004. Nanotechnology: Small matter, many unknowns. Available at: http://www.nanowerk.com/nanotechnology/reports/Nanotechnology_Small_matter_many_unknowns.php (accessed 13 November 2012).

Tiles, M. 2009. Technology and the possibility of global environmental science. *Synthese* 168:433–452.

Toumey, C. 2005. Apostolic succession. *Engineering and Science*, no. 1/2: 16–23.

Yaneva, A. 2005. Scaling up and down: Extraction trials in architectural design. *Social Studies of Science* 35 (6):867–894.

10 Toward a New Ontology of Scientific Vision

Annamaria Carusi and Aud Sissel Hoel

1 Introduction

The emergence of the computational techniques of modeling, simulation, and visualization in all fields of science is producing a new range of visual artifacts that challenge us to reconsider the role of instruments and technologies in scientific observation and representation. One aspect of this is the long-established distinction between qualitative and quantitative methods, which has historically shaped conceptions of scientific methods.[1] In this chapter we discuss the dismantling of the qualitative-quantitative distinction in the practice and instrumentation of computational biology, and the ways in which this prompts us to recognize the limitations of long-established ontological categories.[2]

Disciplinary groupings within the broad domain of computational biology often define themselves according to the instruments and methods they use—i.e., according to whether they are qualitative and observational, or quantitative and mathematical or computational. This distinction is, however, constantly blurred in their practices, which depend upon an impressive array of visual artifacts, including microscopic images, MRI, organ atlases, virtual organs, optical imaging of "real" organs, and visualizations of simulations in the form of movies and stills. Despite the apparent blurring of the quantitative and qualitative in the way these visual artifacts are used, among the scientists there is not an easy acceptance that blurring the distinction is good practice. In the research groups we observed, there is ongoing controversy concerning these methodologies. Training courses for doctoral students in computational biology try to reinforce the idea that these methodologies are distinguishable; they often suggest that the aim of the science is to arrive at a quantitative solution or proof of mathematical models, at which point it is possible to dispense with the qualitative methods that have been used along the way. This holds true even of visualizations, the qualitative appraisal of which is often seen as only a stepping stone to the surer

Carusi and Hoel are equal first authors of this chapter, listed alphabetically.

quantitative analysis. Implicitly, there is an alignment of "good" or "objective" scientific practice with quantitative methods (see also Beaulieu 2002), for example when scientists involved in computational biology and other computational sciences portray themselves as extending the "rigor" of quantitative methods into areas where they were previously difficult to use (e.g., Hey et al. 2009). These observations stand alongside the fact that the methodologies continue to be debated and contested in research meetings and other fora (see also Keller 2002), and that the distinction between qualitative and quantitative is rendered problematic by the very nature of the methods and instruments used and the practices around these. In our discussion we explore different ways of thinking of the interrelationships between the qualitative and the quantitative, and propose that computational biology creates new hybrids of methods which, while appearing to be new, also reveal the instabilities and inadequacies that have always been implicit in the qualitative-quantitative distinction as traditionally conceived.

The hybridization of methods in computational biology is instantiated in the visual artifacts and modes through which much of the science is carried out. Our report of research in the context of computational biology gives an account of this hybridization process, in which new configurations of vision, technologies, and objects (ranging from biological processes to mathematical ideas) arise. In these configurations, the qualitative and the quantitative are intertwined in the very act of observation, indeed even in vision as such. What we argue is that traditional substantivist ontologies, which conceive subjects and objects as pregiven and independently existing entities, fall short of accounting for these intertwinements and, hence, for the nature of scientific vision. For, as we hope to show, there is a sense in which the body measures what it sees; and, conversely, practices of measuring involve a kind of seeing the phenomenon. The case of computational biology allows us to highlight the nature of such intertwining, which nonetheless occurs even in traditional apparently nonhybrid methods. Our efforts to develop a proper understanding of the role and significance of these new configurations leads us to propose an ontological reframing of scientific vision—indeed of vision and perception as such—that has implications beyond the relatively recent practices and techniques we discuss here.

For an ontological framework that may prove helpful for understanding how computational scientific vision operates, we turn to the work of Maurice Merleau-Ponty. His philosophy of embodied perception presented in *Phenomenology of Perception* (1962) is already associated with a strong line of research in science and technology studies (STS) on the role and significance of material practices, touch, and gestures in epistemic work.[3] In this chapter, we focus on Merleau-Ponty's later philosophy, expressed in *The Visible and the Invisible* (1968) and *Nature* (2003). In these works, Merleau-Ponty makes a more radical break with the ontology of subject and object that, on his own account, still haunted his earlier work.[4] In particular we focus on his adoption of the notion of the circuit from the biology of Jakob von Uexküll (1982), and further, on Merleau-Ponty's idea of the body as the "measurant of . . . things" (1968, 152). This perspective brings the qualitative and the quantitative into a new constellation.

The present chapter thus draws on Merleau-Pontian insights to develop a new ontology of scientific vision in relation to the reconfiguration of the qualitative and the quantitative exhibited in emerging computational methods and techniques. This new framework, articulated around the notion of the measuring body, will allow us to recognize the formative and co-constitutive capacities of technologies and the multidimensional ontology to which these capacities give rise.[5] We begin, however, by introducing some of the new artifacts in computational biology that instantiate scientific visuality beyond any ready-made distinction between the qualitative and the quantitative.

2 Blurring Boundaries: Computational Biology

The label "computational biologist" does not define a specific group of scientists, but could refer to anyone who is engaging with and trying to use computational methods across the life sciences, including people who by disciplinary background are physicists, mathematicians, computer scientists, engineers, or experimental biologists. Indeed, computational biology designates a technological methodology for the study of life sciences rather than a scientific area or discipline. It is closely related to systems biology, which is sometimes defined as an *approach* (Kohl and Noble 2009) and sometimes in terms of a new *object of study*: that is, as "the science that deciphers how biological functions arise from the interactions between components of living organisms" (Boogerd et al. 2007, 6).

Computational biology is a program of research that is still seeking to establish itself. This makes the domain particularly relevant for a study of emerging techniques and methods. The development of the program is uneven across the domain of biology and depends on the state of play within specific subdomains with respect to research questions and interdisciplinary relations. In particular, mathematical biology, systems biology and computational biology are very closely related. Computational biology may be seen as the technological side of systems biology, and mathematical modeling is a core technique of both.[6] For this reason, research collaborations often involve both mathematicians and computer scientists; but very importantly, they also involve experimentalists who use mathematical or computational techniques in the wet lab.

The work reported in this paper draws on a longstanding study of a group of computational biologists who have been developing models of multicellular systems (such as cancers) and of whole organs (such as the heart), and of a systems biology lab more closely associated with biochemistry. These groups consist of scientists who came from biochemistry, engineering, mathematics, and computer science, and ranged in rank

from doctoral students to established professors. The study was conducted by partici-
pative observation, interviews, and focus groups, and through analysis of the articles
and other inscriptions produced by the group. Particular attention was given to how
differences in the adoption and use of visualizations in projects working across disciplinary boundaries affected the progress made in the collaboration (see Carusi 2008
and 2011).

In this chapter we draw out the philosophical importance of new computational
methods of modeling, simulating, and visualizing in biology for the understanding
of scientific vision. In particular, we focus on the hybrid practices of a group of computational physiologists developing new approaches to modeling the heart. In the
epistemic context of computational biology, the group's program of research has been
geared toward achieving an adequation between *in vivo/ex vivo* experiment and an *in
silico* model and simulation of a biological process. This can be done only through a
hybridization of methods that, in the computational biologists' own terms, make the
model "gain reality" in relation to experimental data, and make the data gain utility
in relation to the model. Hybridization in the process of model development normally
involves the following steps, presented here in a highly simplified and not necessarily
chronological order:

(1) Obtaining experimental data (relating to factors like concentrations and distributions of chemicals, signals, and speeds) either from publications, or, ideally, from col-
laborating biologists or physiologists who are experimentalists;
(2) Parameter fitting, whereby data are processed in order to obtain the best possible
fit between experiment and model, often bringing into play averages over a number
of experiments;
(3) Constructing mathematical models in the form of ordinary or partial differential
equations (used for continuous rather than discrete processes);
(4) Solving models through computational simulations, involving also numerical
analysis and algorithmic techniques; and
(5) Visualizing the output of the simulation in the form of graphs but also, importantly, in the form of a visual animation that shows, for example, the flow of electrical
currents across the heart under specific conditions.

This process, which involves quantification and mathematization at every step, is
finally rendered in qualitative visual form, but also involves images and visualizations
throughout, as is shown in figure 10.1.

In collaborations where this kind of hybridization occurs, experiments are conducted
with a view to modeling and simulation, and involve a combination of observational
and quantitative techniques. There is a complex interplay between mathematical and
numerical techniques that go into the model and simulation, on the one hand, and on
the other the observation of the visual output of experiments (in the form of images,

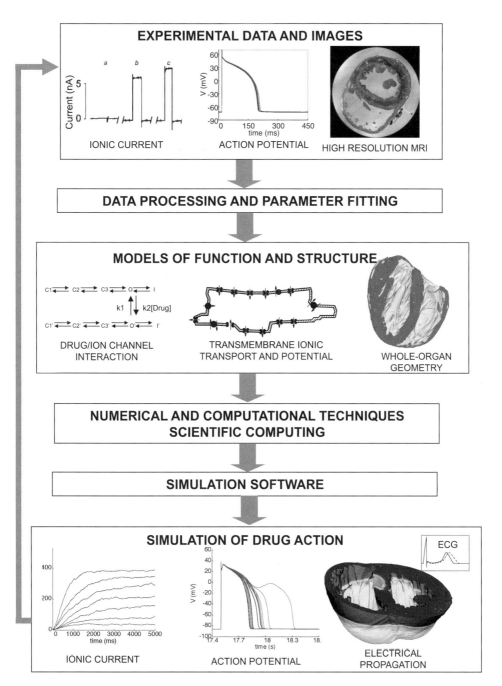

Figure 10.1

A flow chart of the research process of computational physiology, presented in the context of a scientific article whose aim is to demonstrate the computational approach's applicability to drug development. The chart targets an audience of scientists who may not be familiar with the tools and methods of computational science. Each key stage of the process (experiment, model, simulation) is illustrated by means of its typical visual outputs. From Rodriguez et al. (2010), with permission.

graphs, and diagrams) and of simulations (in the form of dynamic visualizations or movies). Transmutations of data from numerical into visual forms and vice versa are at the core of the process: they are the very means to determine, in the computational biologists' own terms, whether the model is a "representation of the physiological process under investigation." This comparison between model output and experimental results therefore is seen, by the scientists, as crucial to the validation of the model/ simulation.

In presentations, workshops, and research sessions with their collaborators, the heart modelers make extensive use of stills and clips from movies of simulations. Although such visualizations are also used for communicative and rhetorical purposes, they are not an optional addition to the research process. The results of modeling and simulation must be visualized in order to be made intelligible, for example by representing complex processes involved in the heart's functioning with key spatiotemporal features. Of course, the visualization itself is the output of a simulation driven by a mathematical model. It is a mathematical formalism that is rendered in concrete perceptual form and described through qualitative adjectives like "spiraling," "smooth," "broken up," and "erratic"; that is, it is a qualitative rendering of a quantitative hypothesis. As noted above, the qualitative comparison with the output of experiments is also a key element of the validation of the simulation result: sometimes, for example, simulation movies or stills are compared with images taken during microscopy sessions or the visual outputs of experiments that take other forms (for example, applying shocks to physical hearts and registering the results). Comparability cannot be taken for granted; it is constructed through the process of calibrating experiments and simulations. For example, the wet-lab experiments are conducted in such a way as to yield parameter values for the construction of the models and simulations; the registering of the results of experiments is evaluated from the perspective of its fit with computational simulations; and the visual outputs of both wet-lab experiments and dry-lab simulations are rendered in a stylistically similar way. This means that the mode of seeing and analyzing the visual field relies on the quantification engendered by computational methods, but that this in turn is shaped by the qualitative visual outputs in an ongoing to-and-fro process.

We will now turn to two key points in this complex process: the first from the experimental stage, and the second from the construction of the mesh for the visualization. Our first example is taken from a systems biology lab, where biochemists are using mathematical models in the context of experiments on bacteria. These experiments are conducted using various types of observational and imaging means, including microscopy. For the microscopists we observed, working in a systems biology paradigm, looking down the microscope at samples had been replaced by looking at a screen on which a digital video of the sample was displayed. In a way reminiscent of the close interrelationship between selection and mathematization described by Lynch

(1990), already at this point of the process digitization lends itself to measurement and quantification. Experimentalists continuously flip between viewing the sample on the screen, displays of measurements, and the outputs of mathematical models using the measurements. These are often juxtaposed on a single screen, or on screens placed side by side. In fact, the whole process of setting up the sample for the experiment is geared toward the modeling that will ensue, and is therefore done so that particular types of measurement are possible. Measurement therefore is informed both by observation and by the models. For example, which concentrations to measure is determined by the expectations of researchers regarding the behavior of the particular samples they are observing, as well as by the observation of the output of the mathematical models into which they are fitting measurements taken during the experiment. In this process, observation is continuously adjusted with reference to the model and vice versa, with measurement playing a mediating role between them.

Our second example is drawn from cardiac electrophysiology and shows another crucial moment in the transmutation of the qualitative and the quantitative in the experimental workflow, this time in the construction of the computational mesh used in the software that drives the simulation and produces the visualization. Where the previous example illustrates the practice of juxtaposing observation and mathematical models, this one shows how the qualitative and quantitative are intertwined in the software programs used for simulation and visualization. The following (in highly summarized form) are the main stages in the production of the meshes which are used for simulation and visualization:

(1) Acquiring images either through histology or through MRI;
(2) Segmenting the MRI data sets into relevant anatomical features;
(3) Tracking contours and surfaces through numerical techniques, and generating surface data through algorithms;
(4) Generating a 3D structure from a 2D structure by adding volume to the surface data, thereby constructing a volumetric mesh.

The mesh is a bridge between the model and the simulation, since it is the means through which the model is processed for the computational simulation. For our purposes, the important point is that the mesh is constructed via a process that includes the whole gamut of techniques from observation and qualitative imaging to a diverse array of mathematical and algorithmic applications. Jointly, these techniques generate both numerical and visual output that is used to analyze the model and its simulation and to compare these to experiments. This process is illustrated in figure 10.2.

The mesh is a way to interweave the qualitative and the quantitative methodologies of computational biology in the computational instrumentation, that is, in the software. There are many factors that may influence the form a mesh takes, including technical factors such as the computational power and the software available, as well

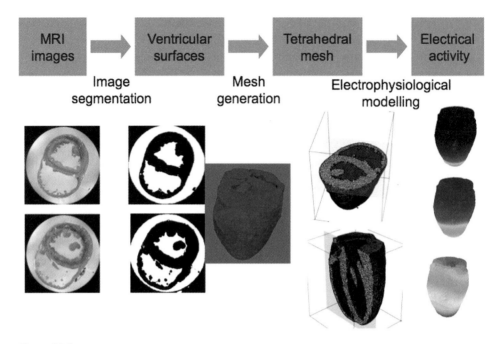

Figure 10.2
The main stages of generating the mesh and ultimately the visualization of the simulation. MRI images are obtained and are segmented by image-processing algorithms to delineate the ventricular surfaces of the heart. These form the tissue boundaries of the anatomically based mesh. Simulations of electrical activity in the heart are run over the mesh, and the output of the simulation is visualized in a movie. From Carusi et al. (2012), with permission.

as social and institutional factors that shape the direction of research that any particular group may take. The crucial point, however, is that even if the mesh is shaped by these factors, it itself acts as a powerful shaper of the visible in this domain, as it is the prime generator of the visualizations that are characteristic of this field. The mesh is the precise site of the transmutation of the visual into the numerical, the continuous into the discrete, the qualitative into the quantitative, and vice versa; in Merleau-Ponty's terms, it is the site of their "crossing-over," one into the other.

Constructing the meshes for the visualizations, including the numerical, algorithmic techniques that go into them, is a major research focus in computational biology, and there is great concern to guarantee "objectivity"—something frequently expressed by the scientists themselves. However, it is also clear that the researchers grapple with questions relating to the epistemic status of these visual outputs. Three frequently mentioned concerns are finding ways of dealing with the distortions of the model that are often required for the visualization; finding the right balance between the level of smoothing required for visual quality and being able to see the uncertainty in underlying data; and avoiding effects that are considered an "artifact of the software." What is undertaken for visual effect has deeper epistemological significance, inviting a reconfiguration of the qualitative and the quantitative and their implicit alignments with subjective perception and objective knowledge.

Thus far we have adhered to the ways the researchers in the domain of computational biology themselves characterize their methods, associating qualitative methods with observation and quantitative methods with statistical, numerical, and mathematical techniques. But, we need to ask, are the new hybridization methods exemplified in computational biology merely blurring the boundaries of this distinction, or do they indicate that there is something wrong with the way this distinction has traditionally been cast?[7] We argue that the blurring of distinctions between qualitative and quantitative methods apparent in this domain in fact challenges our understanding of technologically mediated vision. In philosophical terms, we argue that the substantivist ontology that lies at the basis of established assumptions about how vision links the subject and the object is no longer tenable. Such an ontology is complicit in upholding the distinction between qualitative and quantitative, aligning subjective (doubtful) vision with the subject, and objective (certain) quantification with the object. In order to grasp the new hybridization in a more thoroughgoing way, we need an alternative ontology, the ingredients of which can be found in the later work of Merleau-Ponty.

3   Toward a New Ontology of Scientific Vision

In the broad domain of science studies, efforts to interrogate and reframe traditional ontologies—for example, between subjects and objects or animate and inanimate actors—have provided and continue to provide theoretical and methodological impulses to the field. Several strategies have been deployed to replace traditional essentialist ontologies with ontologies that are relational, taking processes rather than preexisting and preconstituted entities as their starting point.[8] Into this conversation we bring the late work of Merleau-Ponty, which set out to problematize the phenomenological ontology of subjects and objects against the intellectual background of the burgeoning new theories of biology, thermodynamics, physics, and cybernetics of the 1950s.

Merleau-Ponty's later work has received less attention than his account of embodied perception set out in *Phenomenology of Perception*, which first appeared in French in 1945. That work revolutionized the conceptualization of subjecthood and agency in the history of philosophy, providing a compelling account of embodied experience as a counter to the array of disembodied and transcendent selves that the history of Western thought has produced. In STS, the philosophy of Merleau-Ponty has provided support for accounts of scientific inquiry that stress the embodied nature of subjects, the materiality of objects, and the mediating role of instruments.[9] And yet, despite the radical rethinking of the traditional ontology of subjecthood that his account of the subject of perception brought about, Merleau-Ponty himself remained dissatisfied with what he saw as phenomenology's inability to move beyond the subject-object distinction. His work after the *Phenomenology of Perception* consistently aims at ontologically reframing the subject-object distinction in different ways.[10]

Biological metaphors are prominent in both *The Visible and the Invisible* (1968) and the collection of lecture and working notes published under the title *Nature* (2003).[11] The former introduces the notion of "flesh" (to replace the notion of body) as something that is neither of the subject nor of the object, but is rather the principle of interconnectedness and intertwinement that allows subjects and objects to interact. The ontology of the flesh conceives of the body not as a bodily substance but as an incarnated principle of differentiation or articulation: "For if the body is a thing among things, it is so in a stronger and deeper sense than they: in the sense that, we said, it *is of them*, and this means that it detaches itself upon them, and, accordingly, detaches itself from them" (Merleau-Ponty 1968, 137). Throughout his late writings Merleau-Ponty uses the term "flesh" rather broadly; he talks, for instance, about the flesh of vision, about the flesh of the world, and about color or pigment as the flesh of painting. Even if it is the living body that provides the opening to the world and makes possible all further meaning or differentiation, the body is not conceived as the sole differentiator. Symbols and tools have ontological import by virtue of their power to hook into the circuit and transform or modify the metaphysical structure of the flesh. The body and scientific instruments can thus be understood to form a coupled or distributed system. The circuit of organism and environment is *open*, and signs and tools are accorded active roles in the body's ongoing transformations: "Every technique is a 'technique of the body,' illustrating and amplifying the metaphysical structure of our flesh" (Merleau-Ponty 1993, 129).

In his lectures on the topic of nature, held at the Collège de France from 1956 to 1960, Merleau-Ponty draws upon a range of biological theories that challenge causal thinking. These theories seek to provide alternatives to an understanding of biological processes, such as development and evolution, as a series of causal relations between entities that are constituted independently of each other, and are therefore only externally related. Internal relations, Merleau-Ponty insists, hold where the related entities (the *relata*) are what they are only in relation to each other. Crucially, this relational view requires a specific *measure* that serves as the point of view under which the entities in question first become visible. This is where Merleau-Ponty's approach goes beyond a dyadic relationality.[12]

Merleau-Ponty discusses several biologists associated with the new biology of his time who, each in different ways and in a broad range of specific examples, contested mechanistic causal thinking in biology, and emphasized the highly relational and complex nature of biological processes. Here we shall only discuss the influence on Merleau-Ponty's philosophy of Jakob von Uexküll's work on the living organism and its relation to its surroundings, an example that has particular relevance for our claim regarding the role of computational technologies in scientific vision.

Uexküll aimed to account for biological processes, such as embryological and physiological organization, in terms of behavior. This in itself is a shift away from mechanistic causal accounts, since to see something as exhibiting a behavior implies seeing it as always oriented toward something in a targeted way. Uexküll's famous example is that of a tick hanging off a branch until a passing animal emanates enough heat for it to drop down into the animal's fur and burrow its way under the skin. For the tick, body heat is a target that for another organism does not exist. Hence, according to Uexküll, behavior should be accounted for in terms of *meaning* relations rather than causal relations. He framed these meaning relations as occurring within what he termed an "*Umwelt*," that is, a meaningful environment as specified and lived by particular organisms.

According to Uexküll, behavior and *Umwelt* specify or co-create each other, in a series of feedback loops or in a cyclical to-and-fro. The world of the organism is "distilled" by the organism and the possibilities of its behavior within its surroundings. In the case of animals with a more highly developed nervous system, there is a differentiation of sensory inputs to which the animal responds with fine-grained actions; but these differentiated reactions are only possible because the nervous system is formed in response to the surroundings. The notion of the *Umwelt* unites what is normally kept apart, through interrelating the activity that creates the organs and the activity of behavior. There is therefore a reciprocal relation between nature which has created the organism and the organism which creates nature (in the form of its *Umwelt*).

In the last section of *Nature*, Merleau-Ponty echoes Uexküll's account when he writes:

Each part of the situation acts only as part of a whole situation; no element of action has a separate utility in fact. Between the situation and the movement of the animal, there is a relation of meaning which is what the expression *Umwelt* conveys. The *Umwelt* is the world implied by the movement of the animal, and that regulates the animal's movements by its own structure. (Merleau-Ponty 2003, 175)

The *Umwelt*, therefore, is not in front of the organism like a goal or an object. Far from being an inchoate external world confronting the organism, the *Umwelt* is an always already meaningful pattern that the living organism, *qua* living, is embedded in and acts through.

Merleau-Ponty uses these insights from biology to rethink the ontology of perception. As part of this, he emphasizes the continuity between humans and animals and between perception and biological processes. It is important to note, however, that these suggestions of continuity do not amount to a reductive biologism. Like

Uexküll before him, Merleau-Ponty sees as continuous the natural and the cultural or symbolic order. Life itself is conceived in terms of an organizing principle within being, which he refers to as "natural symbolism" (2003, 212). The suggestion is that living organisms operate by a principle of divergence, that is, by a principle that, simultaneously, differentiates and articulates and thus allows a targeted distribution of the surroundings.

The biological input into Merleau-Ponty's theory provides a series of very productive ideas for reframing the ontology of perceptual relations, centering on the notion of the measuring body. In the next section, we will show how Uexküll's idea of the organism and its surroundings forming a circuit is applied to the perceiving or measuring body. We will also elaborate on the notion of the measuring body in its relation to tools and symbolic systems, before drawing out implications for understanding current uses of computational technologies in scientific vision.

4 Perceptual Relations and the Measuring Body

As we have seen with Uexküll and the notion of *Umwelt*, the perceiving body is in a circuit with its surroundings.[13] To be in such a relation implies being open to the surroundings, or to be in a relation of mutual openness and interdependence. The relation of responsiveness between body and surroundings, which in organisms and animals gives rise to the specific structures of the organism or nervous system of the animal, becomes, for the perceiving body, a relation whereby the body always finds itself in a divergence with its surroundings. Think, for example, of the movements of the eyes in order to focus now on this, and now on that: this shifting would not be necessary if the eyes could take in everything at once. Instead, the impetus for the movement comes from a gap which opens as attention and interest shift, a gap which vision seeks to fill by shifting its focus elsewhere. Vision comes from this divergence; it is, so to speak, carved out of the hollows in the texture (the warp and woof) of the surroundings. Vision is therefore something that is neither on the side of the perceiver nor on the side of the surroundings, but something which emerges in the circuit that links them together and specifies each of them in the linking.

The meaningful patterns that result from this dynamic exchange can emerge only against a standard or benchmark for the assessment of what the surroundings yield to the movement and refocusing of the eyes. Merleau-Ponty refers to the body as "the measurement of the world" (2003, 217), which means that the body, in its relationship with the surroundings, divides or articulates these surroundings. This is the reason why the body is not merely a thing among things, but also the very standard according to which things become things for it: "my body is not only one perceived among others, it is the measurant (*mesurant*) of all, *Nullpunkt* of all the dimensions of the world"

(Merleau-Ponty 1968, 248–249). The *Umwelt* is defined by the dimensions cut out for an organism. Each kind of organism opens a specific field of possible actions and entities. Thus understood, the *Umwelt* is a dimension in the sense that a different kind of organism would live in a different field of action and entities. By implication, different organisms are understood to live in different dimensions.

For Merleau-Ponty, another way of saying that the body is a measurement of things is to say that the body is a symbolism (2003, 211). He sees a profound continuity between natural and cultural symbolism through the process of differentiation and articulation. There is a "language before language, which is perception" (2003, 219). It is important to note, however, that symbolism is not understood in representational terms. Even language, for Merleau-Ponty, has to do with perception. Language "reproduces perceptual structures at another level" (2003, 212), since, like perception, it acts as a differentiating medium in the same way as the measuring body does, opening further dimensions and allowing new types and spheres of visibility. Language is conceived in terms of a *measuring* that is productive of a specific *way of seeing* and of perceiving in general. Moreover, language is not the only symbolic system through which the measuring body relates to its surroundings. In other texts, such as *Prose of the World* (1973) and "Eye and Mind" (1993), Merleau-Ponty makes a similar point with regard to painting and, of particular interest for this chapter, to algorithms and mathematical formulations.

At the very point in *Nature* where Merleau-Ponty experiments with a reformulation of vision in informational and computational terms in order to achieve a non-objectivist and noncausal account, a cryptic but highly suggestive note reads: "The modern evolution of mathematics which gets over the dilemma of quality or quantity. Theory of mathematics and of the algorithm to be made a variant of language" (2003, 313). Several points could be made in relation to this note,[14] but for present purposes we focus on the suggestion that, for Merleau-Ponty, mathematics itself is connected to a kind of seeing and to a rethinking of the qualitative-quantitative distinction. Like other symbolic systems, algorithms and mathematical formulations are not mere external means of expression; rather their import is to open new dimensions of vision, and new possibilities of actions and entities. Hence, with the injection of different symbolic systems, the *Umwelt* of the interrogating body becomes multidimensional. When a symbolic system is incorporated into the measuring body, there is a crossing over of qualitative and quantitative, "crossing over" denoting a deeper sense of intertwinement than is evoked by the notions of hybrid and mixed methods. On this new understanding, there is an inner connection between vision and measuring, because the symbolic system modifies the "parameters" of the measuring body, opening a new way of seeing that engages with new aspects and features of the surrounding world.

Drawing on the conceptual understanding offered by Merleau-Ponty and based on the biological thought of Jakob von Uexküll, we propose that computational technologies (especially those used for the type of science in which the techniques of modeling, simulating, and visualizing are closely interrelated with the techniques of experiment and observation) *become injected into the circuits between observer and observed, and become incorporated into the measuring body operative in those contexts*. It is not simply that we can no longer easily differentiate between quantitative and qualitative inputs into a research methodology; rather perception itself appears in a new light, as a measuring informed by the symbolic systems and tools at its disposal. The field of what is seen thus comes into being due to the articulating principles of the mediating symbolic systems.

Hence, when mathematical models and their graphic output are juxtaposed with microscopic observations in the experimental setup, as in our first example described in section 2, this is not simply a case of side-by-side coexistence of two separate modes of taking in information. We have already noted that measuring and observation actively inform each other, but now we can articulate the computational methods of modeling as part of an integrated interrogative setup, consisting of researchers, instruments, screens, and means for gathering and processing data: a distributed space of crossings-over bringing these into internal and co-constitutive relations, in which the measuring body is embedded. In this interrogative setup, factors and features of the interrogated field shift in their salience to scientific vision, and in their availability for possible actions and interaction. The idea of an organizing principle that correlates measuring and vision in a new, internal fashion allows us to rethink the quantitative and the qualitative in perception as well as in the instrumentation of science. This is not the mediation of independent processes, but rather the mutual informing of one by the other, qualitative observation directed by the exigencies of quantification and quantitative processes geared toward their qualitative rendering. The microscope itself is a prime example of a crossing over of measurement and seeing. The digital images and videos of microscopic observations projected onto screens, with the digitization and computational image processing that this implies, are similarly spaces where measurement and seeing are connected in internal and not merely external ways. Therefore, it is not only in setups that involve mathematical modeling that these crossings-over occur: this is a feature of all instrumental settings, even those that incorporate predigital instruments like the analog microscope.

Turning now to the construction of the computational mesh for the visualization of simulations, which was our second example in section 2, we can rearticulate the formative and thus transformative power of instruments—in this case, the software—in the interrogative setups of scientific research. The mesh does not simply render to vision the features of organs and other structures. The entire process which leads to the construction of the mesh is an active intervention in forming and shaping what rises up to vision, and what will then be mapped for the measuring required for the simulation and the visualization. This last step is always a matter of a particular research question, that is, a particular interrogation from a particular perspective. There are no all-purpose meshes; rather, to be an effective instrument, each mesh embodies the research question of a particular instance of modeling and simulation. The mesh opens a specific field of view by carving out and articulating features in accordance with the research question, and making them visible. The mesh mediates, on the one hand, between the actual organs and the simulation, and on the other hand, between the simulation and its visual rendering in movies and animations. However, the mesh also mediates between the simulation and its validation, because of the way in which it brings about comparability between visualizations and biological experiments.

Comparison is key to the process whereby models and simulations are seen to be credible or validated. In an iterative process over many cycles of wet-lab and dry-lab experiments, what is picked out, what is salient to vision in both the visualizations and the experimental outputs, is reshaped constantly, one in answer to the other. This is how the mesh enters into the circuit between the researchers and the interrogated field: the mesh makes a parameterized mathematical model visible, but it also informs the vision whereby outputs are analyzed and interpreted by the researchers in their quest for comparability.

It is important to note that even if there is a circuit here between interrogator and interrogated, it is not a vicious circularity in the sense of empty tautological projections. The constructions involved actually reveal something new about the interrogated field that would not otherwise have been revealed—recall that the measuring body both acts as a standard for articulating the surroundings and answers to those surroundings. The mediation of the circuits of interrogation by the mesh similarly articulates the research field and answers to it, transforming the overall standards of articulation as it does so.

5   Conclusion

We began this chapter by pointing out how computational visualizations and simulations in science challenge the long-established distinction between qualitative and quantitative methods, and by suggesting that the ontological distinction between subject and object that underlies this distinction limits our understanding of the nature of this challenge. Using computational biology as an example, we have shown that computational technologies involve more than a mere blurring of the qualitative-quantitative distinction. Firstly, even in contexts where the microscope is the main experimental and observational instrument, measuring plays an active role in observation, as does observation in measuring. Secondly, in contexts where visual observation is the output of computational modeling, simulating, and visualizing techniques, the quantitative and qualitative are crossed over at each and every step of the process. We have argued that what seems an apparent blurring in fact goes deeper and demands a new ontological approach to technologically mediated vision, which takes into account the reciprocal and co-constitutive relations between vision, technologies, and objects. We have drawn on the notion of the measuring body in a circuit with its surroundings, as developed by Merleau-Ponty and Uexküll, to show that vision itself involves measuring just as measuring involves observation; these processes intermediate each other. This means that the traditional association of the observational with the subjective and qualitative, and the statistical or numerical with the objective and quantitative, is not adequate.

We discussed Merleau-Ponty's notion of the measuring body as a type of distributed symbolism, including natural language as well as mathematical languages like equations and algorithms. Symbolic systems are injected into the circuit of the measuring body and its surroundings, and open up new and different dimensions of action and vision for the body in its context. Developing Merleau-Ponty's ideas further, we have attempted to show that technological mediation opens up a multidimensional ontology. The main lesson from Merleau-Ponty is that the transformations of mediation play a positive role in revealing aspects that could not otherwise be seen or accessed. It is by virtue of incorporating a standard for articulating that the living body provides an opening onto the world. But as Merleau-Ponty makes clear, this opening "is not an opening to everything." It is a "specified opening," which he characterizes as a "dimension," that is, a "point of view under which a variation is possible" (2003, 238). His notion of mediation allows a new understanding of the role of symbols and instruments, including computational techniques, for when symbols and instruments are injected into the circuit, the opening is not covered over but becomes instead further specified. They both amplify and reduce the reach of the perceiving body, allowing new things to come into view, yet always at the cost of introducing new blind spots. The mediation of symbols and instruments modulates the perceiving body's existential relations to its surroundings, giving rise to new perceptual configurations. This creates new landscapes of possible actions and ultimately even of possible interventions.

The computational turn brings to the fore the internal connections between qualitative and quantitative processes, and the new techniques are particularly suited for exploiting these connections. We are not saying that they create an unprecedented mode of vision, but rather that they let us see that our modes of access to phenomena have never in fact been neatly distinguishable into qualitative and quantitative permutations. Considered from the vantage point of the measuring body, the new computational techniques prompt us to realize that what may at first be taken as mere instances of hybridization between preexisting modes actually reveal the inherent hybridity of embodied vision itself. They disturb the prevailing tradition of distinguishing between instruments that allow us to see and instruments that allow us to quantify. By emphasizing the essentially mediated nature of scientific vision, and simultaneously confirming the ontological import of symbols and tools, the notion of the measuring body furnishes us with a new and powerful conceptual understanding of the visual artifacts used for scientific exploration.

Acknowledgements

We would like to thank Claude Imbert, Anne Beaulieu, and Sarah de Rijcke for their helpful comments on an earlier draft of this chapter, and the editors and anonymous reviewers for their constructive input.

Notes

1. See Sepper (1988) for an account of one well-known controversy around scientific method understood as consisting essentially in mathematical measurement or in qualitative observation; Chang (2004) for an account of how the ability to measure has come to be identified with scientific progress. For an account of these different methodologies in biological sciences, see Keller (2002).

2. In recent years, the field of science and technology studies has increasingly turned to questions regarding ontology, which involve a rethinking of subject and object and of associated terms such as agency and passivity. Bruno Latour, John Law, Annemarie Mol, and Steve Woolgar are just some of the leading figures who have contributed to this turn to ontology. See the special issue of *Social Studies of Science* guest-edited by Lezaun and Woolgar (2013).

3. *Phenomenology of Perception* (1962) is considered the classical formulation of Merleau-Ponty's philosophy of embodiment. For a recent paper in STS that draws explicit inspiration from this work, see Myers (2008). Others who have advanced work on embodied perception in STS include Latour (1986), Goodwin (1994), Prentice (2005), and Suchman (2007).

4. In the working notes published in *The Visible and the Invisible*, Merleau-Ponty remarks that the problems he encountered in *Phenomenology of Perception* were due to the fact that this work still in part depended on the philosophy of consciousness (Merleau-Ponty 1968, 183).

5. Among the current attempts to reframe ontology, the one that resonates most strongly with our approach is Karen Barad's agential realism, which is a version of a performative ontology. While we cannot do full justice here to the differences and similarities, it is important to note that Barad's agential realism draws on the philosophical physics of Niels Bohr and is explicitly nonphenomenological, as she herself emphasizes (2007, 429, fn18). As such, it differs from approaches drawing on Merleau-Ponty.

6. For history and further specifications see Varenne (2010) and Powell et al. (2007).

7. We propose that this transformation goes deeper than the linguistic hybridization associated with interdisciplinary collaboration. The latter has been articulated by Peter Galison (1997) as the construction of a "trading zone" for communication across disciplines in a type of pidgin or creole. What we have in mind here is a blurring of distinctions that challenges our very understanding of technologically mediated vision.

8. The most important of these approaches are: (1) actor-network theory, with its emphasis on relational networks between human and nonhuman actants (Latour 2005; cf. Law and Hassard 1999); (2) performative theories, with their emphasis of nonrepresentational accounts, stressing the nature of knowledge as something that is enacted rather than represented (Pickering 1995; Barad 2007); and (3) process theories, drawing on Alfred North Whitehead, with their emphasis on processes whereby entities are constituted in relation to each other (Stengers 2008). It is important to note that these approaches are not separate and distinct, since they overlap in important respects in their concerns and strategies.

9. Foremost among these, and possibly staying closest to Merleau-Ponty's philosophical program, is postphenomenology, put forward by Don Ihde and Peter-Paul Verbeek, which distinguishes between the different kinds of mediation between subjects and objects effected by different types of instruments, including hybrid and composite relations (Verbeek 2008). A phenomenological analysis of visualizations is undertaken by Araya (2003). The Merleau-Pontian perspective has been explicitly related to the life sciences by Myers (2008), who discusses how models of molecules are interrelated with the embodied practices of researchers.

10. Further comments by Merleau-Ponty that position his new work in relation to *Phenomenology of Perception* are found in Merleau-Ponty (1968, 229 and 253). Also see Saint-Aubert (2008, 7–40).

11. Merleau-Ponty is unusual among philosophers in his willingness to draw upon the scientific theories of his time. In his early work he drew on psychology and linguistics, and in his later work he was highly attuned to thermodynamics, physics, chemistry, and cybernetics, among other fields. His critical engagement with cybernetics is of continuing relevance to current debates about the nature of computation and its relation with perception, cognition, and agency. Merleau-Ponty recognized the challenge presented by computation to accepted ways of thinking about meaning and perception, but was also critical of the way cybernetics cast these problems in "machine" terms. In his turn to biology, Merleau-Ponty explicitly referred to cybernetics to highlight the impossibility of describing "Nature" in "pure" terms, unadulterated by artifice: "It is not possible to speak of Nature without speaking of cybernetics. Maybe this is only an ultrafinalism without mechanism, but we cannot think Nature without taking account to ourselves that our idea of Nature is impregnated with artifice" (2003, 86).

12. See Hoel (2011 and 2012) for a discussion of dyadic relationality versus triadic or differential relationality.

13. Merleau-Ponty also uses the term *Ineinander*, or the "one in the otherness," to describe the circuit (2003, 215).

14. The authors are developing this precise point concerning the relation between perception, language, and algorithms in a forthcoming paper titled "The Measuring Body."

References

Araya, Augustin A. 2003. The hidden side of visualization. *Techné* 7(2). Available at: http://scholar.lib.vt.edu/ejournals/SPT/v7n2/araya.html (accessed 2 January 2013).

Barad, Karen. 2007. *Meeting the Universe Halfway: Quantum Physics and the Entanglement of Matter and Meaning*. Durham: Duke University Press.

Beaulieu, Anne. 2002. Images are not the (only) truth: Brain mapping, visual knowledge, and iconoclasm. *Science, Technology and Human Values* 27:53–86.

Boogerd, Fred, Frank J. Bruggeman, Jan-Hendrik S. Hofmeyr, and H. V. Westerhoff, eds. 2007. *Systems Biology: Philosophical Foundations*. Amsterdam: Elsevier.

Carusi, Annamaria. 2008. Scientific visualizations and aesthetic grounds for trust. *Ethics and Information Technology* 10:243–254.

Carusi, Annamaria. 2011. Computational biology and the limits of shared vision. *Perspectives on Science* 19 (3):300–336.

Carusi, Annamaria, Kevin Burrage, and Blanca Rodriguez. 2012. Bridging experiments, models and simulations: An integrative approach to validation in computational cardiac electrophysiology. *American Journal of Physiology. Heart and Circulatory Physiology* 303 (2):H144–H155.

Chang, Hasok. 2004. *Inventing Temperature: Measurement and Scientific Progress*. New York: Oxford University Press.

Galison, Peter. 1997. *Image and Logic: A Material Culture of Microphysics*. Chicago: University of Chicago Press.

Goodwin, Charles. 1994. Professional vision. *American Anthropologist* 96 (3):606–633.

Hey, Tony, Stuart Tansley, and Kristin Tolle, eds. 2009. *The Fourth Paradigm: Data-Intensive Scientific Discovery*. Redmond, WA: Microsoft Research.

Hoel, Aud Sissel. 2011. Thinking "difference" differently: Cassirer versus Derrida on symbolic mediation. *Synthese* 179 (1):75–91.

Hoel, Aud Sissel. 2012. Technics of thinking. In *Ernst Cassirer on Form and Technology: Contemporary Readings*, ed. A. S. Hoel and I. Folkvord. Basingstoke, UK: Palgrave Macmillan.

Keller, Evelyn Fox. 2002. *Making Sense of Life: Explaining Biological Development with Models, Metaphors, and Machines*. Cambridge, MA: Harvard University Press.

Kohl, Peter, and Denis Noble. 2009. Systems biology and the virtual physiological human. *Molecular Systems Biology* 5 (292). doi:10.1038/msb.2009.51.

Latour, Bruno. 1986. Visualization and cognition: Thinking with eyes and hands. *Knowledge in Society* 6:1–40.

Latour, Bruno. 2005. *Reassembling the Social: An Introduction to Actor-Network-Theory*. Oxford: Oxford University Press.

Law, John, and John Hassard. 1999. *Actor Network Theory and After*. Oxford: Blackwell Publishing.

Lezaun, Javier, and Steve Woolgar, eds. 2013. A turn to ontology in social studies of science? Special issue of *Social Studies of Science* 43.

Lynch, Michael. 1990. The externalized retina: Selection and mathematization in the visual documentation of objects in the life sciences. In *Representation in Scientific Practice*, ed. M. Lynch and S. Woolgar, 153–186. Cambridge, MA: MIT Press.

Merleau-Ponty, Maurice. 1962. *Phenomenology of Perception*. Trans. C. Smith. London: Routledge and Kegan Paul.

Merleau-Ponty, Maurice. 1968. *The Visible and the Invisible*. Trans. A. Lingis. Evanston: Northwestern University Press.

Merleau-Ponty, Maurice. 1973. *The Prose of the World*. Trans. J. O'Neill. Evanston: Northwestern University Press.

Merleau-Ponty, Maurice. 1993. Eye and mind. In *The Merleau-Ponty Aesthetics Reader: Philosophy and Painting*, ed. G. A. Johnson, 121–149. Evanston: Northwestern University Press.

Merleau-Ponty, Maurice. 2003. *Nature: Course Notes from the Collège de France*. Compiled and with notes by D. Séglard; trans. R. Vallier. Evanston: Northwestern University Press.

Myers, Natasha. 2008. Molecular embodiments and the body-work of modeling in protein crystallography. *Social Studies of Science* 38 (2):163–199.

Pickering, Andrew. 1995. *The Mangle of Practice: Time, Agency and Science*. Chicago: University of Chicago Press.

Powell, Alexander, Maureen A. O'Malley, Staffan Müller-Wille, Jane Calvert, and John Dupré. 2007. Disciplinary baptisms: A comparison of the naming stories of genetics, molecular biology, genomics and systems biology. *History and Philosophy of the Life Sciences* 29 (1):5–32.

Prentice, Rachel. 2005. The anatomy of a surgical simulation: The mutual articulation of bodies in and through the machine. *Social Studies of Science* 35 (6):837–866.

Rodriguez, Blanca, Kevin Burrage, David Gavaghan, Vincente Grau, Peter Kohl, and Dennis Noble. 2010. The systems biology approach to drug development: Application to toxicity assessment of cardiac drugs. *Clinical Pharmacology and Therapeutics* 88 (1):130–134.

Saint-Aubert, Emmanuel de. 2008. *Maurice Merleau-Ponty*. Paris: Hermann.

Sepper, Dennis L. 1988. *Goethe contra Newton: Polemics and the Project for a New Science of Colour*. Cambridge: Cambridge University Press.

Stengers, Isabel. 2008. A constructivist reading of process and reality. *Theory, Culture and Society* 25:91–110.

Suchman, Lucy A. 2007. *Human-Machine Reconfigurations: Plans and Situated Actions*. 2nd ed. Cambridge: Cambridge University Press.

Uexküll, Jacob von. 1982 [1940]. Theory of Meaning. *Semiotica* 42 (1):1–24.

Varenne, Franck. 2010. *Formaliser le vivant: Lois, théories, modèles*. Paris: Hermann.

Verbeek, Peter-Paul. 2008. Cyborg intentionality: Rethinking the phenomenology of human-technology relations. *Phenomenology and the Cognitive Sciences* 7:387–395.

11   Essential Tensions and Representational Strategies

Cyrus C. M. Mody

In advancing themselves and their work, scientists balance between innovation and tradition, between building upon and breaking down the current state of knowledge. This balancing act is so central to research that Thomas Kuhn (1977) called it the "essential tension" of science. Yet it is not self-evident whether a particular argument or piece of evidence should be seen as conventional or iconoclastic. Many contributions could be assimilated easily to what has gone before or could be taken as anomalous and disruptive, depending on how they are presented and interpreted.

If a scientist prefers colleagues to interpret his or her work as either conventional or iconoclastic, he or she can *represent* the data and conclusions in a way that guides audiences toward the preferred interpretation. In presenting their data, scientists choose from a variety of representational conventions (or invent their own idioms). The choice of conventions helps scientists signal which disciplinary and institutional frameworks they would like to embed their work in, and helps audiences familiar with those frameworks understand which aspects of that research should be seen as novel and which should be seen as conforming to present knowledge.

1   Nanotechnology, Computerization, and Normal versus Revolutionary

This chapter explores the complex relationship between representational conventions and representations of (un)conventionality through a case study of scanning probe microscopy.[1] Probe microscopy is a good case for several reasons. First, probe microscopes—especially the most common variants, scanning tunneling microscopes (STM) and atomic force microscopes (AFM)—are often described in popular, official, and even analytical accounts as the foundational tools of nanotechnology.[2] According to the US National Nanotechnology Initiative, nanotechnology is "the understanding and control of matter at dimensions between approximately 1 and 100 nanometers, where *unique* phenomena enable *novel* applications" (my emphasis). Officially, then, novelty and disruption demarcate nanotechnology. Yet many scientists conducting nanotechnology research publicly worry that their field's novelty has been overhyped, encouraging needless regulation, public backlash, or disappointed funders.[3] Nanoscientists face difficult, context-specific choices in representing their work as novel or conventional. These choices are made especially clear in probe microscopy, which has generated an astonishingly diverse and contested set of visual (and nonvisual) conventions for presenting data.

The underlying source of such representational diversity provides a second reason to examine probe microscopy. There is a persistent and common belief among many practicing probe microscopists, as well as among historians and sociologists interested in nanotechnology, that these tools would not have been as representationally flexible or powerful if not for the invention of cheap, fast personal computers.[4] Though I will contest this belief somewhat, it contains a grain of truth, rooted in the technical means through which STMs and AFMs generate images. Probe microscopes work by bringing a small solid probe close to the surface of a sample and then scanning (or "rastering") the probe across the sample in two dimensions. Your eyes are "rastering" across this page in much the same way: you are moving them back and forth in one direction, while incrementally moving them line by line in the perpendicular direction. In most probe microscopes, the strength of interaction between probe and surface is sampled at regular intervals during each scan.[5] Those data then go into a numerical matrix containing $N \times M$ points (where N is the number of lines and M is the number of data points per line). The raw data in that matrix are usually converted into a pictorial image, although other representations (graphs, sounds, haptic feedback, etc.) are not uncommon.

That conversion from matrix to picture is, today, almost always done by a computer program. Such programs heavily process the raw data to (in microscopists' terms) "correct" for flaws and errors and to "render" the image more appealing or informative. Such enhancements to STM and AFM images have troubled some philosophers, who view them as an indication that nanotechnologists have been careless and indifferent about the truth value of their representations (see, for instance, Pitt 2006, 131–141). However, probe microscopists themselves have passionately debated about which kinds of image processing are appropriate. What's more, while computers have augmented the diversity of ways of presenting probe microscope data, they were not initially viewed as indispensable to presenting such data. Indeed, some very successful early microscopists were ambivalent about this use of computers. Probe microscopists' use, disuse, and misuse of computers for image processing were closely related to the institutional and disciplinary contexts in which they were (or wished to be) embedded. Microscopists' preferences regarding computer-generated images were also shaped, in complex ways, by how they wanted their work to be interpreted. Computers sometimes facilitated the ability to represent the technique as novel and disruptive, and sometimes made it easier to fit the technique into existing frameworks.

Finally, probe microscopy is an interesting case because it has been used for both routine and novel activities. At one extreme, AFMs are used in semiconductor chip factories (known in the industry as "fabs") to inspect silicon wafers to ensure that manufacturing steps are completed successfully. Though information generated in this way *could* be used to learn something new about semiconductor physics, the immediate purpose is to know whether to continue processing a wafer or to discard it. At the other extreme, some STMs and AFMs have gained reputations as revolutionary tools. For instance, low-temperature STMs have been used to manipulate, and generate spectroscopic data from, individual molecules. Some variants of the STM and AFM have imaged (and even manipulated) *sub*-atomic features (see figure 11.8 and Moon et al.).

Probe microscopists' choices to depict their *data* in conventional or unorthodox ways do not always align with their desire to have colleagues interpret their *findings* as conventional or unorthodox. At times, microscopists represent anomalous results in conventional ways, perhaps to reassure their audiences that their methods are robust. At other times, they defy the pictorial conventions with which their colleagues are familiar, though not necessarily in order to advance some revolutionary claim— sometimes unconventional representations are simply meant to draw attention to the microscopist, or to some feature of his or her microscope. In choosing how to represent their data, and how they would like those representations to be interpreted, microscopists continually look to the institutional and disciplinary networks that generate audiences, produce knowledge, supply materials, raise research questions, develop experimental technologies, and establish representational conventions that can be followed or broken.

2   From Graphs to Images

Though there were precursors to the STM and AFM, a probe microscopy community did not form until the invention of the scanning tunneling microscope in 1981. As I've argued elsewhere, the technical expertise of the STM's inventors was matched by their skill in cultivating audiences to pay attention to, and replicate, their work.[6] The inventors of the STM sometimes followed familiar representational conventions and sometimes employed playful, novel, and unusually creative representational forms.

This mixed representational strategy grew, in part, from the organizational and disciplinary context in which the STM originated. The idea for the STM was first broached in 1978, when Heinrich Rohrer, a senior researcher at the IBM laboratory in Zurich, was looking for a new project. At the time, IBM was investing more than $10 million per year (in US dollars at the time) to develop a computer based on superconducting (rather than traditional semiconducting) materials. Rohrer himself was not formally part of the superconducting computer project, but he wanted his new project to contribute both to fundamental research and to aid his colleagues in the superconducting computer effort. Rohrer and a new hire, Gerd Binnig, identified highly localized tunneling spectroscopy as a project that might fit those requirements.

To pursue localized tunneling spectroscopy, Binnig and a technician, Christoph Gerber, began building an apparatus to bring a sharp metal probe a few angstroms from a sample while placing a voltage between probe and sample. In theory, electrons would then quantum-mechanically "tunnel" between probe and sample. As far as the Zurich team knew, no one had ever successfully measured this "metal-vacuum-metal tunnel current" before. Confirming theoretical predictions concerning the phenomenon would be a nice piece of basic research. At the same time, Rohrer foresaw that this apparatus could inspect the thin films used to make superconducting logic chips and thereby improve those chips' manufacturability.

Binnig soon suggested that the vacuum tunneling apparatus could be elaborated into a general-purpose *microscope* if the probe scanned over the sample. Thus, the vacuum tunneling project became a "scanning tunneling microscope" project. However, the STM presented greater technical difficulties than a simple tunneling apparatus.[7] Rohrer's colleagues told him, "You are totally crazy—but if it works, you'll get the Nobel" (quoted in Fisher 1989). Perhaps to assuage these colleagues, Binnig and Gerber finished building the stationary vacuum tunneling apparatus and published a graph that paired their experiment's "vacuum tunneling signature" with the theoretical prediction for that signature (Binnig et al. 1982). In presenting their stationary vacuum tunneling data, the Zurich team hewed closely to conventions followed by tunneling researchers, perhaps to show their experiment was *not* "totally crazy."

Next, Binnig and Gerber built a true scanning tunneling microscope. At the time, the Zurich team believed this was a completely novel experiment. However, as the first STM results neared publication, they became aware of a precursor known as the Topografiner. In fact, when the Zurich team first applied for a US patent, the patent examiner struck several claims on the grounds that the STM too closely resembled the Topografiner. Thus, the STM team faced conflicting constraints. Gaining the patent (and probably realizing their own ambitions for establishing the microscope's novel capabilities) made it desirable to present the STM as a radical break (for a similar story about patenting, see Myers 1995). Conversely, the need to reassure audiences that the STM was reliable and credible encouraged them to portray the microscope as conventional and nondisruptive.

3 "The Former for Credibility, the Latter for Analysis and Discussion"

This balancing act pervaded the Zurich team's early STM publications, especially those presenting images of "surface reconstructions." In looking for materials to examine with the STM, Binnig and Rohrer were directed toward the topic of reconstructions that occur in semiconductor (and some metal) crystals exposed to extreme vacuum.[8] Inside a semiconductor crystal, atoms share electrons with their neighbors in all three dimensions. At the surface, however, the neighbors above the outermost atoms are missing. The atoms at the surface will, with experimental coaxing, rearrange ("reconstruct") to make up for these "dangling bonds."

Surface scientists had known about surface reconstructions since the 1950s through indirect approaches, especially low-energy electron diffraction (LEED). LEED works by directing a beam of electrons through the top few atomic layers of a crystal; the electrons interfere with each other as they pass through the crystalline lattice before shining onto a fluorescent screen, generating a pattern of bright spots on a dark field. It is possible to mathematically transform a model of a reconstruction to arrive at its LEED pattern, but doing the inverse transform of a LEED pattern does not give a unique solution—several different arrangements of atoms could produce the same pattern.

Thus, a microscope that could identify where the outermost atoms were located would help decide which models were correct. One reconstruction known as the silicon(111) 7 × 7 had a *very* large number of suggested models and therefore seemed particularly amenable to STM. At the time, however, it was not clear that STM could resolve single atoms in a reconstruction. Initially, Gerd Binnig was somewhat alone in believing that virtually all the tunneling current would pass through the outermost atom of the tip, thereby allowing it to resolve single atoms.

As they gained familiarity with the 7 × 7 and the STM, the Zurich team became more confident in Binnig's view until, in late 1982, they started to generate images that they believed showed the outermost atoms of the 7 × 7. This was, by all accounts, a moment of elation.[9] Yet it was unclear how to communicate how momentous their discovery was, because the first STMs did not generate an "image" in any normal sense. Instead, an oscilloscope or chart recorder displayed one line scan at a time. The Zurich team learned to "see" an image by recognizing patterns recurring from line to line. This would be rather like your reading each line of text on this page one by one, in isolation, and then forming a mental image of the page by remembering what all the individual lines looked like. A small, dedicated band of researchers might learn such a skill, but it would be difficult to ask a larger audience of people uncommitted to the STM to invest the time and mental effort needed to interpret images in this way.

Thus, to communicate their 7 × 7 results, the Zurich team needed to explain how the STM worked and demonstrate that the instrument was calibrated correctly so that it could resolve individual atoms. They needed to translate their 7 × 7 data into a form their audience could readily understand. Finally, they wanted to show that their 7 × 7 data could be accommodated within the existing framework of surface science, so that their findings would be recognized as an important contribution to one of that field's longest-standing problems.

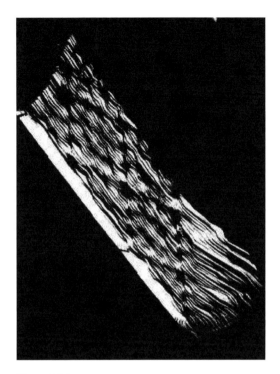

Figure 11.1
Binnig and Rohrer's first published image of the silicon 7×7. The image is actually a photo of a three-dimensional model. From Binnig et al. (1983b).

They therefore presented two radically different versions of their 7×7 data, as Binnig and Rohrer (1987) put it in their Nobel lecture, "the former for credibility, the latter for analysis and discussion." The "former" refers to an extraordinary representation that remains one of the most famous STM images (figure 11.1) (Binnig et al. 1983a, 120).

To make this image, Christoph Gerber took a sequential series of chart recorder traces of line scans of the 7×7 (similar to those in figure 11.2). He glued the chart recorder strips to pieces of cardboard, then cut along the traces to give each trace a slight three-dimensional presence. Finally, he glued the whole sequence of cardboard strips together to form a stack, and nailed the stack to a piece of Plexiglas.[10] The first published STM image of the 7×7 was a photograph of this cardboard sculpture—with the nails securing the cardboard to the Plexiglas visible on the side!

The cardboard rendering successfully translated STM line traces into a form a wider audience could appreciate. It conveyed to that audience how the STM worked: the sculpture's tactile, "homemade" quality spoke to the tactile, homemade quality of the instrument. The photo of the cardboard 7×7 emphasized the STM's (and the STMers')

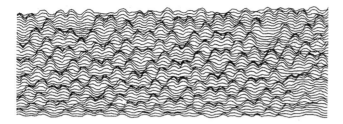

Figure 11.2
Image of the silicon 7×7 made by juxtaposing a series of line scans on a chart recorder. From Demuth et al. (1986). This is the kind of data which Christoph Gerber photocopied (one copy per line), glued onto cardboard, cut (one strip per line), and then glued together into a stack to make the model in figure 11.1. Reprinted with permission of *IBM Journal of Research and Development.*

revolutionary character. No other instrument could generate such data, and few other groups would have represented their data in such a creative way. Yet, if the unconventionality of the 7×7 sculpture made it appealing, it also made it difficult to accommodate the Zurich team's work to the framework of surface science—especially since the STMers were not themselves surface scientists. Thus, in the same article in which the cardboard rendering appeared, the STMers offered another representation in the pictorial idiom of surface science "for analysis and discussion."

That is, for their second representation of the 7×7, the STMers looked to the literature to see how surface scientists usually depicted surface reconstructions. Since surface scientists had no way to image the atomic structure of surface reconstructions in real space, the only real-space representations of such reconstructions prior to the STM were models based on information from indirect characterization techniques. Such models were usually depicted in publications as schematic two-dimensional diagrams resembling the three-dimensional ball-and-stick models that most surface scientists even today keep in their offices. To speak to surface scientists, therefore, the Zurich team presented a model of the 7×7 in a form that closely resembled conventional two-dimensional pictograms (see figure 11.3). They (literally) re-presented the 7×7 less as a tangible, material object and more like an abstract model of that object.

Neither of Binnig and Rohrer's representations of the 7×7 is "more true" in any obvious way. Each robustly conveys some characteristics of the 7×7 and the STM (and distorts others). But one representation seems to have been aimed at assimilating to a preexisting body of knowledge, while the other hinted at more exotic possibilities.

4   Corporate Surface Science and the Ecology of Images

The Zurich team's initial 7×7 results won them instant acclaim, and Binnig and Rohrer were awarded the Nobel Prize in Physics just three years later. There were, however, a few naysayers, some of whom referenced the team's representational choices.

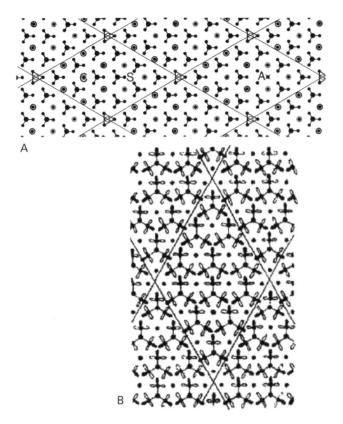

Figure 11.3

(A) Binnig and Rohrer's proposed model of the silicon 7×7 structure, much more in accordance with the pictorial language of surface science than their cardboard rendering of the 7×7 surface. From Binnig et al. (1983b). (B) A classic surface science schematic model of the 7×7 reconstruction, reprinted from Snyder (1984) with permission from Elsevier.

For instance, some skeptics apparently claimed that the STMers had fabricated their data on a computer—to which their rough-hewn cardboard rendering was an effective riposte. Alternatively, some skeptics insisted that the STM cardboard image of the 7×7 could not be real since, after all, there were nails visible in it![11] These naysayers could be dismissed partly because the STMers' more analytical representation of the 7×7, in the pictorial language of surface science, was acknowledged as a major contribution.

The Zurich team had not, however, "solved" the 7×7 reconstruction. Their images were interpreted as definitively establishing the locations of the outermost "adatoms" of the 7×7 but not the locations of atoms in lower layers. The Zurich team's images did allow surface scientists to discard many proposed solutions, and aspects of the model offered by the team were incorporated into later models. The STMers' model, however, was not the one finally accepted a few years later.[12]

Most surface scientists therefore saw the STM as a potentially revolutionary tool, yet one requiring extensive adaptation and interpretation. Both surface scientists and the Zurich STMers raised important questions that needed to be better understood before surface scientists could fully embrace the STM. What did the "bumps" in the cardboard rendering mean? Did they correspond to the locations of atoms, or did they require a more complicated interpretation? Why did the STM yield very different images of a surface at different voltages between probe and sample? Were the surface features seen at one voltage "more real" than features seen at other voltages? Contrary to the worries of some philosophers, STMers and surface scientists hotly debated these questions and eventually converged on the view that STM images should not be simplistically interpreted in a realist way.

That convergence was achieved largely by the generation of young researchers working in corporate laboratories who replicated the STM in the mid-1980s. At the time, corporate research laboratories—especially Bell Laboratories and IBM's laboratories in Zurich and California and Yorktown Heights, New York—played an outsized role in semiconductor surface science (the subfield most interested in the silicon 7 × 7 reconstruction and therefore initially most receptive to the STM). These organizations were strongly associated with leading-edge computing, and therefore most of their STM groups had access to advanced computing equipment. Some IBM Yorktown groups even reportedly faced pressure from management to adopt the latest versions of the IBM PC to control their STMs and generate images.[13]

Corporate surface science STMers used computers to create a common visual style that, for a time, was somewhat distinct from the visual style of other STM groups. That shared visual style sprang, in part, from the pressure corporate STMers faced to produce results quickly. Some were postdocs whose fellowships ended after a year or two, while others were very junior staff scientists whose career prospects depended on publishing articles quickly in high-quality journals. Surface scientists at Bell Labs and IBM Yorktown faced intense competition, both within and beyond their organizations. These researchers did not have the time or manpower to create exquisite three-dimensional cardboard renderings of every publishable result. Nor did they have time to learn to mentally reassemble images line by line as the Zurich team had.

Instead, they used their access to powerful computers to become more efficient producers and interpreters of STM images. They took the line-by-line traces produced by their STMs and fed them through software that interpolated between line scans and corrected for obvious scan errors. This produced images seemingly from a vantage point directly above the sample—a perspective that allowed images to be directly compared with models derived by surface theorists (see figure 11.4). Indeed, some corporate surface science STMers generated simulations of what an STM image of a

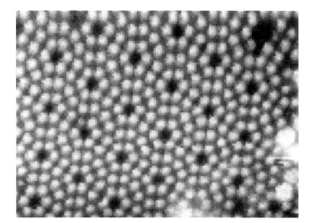

Figure 11.4
Image of the silicon 7×7 made by digitizing STM data and then using software to fill in space be-tween line scans. From Demuth et al. (1986). Note how much easier this version is to "read" than figure 11.1 (which was made by the same group and represents the same surface). Note also how figure 11.4 represents the data from a top-down vantage, whereas figure 11.1 is more from a three-quarters vantage point. Reprinted with permission of *IBM Journal of Research and Development*.

surface would look like if that surface had the atomic structure predicted by various models—allowing them to compare the simulations for each model with actual STM images and identify which models fit best.

Corporate surface science STMers' visual style was also shaped by their complex relationship with practitioners of LEED and various surface spectroscopic techniques.[14] STMers needed to accommodate their results to those from other techniques, yet the practitioners of those techniques sometimes regarded STMers as competitors. Many (though not all) of LEED's applications, for instance, were overtaken by STM in the early 1990s. In the 1980s, STMers faced criticisms from practitioners of other tech-niques—especially those who were senior managers at IBM and Bell Labs—regarding interpretation of STM images. For instance, some LEED specialists claimed the STM could only characterize small, nongeneralizable areas of a surface. In response, some IBM Yorktown and Bell Labs surface science STMers occasionally Fourier-transformed their STM images into frequency space patterns (figure 11.5). Doing so offered little information about their samples, but it did make their images slightly more compa-rable with LEED images of the same surfaces.

Practitioners of other surface science techniques also eagerly seized on STMers' missteps. For instance, two IBM groups published almost indistinguishable images of the same surface at the same time but with contradictory interpretations. Evidently, STMers converged on a pictorial idiom faster than they converged on an interpretive

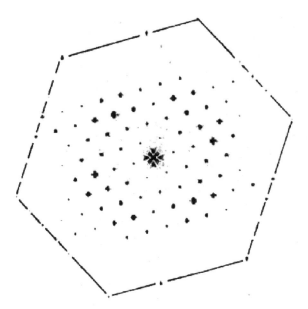

Figure 11.5
Fourier-transformed STM image. One way for STMers to insinuate their instrument into surface science was to Fourier-transform their images into inverse-space representations similar, though not equivalent, to LEED images of the same surfaces (since LEED is an inverse-space tool, though one that averages over a larger area, and a somewhat lower depth, than the STM). From Demuth et al. (1988). Reprinted with permission from John Wiley & Sons, Inc.

framework. Both interpretations were contested by surface scientists outside IBM; those skeptics could only establish that a third model was correct, however, by grudgingly adopting the STM themselves.[15] As would recur many times in probe microscopy, the combination of striking images and questionable interpretations only increased the popularity of the technique.

Corporate surface science STMers were at their most aesthetically creative when devising representations of spectroscopic data. "Scanning tunneling spectroscopy" involved imaging a surface at more than one voltage between probe and sample, then combining the resulting images into a single representation. For instance, one IBM group scanned gallium arsenide at two different voltages—one tuned to electron energy states associated with gallium atoms at the surface, the other tuned to those of arsenic atoms. They combined the resulting data sets into an image showing one scan (in green) overlaid on another (in red), indicating the alternation of the two kinds of atoms. That image was later republished (in pink and blue) as a cover of *Physics Today* (figure 11.6).

By then, the STM was becoming accepted in surface science. As a result, the pictorial idiom of corporate surface science STM images gradually became more diverse and

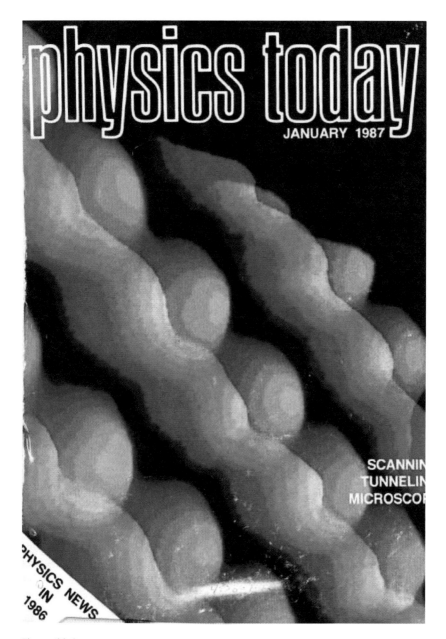

Figure 11.6
STM image of gallium arsenide made by Joseph Stroscio and Randall Feenstra at the IBM laboratory in Yorktown Heights, New York. Cover reprinted with permission from *Physics Today* 40, no. 1 (January 1987). © 1987 American Institute of Physics.

colorful. As the practitioners of corporate surface science STM became more professionally established, a few moved toward more aestheticized representations of their data, in which the mode of representation drew almost as much attention as the surfaces being represented. We will see later that a few high-profile members of the first generation of surface science STMers raised their profiles even higher by shifting their STM images away from the pictorial conventions of surface science and toward the worlds of art and public relations.

5 The Dispensability of Computing

At the same time that young researchers at IBM and Bell Labs were bringing the STM into surface science, the STM's inventors were making connections outside IBM and surface science. The original Zurich STMers were at the periphery of IBM's research empire, and none of them were surface scientists. Their writings and interviews over the years evince some impatience with the constraints of corporate surface science's established body of knowledge.

It is unsurprising, then, that the original Zurich team did not fully adopt the pictorial idiom of early Yorktown and Bell Labs surface science STMers. Indeed, Binnig and Rohrer sometimes made representational choices that befuddled their Yorktown and Bell Labs colleagues—none more so than their avoidance of computer control and image processing. As several early Yorktown and Bell Labs surface science STMers recount, Rohrer, especially, encouraged new STMers to eschew computers, and claimed the Zurich team made their greatest discoveries partly because they did not use computers.[16]

Today, Rohrer's assertion might seem odd or even Luddite. However, the need for computers in tunneling microscopy was not a given in the mid-1980s. Most STMers at that time used large racks of analog electronics to tune their microscopes. Even many STMers who relied in part on computers and digital components still considered sophisticated analog circuitry indispensable. This was partly because digital circuits of the day were seen as too slow and noisy for some aspects of STM operation. Many early STMers also viewed analog electronics as more transparently intelligible and capable of subtler control than digital electronics.[17] Manufacturers of digital components may not fully (or accurately) give details of their operation, and digital components, by definition, have gaps between the values at which they can be set. As digital electronics have improved, those gaps have become smaller, but in the 1980s some STM builders wanted assurance that their microscopes' best operating parameters did not fall in those gaps.

The complexities of analog and digital control are nicely demonstrated by two California academic groups that became especially productive of new designs and results, and (not coincidentally) became especially friendly with the original Zurich team.

These were Paul Hansma's group at the University of California at Santa Barbara and Calvin Quate's at Stanford. In the early 2000s, when Binnig and Rohrer wanted to recognize contributions to the technique they had invented, they presented lifetime achievement awards to Quate and Hansma and nobody else. The award itself was a cast-metal version of the cardboard 7×7 sculpture!

Elsewhere I have referred to researchers affiliated with Quate's, Hansma's, and Binnig's groups as the "Zurich-California network" (Mody 2011, chapter 4). People, ideas, designs, and tools traveled among the nodes in this network in ways that pushed probe microscopy beyond surface science. Certain habits for representing data were also shared among the members of this network. For instance, these groups experimented with ways of constructing and circulating images without the aid of a computer. One common method was to display line traces on a storage oscilloscope capable of showing each line slightly offset from the last. Oscilloscope displays, though, are ephemeral and difficult to transport—the opposite of Bruno Latour's (1987) "immutable mobiles." In Hansma's group (and probably others), therefore, microscope operators took Polaroids of oscilloscope displays and glued the Polaroids into notebooks and dissertations and published copies of them in journal articles.

Most members of the Zurich-California network did not abjure computer-generated images, however. Indeed, some members of this network—such as Arturo Baro and Nico García's team at the Autonomous University of Madrid—invented software for displaying STM images at about the same time as the Bell Labs and IBM Yorktown groups did so. The difference is that the Zurich-California groups did not *automatically* resort to computer-generated images. Instead, they experimented with techniques for representing their data via analog, digital, and hybrid means.

For instance, Calvin Quate became interested in transferring images onto videotapes, and he acquired a device (known by Quate and Hansma group members as "the Arlunya" after its Australian manufacturer) that purportedly could do so. This equipment was bothersome enough that Quate gave it away to Hansma, whose postdoc Othmar Marti (formerly a graduate student in Zurich working for Rohrer) somehow mastered it well enough that videocassettes became a standard data storage device in the Hansma lab.[18] Hansma group members sometimes used a microphone to narrate onto the tape what they were doing as they took data.

In the 1980s, the Zurich-California groups were relatively experimental, not only in the technologies they used to generate images but also in the pictorial idioms employed in those images. Both Binnig and Hansma are amateur artists (painting and photography, respectively), and in their Nobel lecture Binnig and Rohrer explicitly compared their cardboard rendering of the 7×7 to a sculpture by Ruedi Rempfler. Binnig later wrote a curiously illustrated book about creativity, in which he posited that the universe itself is a creative entity and that the key to discovery is undisciplined exploration unfettered by prior knowledge. In the 1980s, members of the Zurich-California groups distinguished themselves in part by generating painterly STM and AFM images that were brightly colored, subtly shaded, and/or intricately worked. For instance, Erich Stoll, one of Binnig and Rohrer's colleagues at Zurich, presented a pink, blue, and yellow color scheme for STM images at one of the first probe microscopy conferences in 1985, which somewhat scandalized surface science STMers.[19]

The more important distinguishing feature of the Zurich-California groups was that they formed collaborations with researchers from a very disparate range of fields: mineralogy, molecular biology, materials science, genetics, and electrical engineering, among others. These interdisciplinary (and usually interorganizational) collaborations influenced their representational choices to the same degree that corporate surface scientists' disciplinarity and organizational affiliations had. Members of the Zurich-California groups needed to attract the attention of researchers across a variety of fields to the point where practitioners from those fields would adopt probe microscopy and adapt it to their disciplinary conventions. That, after all, was what the Zurich team had done with surface science: their images of the 7×7 attracted attention and drew surface scientists into building their own STMs. Other disciplines were slower to join in, and the Zurich-California teams needed to experiment—with instrument designs *and* representational choices—to repeat the success of the 7×7.

Of course, where there is experimentation there can be failure. Perhaps the most notable failure was an STM image of DNA on the 19 July 1990 cover of *Nature*. This image, from a Caltech group on the fringes of the Zurich-California network, generated tremendous excitement during the early years of the Human Genome Project because it hinted that the STM might be capable of sequencing DNA. However, both surface science STMers and more central members of the Zurich-California network (especially Gerd Binnig and a Hansma collaborator, Stuart Lindsay) contested the Caltech *Nature* cover and images like it made by other groups.[20] Their most potent critique was that similar images could be made even when there was no DNA to be imaged—graphite and other substrate materials could "mimic" the behavior of DNA under an STM tip. Binnig reportedly would make that point by giving conference talks in which he showed beautiful STM images of DNA, only to reveal in the end that some of those images were fakes!

The DNA case is an extreme example of something many members of the Zurich-California groups candidly admit: there was so much demand from journals in the late 1980s for pretty STM and AFM pictures that groups did not always have time to ensure that their results were irreproachable. Occasionally, their interpretations of images were rushed or simplistic. Members of these groups argue that even incorrect interpretations advanced the field by drawing other practitioners into probe microscopy. The DNA episode shows the limits of that approach, though. When the rush to make big claims for probe microscope images threatened to tarnish the reputation of the technique, members of the Zurich-California groups were willing to intervene to enforce quality control.

6 The Commercialization of Representation

The delicate balance between drawing practitioners in with striking images and bold interpretations and guarding against *overly* bold claims and images became more complicated after the advent of commercial probe microscopes, which from 1987 onward were marketed by start-up companies affiliated with the Zurich-California network. That year marked the founding of Digital Instruments (DI) of Santa Barbara, which had close to a fifty percent share of the probe microscopy market through the end of the 1990s.

As the name implies, DI's founder, Virgil Elings, wanted to sell microscopes with digital controllers and computer-generated images. His past experience with building instruments led him to believe that digital control offered greater flexibility in the face of uncertainty about an instrument's future applications and design. As Elings tells it, Hansma—who was then Elings's departmental colleague at UC Santa Barbara—was initially skeptical that an all-digital controller would achieve the subtlety of his analog instruments.[21] Oddly enough, DI's first product (because of a rush to achieve first-mover status) was an *analog* instrument that displayed its output with an oscilloscope. However, DI's second, digital microscope quickly went on the market and sold well enough to double the size of the probe microscopy community in three years.

Elings and Hansma came to an accommodation that allowed people, ideas, designs, and software to circulate between UC Santa Barbara and DI. One area of mutual benefit lay in devising ways to represent microscope data. Some of the code for displaying and processing images on DI microscopes was written by current (or recently graduated) Hansma students. DI also developed its own code, to which members of the Hansma group had access because they were given free microscopes with which to develop applications that DI could commercialize. Using those donated microscopes and image-processing algorithms, Hansma group members and DI employees coauthored articles with images featured on the covers of prestige journals. For a time in the early 1990s DI ran an ad campaign with the slogan "We Have Science Covered," which showed a DI microscope surrounded by issues of *Science* with cover images generated on DI machines. Most of those covers came from articles coauthored by DI employees and UCSB researchers.

DI and its competitors were so successful that, by the early 1990s, substantial majorities of probe microscopists were buying rather than building their machines. For most users, there was little incentive to build a microscope, because a commercial instrument sufficed to characterize materials of interest to their disciplinary colleagues. Most users did not have the time, skill, or esoteric requirements that made building an instrument advantageous. That demographic change in the probe microscopy community was linked to a shift in how probe microscope data were represented: probe microscope images become more standardized, and commercial microscopes established pictorial conventions most users were content to accept.

After commercial microscopes achieved a certain market size and maturity, building one's own microscope became a niche activity—rather like building one's own car after the introduction of the Model T (Lucsko 2008). Some researchers continued building microscopes for experiments that required specialized capabilities which only a small number of other users would want and which therefore would never constitute a large enough market for a commercial manufacturer to cater to. Others built specialized parts to add on to commercial microscopes, sometimes in the hope that DI or another company would commercialize those parts.

Most probe microscopists, however, used DI products with few modifications. These users were, in general, even less interested in experimenting with ways of representing their data than they were in modifying their microscopes. Most DI customers simply relied on the default graphics package, which output every image with a distinctive golden-brown color. Users were not, of course, restricted to the default graphics settings on DI's microscopes. DI's software allowed a variety of perspectives and color schemes. Other companies (some specifically focused on probe microscopy) sold graphics packages that processed image files made with a DI microscope that enabled even more variation. However, most users weren't interested in such options. AFM images in a golden-brown color scheme and viewed from a top-down perspective flooded journals in the 1990s, making DI's default settings the standard "look" of probe microscopy (see figure 11.7).

Trevor Pinch and Frank Trocco (2002) observe a similar phenomenon in their study of the Moog synthesizer. The original Moogs, like the first generations of STMs and AFMs, had bulky analog electronics and were custom-modified by users. Early Moog users developed unique sounds by physically tweaking different combinations of electronic "patches"—in the same way that early builders of STMs physically tweaked their analog electronics for subtle control of operating parameters. To carry the analogy further, the Moog became widely known through best-selling performances by early virtuosi—such as Wendy Carlos, the performer of the popular "Switched-on Bach" recordings, and Keith Emerson of the rock band Emerson, Lake, and Palmer—in the same way that STM became widely known through unique, virtuoso performances such as the cardboard rendering of the 7 × 7. Virtuoso performances on both STM and Moog created demand for commercial instruments from less virtuosic users. With commercial microscopes, of course, customers were often virtuosi in other practices than microscope operation—in specimen preparation, for instance. Commercial microscopes, especially DI's NanoScope line, were tailored to the needs of such

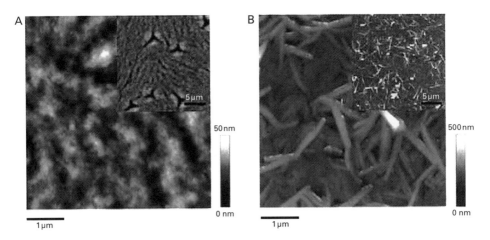

Figure 11.7
Stereotypical golden-brown AFM image made with a Digital Instruments NanoScope. The sample is a liquid crystal thin film spin-coated onto a silicon wafer. From Schmidt-Mende et al. (2001). Reprinted with permission from the American Academy of Arts and Sciences.

"average users" (as DI engineers referred to them), because such microscopes incorporated enough of the functionality of home-built microscopes to be useful, but not so much as to be daunting or unreliable.

The NanoScope was to a large extent a "black box" that discouraged "tampering" by customers.[22] However, most users did not even tamper with off-the-shelf instruments to the extent allowed by the technology. Again, this resonates with Pinch and Trocco's finding that even though the Mini-Moog allowed some customization of sounds, most users never departed from its preset patches. The sounds that came with the Mini-Moog from the factory became the distinctive sounds of that instrument. Likewise, the default golden-brown images (with a perspective vantage point directly above the sample) that a NanoScope generated became the standard look of probe microscope images. To speculate somewhat, for the average user there may have been real advantages in sticking with the default color scheme. Having a NanoScope was useful partly in that customers could publish AFM images of their samples that were instantly recognizable by journal readers *as AFM images*.

To reiterate, "average users" can only be seen as "average" in terms of their relation to available probe microscope technology. A researcher could easily be an "average" user of a NanoScope, in not demanding special capabilities or modifications, while also being a revolutionary contributor to his or her home discipline. By contrast, some customers who made extreme demands on their AFMs were not engaged in original research. Semiconductor manufacturers, in particular, were willing to pay top dollar for specialized, ultrareliable microscopes for process line inspection—a task far from cutting-edge science.

7 Art for Science's Sake

At the other extreme were researchers who continued building specialized microscopes even after commercial microscopes had become available. These virtuosi commanded the respect of peers in their own disciplines, and across the interdisciplinary probe microscopy community, for both their microscope designs and the data yielded by those microscopes. By the 1990s, a few such virtuosi also distinguished themselves by the novel, and heavily aestheticized, ways in which they displayed their data.

By "heavily aestheticized," I mean that their representations overtly incorporated a large amount of image processing that drew on pictorial conventions usually associated with genres of art such as landscape painting, and/or that pushed beyond visual representation as a means for conveying microscope data. In a world where golden-brown, top-down DI NanoScope images were becoming ubiquitous, people and organizations that built microscopes—whether for themselves or for the market—resorted to novel representational forms to emphasize the special capabilities of their instruments. Conference-goers or readers of journal articles might have been aware of those special capabilities anyway, but only if they paid close attention to the data being presented. By opting for eye-catching representational forms, instrument *builders* could distinguish themselves from instrument *buyers* instantly, even to casual observers. Figure 11.8 offers a good example; it shows subatomic features of an electron gas created and then imaged with an STM built by Hari Manoharan, a former postdoc in the IBM group of Don Eigler. Eigler and his former postdocs have specialized in generating these kinds of striking, heavily rendered, "photorealistic" images of atoms and subatomic features. As several observers have noted (Ruivenkamp and Rip, this volume; Toumey 2009; De Ridder-Vignone and Lynch 2012) such representations are *unconventional* in departing from the pictorial idiom that their audience is used to seeing in scientific journals, but *conventional* in drawing on pictorial idioms common in popular media and art (e.g., photorealism).

For makers of commercial microscopes, meanwhile, novel ways of representing data offered a powerful form of advertising. For instance, Asylum Research, a DI spin-off, incorporated a way to convert their product's data into *auditory* form, partly to help the company's engineers debug their products and partly to help salespeople better explain those products to potential customers at trade shows.[23] In the early 2000s, Asylum representatives at trade shows were also bragging about bringing programmers from the video game industry to improve the images made with their products;

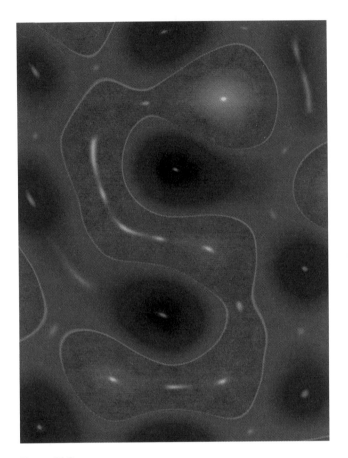

Figure 11.8
STM image of an electron gas present on a copper surface, modified by the presence of adsorbed carbon monoxide molecules in such a way that the electron waves interfere with each other to form a letter "S" when imaged by the STM tip. Image made by Hari Manoharan's group at Stanford University (hence the "S" for Stanford) and reported in Moon et al. (2009). Cover of *Nature Nanotechnology*, volume 4, number 3 (March 2009), reprinted with permission of Nature Publishing Group.

posters showing off their handiwork would be given away at Asylum's booths. Even Veeco, a semiconductor equipment firm that merged with DI in 1998, sponsored an annual calendar contest that crowd-sourced the production of heavily aestheticized probe microscope images to use as advertising.

A few individual probe microscopists "advertised" themselves with novel representations in a similar way. For instance, Don Eigler audibilized his STM's output much as Asylum's engineers did. Descriptions of Eigler listening to atoms have appeared in many popular articles, as have items from his "gallery" of STM images that consciously imitate various artistic genres: portraits (of a molecular "man"), landscapes, seascapes, calligraphy (atoms positioned to spell words in Japanese kanji as well as Roman letters), etc. (see Toumey 2009). Like Asylum's representatives, Eigler emphasized that the image-rendering software he developed shared ties with the software used in special effects and video game industries.

Eigler's novel forms of representation blur the distinction between individual and organization. His most famous STM image (Eigler and Schweizer 1990), for instance, depicts 35 xenon atoms positioned to spell out I-B-M—both in homage to and as a billboard for the company that made his research possible. Indeed, one hallmark of probe microscopy virtuosi is that they produce images that blend science, commerce, and art. For instance, Eric Heller, a Harvard theorist who sometimes works with Eigler, has a reputation for producing appealing representations of his collaborators' experimental findings that journals often use as cover images. On the side, Heller then markets those images as "digital fine art."

This powerful mix of science, commerce, and art has allowed some virtuosi to dramatically transcend the organizational and disciplinary frameworks in the specialized fields in which they started. For instance, Wolfgang Heckl—one of Gerd Binnig's former postdocs and closest protégés—has parlayed his virtuosic probe microscopy research into the directorship of one of Germany's most visible cultural institutions, the Deutsches Museum (where the cardboard 7×7 is proudly displayed). Similarly, Jim Gimzewski, a former IBM surface scientist who was one of the first in the Zurich lab to replicate the STM, has constructed various playful molecular "sculptures" (such as an "abacus" made from fullerenes), presented in the form of exquisite STM and AFM images. He even has created a new field of "sonocytology" based on the "screams" of bacteria audibilized through an AFM (Roosth 2009). For this representational experimentation, Gimzewski has gained recognition from *Wired* magazine, and his collaborations with artists are displayed at high-profile venues such as the Los Angeles County Museum of Art.

In the end, probe microscopy is stronger for containing the full spectrum of contributors from virtuosi to "average" users. On the one hand, instruments that remain only in the hands of virtuosi can become so idiosyncratic that the wider scientific community pays little attention to them. Binnig and Rohrer themselves believed this was a danger for STM. On the other hand, instruments that are only for "average users" can stagnate and be shifted into "service departments" or introductory classrooms where they have little contact with original research.

Likewise, probe microscopy is probably stronger for offering its practitioners so many ways to represent their data. In some cases, the virtuosi are clearly on to something: there are some phenomena where conventional pictorial idioms—or even visual representation itself—are not sufficient to convey the subtleties of the phenomena in question. Even the average users know this—which is why the image-rendering techniques used by the virtuosi usually appear on the market a few years later.

But at other times, average users—and the microscope manufacturers who cater to them—have it right. Conventions, idioms, a standardized "look" of a class of instrumentation—these things allow interpretation, dissemination, and debate to happen more quickly than if every microscope image were a finely crafted, unique gem. Even the virtuosi know this—which is why, for some audiences and in some situations, they are happy to convey results with a drab graph or a numerical chart rather than with a splashy image.

The representational flexibility that has grown up around probe microscopy (and many other instruments) goes a long way toward explaining why Kuhn's "essential tension" usually doesn't lead to much tension or conflict. As many observers (e.g., Galison 1997) have noted, science is disaggregated in a way that Kuhn struggled to acknowledge; revolutions in theory occur during periods of "normalcy" in instrumentation and experiment and vice versa. Similarly, instrumentation is disaggregated from representation. As probe microscopists have shown, instrument users can tune their representational options very finely so as to make the tensions of science less than essential: revolutionary findings can be softened through use of bland pictorial conventions, or they can be made spectacular through daring representational innovations; quotidian findings of normal science can be made fresh through heavily aestheticized representations, or they can be made easily mergeable into a body of knowledge by being presented in a standardized idiom. The choice of representational tack usually has more to do with the scientists' immediate audience, career aspirations, and organizational and disciplinary context than with their data's essentially conventional or disruptive nature.

Notes

1. This chapter builds most closely on Mody (2004). Other works that discuss probe microscope images and/as representations of novelty include: Hessenbruch (2004, 135–144); Hennig (2006); Hanson (2004); Granek and Hon (2008, 101–125); and Milburn (2005).

2. This view is examined in Baird and Shew (2004, 145–156).

3. See, among others, Kehrt and Schüßler (2009).

4. By far the most cogent expression of this view is Johnson (2006).

5. The kind of probe-sample interaction depends on the type of microscope. In the STM, the measured interaction is the "tunnel current" of electrons moving between probe and sample. In the AFM, it is the van der Waals, capillary, and other interatomic forces between probe and sample.

6. Mody (2011). References regarding the general history of the STM and AFM are provided there. For a brief overview, see Binnig and Rohrer (1987).

7. In particular, scanning the probe makes it more difficult to dampen vibrations that uncontrollably move the probe relative to the sample and blur the image generated.

8. A deeper description of surface reconstructions, the STM, and the silicon 7×7 can be found in Mody and Lynch (2010).

9. Christoph Gerber, personal communication.

10. The effect is very similar to Frank Gehry's "Easy Edges" line of cardboard furniture, first introduced in the late 1960s.

11. This paragraph draws on conversations with Arne Hessenbruch.

12. Charles Duke, personal communication.

13. Interview with Joe Griffith. The importance of computing in the surface science revolution of the 1970s is discussed in Duke (1984).

14. Much of this chapter resembles Bernike Pasveer's (1989) arguments about early X-ray technology, especially my points about organizational and disciplinary contexts and about experimentation in making the STM comparable to other techniques.

15. Interview with Stan Williams and personal communication from Jun Nogami.

16. Interviews with Joe Griffith and Bob Hamers.

17. A point explored in Haring (2007).

18. Interviews with Scot Gould and Craig Prater.

19. Analyzed further in Hennig (2006).

20. This paragraph draws on interviews with Tom Beebe, Paul Hansma, Stuart Lindsay, Bob Wilson, and Bob Hamers.

21. Interview and personal communication with Virgil Elings and Paul Hansma. Other interviews relevant to this section are listed in Mody (2006).

22. The idea of the "black box" is discussed in Latour (1987).

23. Explored further in Mody (2005).

References

Baird, Davis, and Ashley Shew. 2004. Probing the history of scanning tunneling microscopy. In *Discovering the Nanoscale*, ed. Davis Baird, Alfred Nordmann, and Joachim Schummer, 145–156. Amsterdam: IOS Press.

Binnig, Gerd, and Heinrich Rohrer. 1987. Scanning tunneling microscopy—from birth to adolescence. *Reviews of Modern Physics* 59 (3):615–625.

Binnig, Gerd, Heinrich Rohrer, Christoph Gerber, and Eduard Weibel. 1982. Vacuum tunneling. *Physica B+C* 109–110: 2075–2077.

Binnig, Gerd, Heinrich Rohrer, Christoph Gerber, and Eduard Weibel. 1983a. 7 × 7 Reconstruction on Si(111) resolved in real space. *Physical Review Letters* 50 (2):120.

Binnig, Gerd, Heinrich Rohrer, Christoph Gerber, and Eduard Weibel. 1983b. Revisiting the 7r7 reconstruction of the Si(111). *Surface Science* 157:L373–L378.

Demuth, J. E., R. J. Hamers, R. M. Tromp, and M. E. Welland. 1986. A scanning tunneling microscope for surface science studies. *IBM Journal of Research and Development* 30 (4): 396–402.

Demuth, J. E., U. Koehler, and R. J. Hamers. 1988. The STM learning curve and where it may take us. *Journal of Microscopy* 152:299–316.

De Ridder-Vignone, Kathryn, and Michael Lynch. 2012. Images and imaginations: An exploration of nanotechnology image galleries. *Leonardo* 45:447–454.

Duke, C. B. 1984. Atoms and electrons at surfaces: A modern scientific revolution. *Journal of Vacuum Science and Technology* 2 (2):139–143.

Eigler, D. M., and E. K. Schweizer. 1990. Positioning single atoms with a scanning tunneling microscope. *Nature* 344 (5 April): 524–526.

Fisher, Arthur. 1989. Seeing atoms. *Popular Science* (April): 102–107.

Galison, Peter. 1997. *Image and Logic*. Chicago: University of Chicago Press.

Granek, Galina, and Giora Hon. 2008. Searching for asses, finding a kingdom: The story of the invention of the scanning tunneling microscope (STM). *Annals of Science* 65:101–125.

Hanson, Valerie Louise. 2004. Haptic visions: Rhetorics, subjectivities and visualization technologies in the case of the scanning tunneling microscope. Ph.D. diss., Pennsylvania State University.

Haring, Kristen. 2007. *Ham Radio's Technical Culture*. Cambridge, MA: MIT Press.

Hennig, Jochen. 2006. Changes in the design of scanning tunneling microscopic images from 1980 to 1990. In *Nanotechnology Challenges*, ed. Joachim Schummer and Davis Baird, 143–163. Singapore: World Scientific.

Hessenbruch, Arne. 2004. Nanotechnology and the negotiation of novelty. In *Discovering the Nanoscale*, ed. Davis Baird, Alfred Nordmann, and Joachim Schummer, 135–144. Lansdale, PA: IOS Press.

Johnson, Ann. 2006. The shape of molecules to come. In *Simulations: Pragmatic Constructions of Reality*. Sociology of the Sciences Yearbook 25, ed. Johannes Lenhard, Gunther Küppers, and Terry Shinn, 25–39. Dordrecht: Springer.

Kehrt, Christian, and Peter Schüßler. 2009. Nanoscience is 100 years old. The defensive appropriation of the nanotechnology discourse within the disciplinary boundaries of crystallography. In *Governing Future Technologies: Nanotechnology and the Rise of an Assessment Regime*, ed. M. Kaiser, M. Kurath, S. Maasen, and C. Rehmann-Sutter, 37–53. Dordrecht: Springer.

Kuhn, Thomas S. 1977. *The Essential Tension: Selected Studies in Scientific Tradition and Change*. Chicago: University of Chicago Press.

Latour, Bruno. 1987. *Science in Action: How to Follow Scientists and Engineers through Society*. Cambridge, MA: Harvard University Press.

Lucsko, David. 2008. *The Business of Speed: The Hot Rod Industry in America, 1915–1990*. Baltimore: Johns Hopkins University Press.

Milburn, Colin N. 2005. Nanovision: Engineering the future. Ph.D. diss., Harvard University.

Mody, Cyrus C. M. 2004. How probe microscopists became nanotechnologists. In *Discovering the Nanoscale*, ed. Davis Baird, Alfred Nordmann, and Joachim Schummer, 119–134. Lansdale, PA: IOS Press.

Mody, Cyrus C. M. 2005. The sounds of science: Listening to laboratory practice. *Science, Technology and Human Values* 30:175–198.

Mody, Cyrus C. M. 2006. Corporations, universities, and instrumental communities: Commercializing probe microscopy, 1981–1996. *Technology and Culture* 47 (1):56–80.

Mody, Cyrus C. M. 2011. *Instrumental Community: Probe Microscopy and the Path to Nanotechnology, 1960–2000*. Cambridge, MA: MIT Press.

Mody, Cyrus C. M., and Michael Lynch. 2010. Test objects and other epistemic things: A history of a nanoscale object. *British Journal for the History of Science* 43:423–458.

Moon, Christopher R., et al. 2009. Quantum holographic encoding in a two-dimensional electron gas. *Nature Nanotechnology* 4:167–172.

Myers, Greg. 1995. From discovery to invention: The writing and rewriting of two patents. *Social Studies of Science* 25:57–105.

Pasveer, Bernike. 1989. Knowledge of shadows: The introduction of X-ray images in medicine. *Sociology of Health and Illness* 11:360–381.

Pinch, Trevor J., and Frank Trocco. 2002. *Analog Days: The Invention and Impact of the Moog Synthesizer*. Cambridge, MA: Harvard University Press.

Pitt, Joseph. 2006. When is an image not an image. In *Nanotechnology Challenges: Implications for Philosophy, Ethics, and Society*, ed. Joachim Schummer and Davis Baird, 131–141. Singapore: World Scientific.

Roosth, Sophia. 2009. Screaming yeast: Sonocytology, cytoplasmic milieus, and cellular subjectivities. *Critical Inquiry* 35:332–350.

Schmidt-Mende, L., A. Fechtenkötter, K. Müllen, E. Moons, R. H. Friend, and J. D. MacKenzie. 2001. Self-organized discotic liquid crystals for high-efficiency organic photovoltaics. *Science* 293:1119–1122.

Snyder, L. C. 1984. Modified milk-stool on wurtzite layer model for Si(111) 7r7 surface reconstruction. *Surface Science* 140:101–107.

Toumey, Chris. 2009. Truth and beauty at the nanoscale. *Leonardo* 42 (2):151–155.

12   In Images We Trust? Representation and Objectivity in the Digital Age

Emma K. Frow

This chapter draws on a contemporary debate in scientific publishing to address ideas about objectivity, trust, and practices of representation in the digital age. Over the last 10–15 years, digital images have become a ubiquitous feature of scientific journal articles, not least because during this period most journal publishers have implemented electronic submission systems for manuscript text files and their accompanying images (e.g., *Nature* 2000; Rossner 2002). Whether or not researchers acquired their original image data in digital format, the process of preparing images for publication now requires them to be rendered digitally; a wide and growing array of imaging equipment and software tools allows capture, processing, and adjustment of digital images. But accompanying this increasing availability and use of digital imaging technologies is a certain crisis of trust in the published image, particularly in subdisciplines of biology such as cell biology, molecular biology, and genetics.[1] In recent years, editors of several high-profile science journals, including *Science*, *Nature*, and *PLoS Biology*, have been expressing concern that the ability to use image-processing software (particularly Photoshop) facilitates the production of aesthetically attractive but scientifically misleading images. In response to such concerns, these and other journals have been establishing guidelines for digital image processing.[2] Some are also performing random spot checks of submitted images; others have even hired forensic experts to systematically scrutinize all images accepted for publication (e.g., Pearson 2005; Couzin 2006). These interventions are discussed in terms of trying to restore trust in the published image: "finding ways to regain our trust in scientific images is a goal on which we can all agree" (*Nature* 2006a, 892).

This crisis of trust is presented as a matter of significant concern within the scientific community, but need not be seen as an inevitable consequence of introducing new image-processing technology. How might we articulate and understand the relationship between digital image processing and the trustworthiness of images? The science studies community began to raise such questions some 20 years ago. In his 1991 paper "Science in the Age of Mechanical Reproduction," Lynch associates digital image processing with the possibility of increasing trust in images under certain conditions:

The moral and epistemic qualities assigned to an "original" representation may undergo yet another transformation. The "raw data" collected by taking a photograph or micrograph can be "improved upon," not by re-drawing or re-touching the original, but by subjecting it to mathematical transformations that subtract noise or otherwise create a mechanically "improved" or "enhanced" image. Assuming that the software functions are justified by accepted mathematics and physics, the processed image can be held to provide more "trustworthy" or "authentic" evidence than the raw data. In a digital image, many of the surplus details in a raw frame become the "noise" to be eliminated rather than guarantors of the authenticity of the product. (Lynch 1991a, 221)

In discussing Lynch's article, Taylor and Blum also suggest a link between digital image processing and the trustworthiness of an image, proposing that "just as historically the advent of photography promised an escape from the 'fallibility' of drawing, so today the availability of computer-generated imagery may be transferring the mantle of trustworthiness from the passive lens and film to the interactive program" (Taylor and Blum 1991, 126). Intriguingly, journal editors do not seem quick to embrace the suggestion that digitally enhanced or manipulated images might be seen as more trustworthy than raw image data. Instead, a rather different view prevails: editorial statements such as "slightly dirty images reflect the real world" (*Nature* 2006a, 892) associate trust in an image with the perception that it has not been enhanced or adjusted to any great extent.

At first glance these statements might suggest a mismatch between editorial concerns and understandings of images derived from science studies. A large body of established research in science and technology studies (STS) advances the view that published images are carefully arranged, crafted, or even "designed."[3] In this context, digital image-processing technologies could be treated as another set of tools in the design process, ones that in principle might offer possibilities for further refinement, isolation, and presentation of the phenomena under investigation. The ability to highlight some data at the expense of others—identified by journals as problematic with respect to digital image processing—is arguably a necessary component of scientific inquiry and the production of *any* image, not just digitally processed ones. However, journal editors seem to view the spread of digital image processing as a possible threat to the integrity and objectivity of visual representation. Their concerns are being raised at a time when digital imaging tools are becoming ever more embedded in experimental practices across the natural and medical sciences.[4]

This chapter is concerned with the question of why digital image processing might be seen to threaten trust in the published image, particularly in molecular biosciences. Drawing on recent journal guidelines and commentaries concerning digital image processing, as well as STS scholarship on representation in scientific practice, it explores how and where trust is invested in practices of image making.[5] Images have long had an important role in communicating scientific findings; why should there be particular concern about their trustworthiness now? Is there something fundamentally different (and troubling) about digital images as compared with analog representations? For journals, metaphysical questions concerning the nature of digital images (e.g., Mitchell 1992; Lister 2004) are by and large overshadowed by more practical considerations relating to the role images have and how they are used in research articles. Digital technologies might offer new possibilities for images to be adjusted, circulated, and scrutinized, but the central role of a published image—to allow readers to "see for themselves" and evaluate what a phenomenon "looks like"—remains unchanged.[6] Furthermore, this role must somehow be sustained across changing configurations of technologies and scientific practices. The digital realm does offer new possibilities for intervening in the relationship between the "surface" of an image that is made available to the viewer on the printed page or computer screen, and the "raw data" underlying that image—interventions that can be imperceptible to the reader. In extreme cases, readers might be shown images that are not based on any underlying experimental data, but rather are composed in front of a computer screen.[7] The journal guidelines can be read as an intervention that attempts to strengthen the correspondence between image and data, and to improve the traceability or the historicity of an image by shaping author practices of record keeping and image adjustment. By establishing norms of practice that strive to anchor "surface" representations to underlying data, the guidelines associate image-making practices with ideas about accountability and trust.

1   The Guidelines

The key journals considered in this chapter are *Science*, the *Journal of Cell Biology* (*JCB*), and the *Nature* family of journals.[8] Editors at *JCB* (published by the Rockefeller University Press) have been among the more proactive in establishing image-processing guidelines. Their summary recommendations are as follows:

No specific feature within an image may be enhanced, obscured, moved, removed, or introduced. The grouping of images from different parts of the same gel, or from different gels, fields, or exposures must be made explicit by the arrangement of the figure (e.g., using dividing lines) and in the text of the figure legend. Adjustments of brightness, contrast, or color balance are acceptable if they are applied to the whole image and as long as they do not obscure or eliminate any information present in the original. Nonlinear adjustments (e.g., changes to gamma settings) must be disclosed in the figure legend. (Rossner and Yamada 2004, 12)

This guidance has been used as a starting point for several other journals, as well as by the Council of Science Editors in their 2009 *White Paper on Promoting Integrity in Scientific Journal Publications*. Across the different journals' image-processing guidelines, four general points consistently emerge.[9] The first relates to the composition of images and the scope of acceptable manipulations. Any adjustments made using digital processing software must be applied to the whole image, not selectively to discrete parts of the image. Furthermore, no global adjustment should be made if it hides or removes information present in the original image—for example, "contrast should not be adjusted so that data disappear."[10] Practices such as adjusting contrast levels on images of gels or blots in order to eliminate faint (and, perhaps according to the author, spurious or irrelevant) bands are therefore not acceptable (see Rossner and Yamada 2004, 13). Nor, according to this guideline, should authors remove discrete artifacts from an image, for example a speck of dust, or a cosmetic defect arising from a known imperfection with the scientific equipment being used (for examples of such adjustments, see Lynch and Edgerton 1988, 205–209; Frankel 2002, 274–275).

The second point also relates to the composition of images, and focuses on manipulations involving the rearrangement or grouping of images. According to journal guidelines, producing composite images by cutting and pasting together (selected portions of) images can be acceptable provided that the various subimages are clearly delineated on the figure, and that the composite nature of the figure is described in the accompanying figure legend.[11] Consistent with this, a third point emphasized in the journal guidelines is transparency in process. Just as the arrangement of composite images should be accounted for in the relevant figure legend, so too should other aspects of image preparation be detailed. For example, the *Nature* journal guidelines stipulate that "authors should list all image acquisition tools and image processing software packages used," and should "document key image-gathering settings and processing manipulations."[12] For images obtained using microscopy, such information would include the make and model of the microscope and lens used, together with a list of the instrument settings used for image capture, a description of the experimental sample, and details of any postacquisition adjustments. The fourth key issue mentioned in most of the journal guidelines is a requirement to keep all original data relating to images. This stipulation applies less to the manipulation of digital images themselves than to practices of data storage and record keeping by scientists.

The guidelines as outlined above offer fairly general direction—necessarily so, for they must be broadly applicable to an ever-expanding diversity in assemblages of materials, software, and practices that can be used for image making.[13] Furthermore, researchers trained in different disciplines may abide by different conventions for image making (e.g., Ruivenkamp and Rip 2010, 12). The journals *Science* and *Nature* publish research from across the spectrum of natural science disciplines, and so their publication guidelines must be applicable to the scientific community as a whole. However, based on the concrete examples provided in both the guidelines and their associated commentaries, journal editors are concerned primarily with images from cellular and molecular biology, and specifically single out digital photographs of electrophoretic gels, immunolabeled blots, and microscopy data (e.g., *Nature* 2006b, 203). Furthermore, Photoshop is the only image-processing software consistently mentioned across the journal guidelines and editorials. In what follows, I explore how these guidelines relate to specific concerns regarding digital photographs and practices of image making using image-processing software (particularly Photoshop) to understand why there might be such concern with restoring trust.

2 Ethics and the Composition of Digital Images

Journal editors acknowledge the impossibility, and indeed the undesirability, of banning all image manipulations. As suggested in a *Nature* news feature, "no one wants to ban image manipulation outright. In cell biology experiments, for example, researchers often have to adjust the relative intensities of red, green and blue fluorescent markers in order to show all three in a single image" (Pearson 2005, 953). Sophisticated equipment and data-processing algorithms can be necessary to derive meaningful signals and "make visible" certain types of phenomena in the first place.[14] Furthermore, the requirement to submit images to journals in digital form means that a certain amount of labeling, formatting, and processing becomes inevitable. For some types of images (e.g., microscopy), image data are now typically acquired in digital format from the outset, whereas in other cases analog data are converted into digital representations through scanning or photography (e.g., a digital photograph of a western blot). In this context, the journal guidelines are an attempt not to outlaw image processing, but to define limits or "draw a line" regarding acceptable and unacceptable practices of image adjustment (Frow 2012).

The digital realm provides a dynamic and interactive interface for image manipulation, one that affords authors great—perhaps unprecedented—control over the detailed composition and appearance of the final image.[15] Journal editors identify this as potentially problematic, expressing concern about the temptation facing researchers to "beautify" (or in extreme cases, to fabricate) images. Factors they identify as contributing to this temptation include the availability and ease of using image-processing software such as Photoshop (*Nature* 2006a) and the pressure on authors to publish their work in high-impact-factor journals (*Nature* 2006c; Franzen et al. 2007). With the composition and content of digital images more easily manipulable, an author's behavior during postacquisition image processing can have an increasingly prominent role in shaping the final appearance of the image. Journal editors acknowledge that scientists' inclination to beautify their images is not new—"pretty pictures" have always been admired and strived for—but they argue that the ease of doing so has increased: "It is doubtful that scientists were more angelic then than now. It is more likely that, when it came to image manipulation, they wouldn't because they couldn't" (*Nature* 2006a, 891–892). The journal guidelines can be seen as an intervention designed to establish norms for digital image processing, in the absence of a perceived need for such explicit rules in the era of analog images.

Some types of transformations that authors might apply to digital images are qualitatively similar to the adjustments routinely performed on analog data. These include, for example, cropping or cutting and pasting data to produce composite images, and adjusting color or brightness and contrast levels (Knorr-Cetina and Amann 1990, 279–280). It is these types of image adjustments that receive most attention in the recent journal guidelines and their associated commentaries, perhaps precisely because these are the types of manipulations that can readily be performed according to more "traditional" methods.[16] Analog and digital images that have been subjected, for example, to contrast enhancement can end up looking indistinguishable on the printed page. Notably, even the "bricolage"-like vocabulary used to describe such adjustments is similar for digital and analog images.[17] However, the actual transformational procedures in question are different. Most obviously, the physical site at which the adjustments take place is different (a developing studio or laboratory benchtop for analog images, a computer interface for digital images). The adjustment of analog images involves manipulation of material artifacts (for example, cutting photographs or subjecting negatives to variations in the developing process). Adjusting digital images involves applying mathematical transformations to the underlying pixel matrix, and allows the relationship between pixels to be manipulated in highly selective and nonlinear ways if desired.

The journal guidelines that relate to image composition effectively restrict the scope for differences in how digital and analog images might be processed. Editors seem to use practices of analog image-making as a reference point or yardstick against which digital technologies are compared. Their examples and concerns relate predominantly to scientific disciplines in which analog representations have long been commonplace, such as cell and molecular biology, as opposed to newer fields like nanotechnology where digital imaging has been used from the outset (see Ruivenkamp and Rip 2010). The requirement for adjustments to be applied across an entire image rather than to selected portions of it acknowledges the fact that it is much easier to selectively alter parts of an image using digital media (for example, to add or move signals, or to remove artifacts and unwanted data); such manipulations are certainly possible with analog photographs, but are typically more difficult and time-consuming. Effectively, this stipulation limits the nature of adjustments for digital image processing to what is possible with analog images. Moreover, some of the guidelines for digital image processing are stricter than what journals have typically permitted for analog equivalents. For example, when discussing the preparation of figures containing blots and gels, the editors of *JCB* would prefer authors to "perform multiple exposures to get the bands at the density you want, without having to overadjust digitally the brightness and contrast of the scanned image" (Rossner and Yamada 2004, 13). Here, contrast adjustments of digital images on-screen are less welcome than the equivalent manipulations of analog photographs, a point acknowledged by the journal editors: "It may be argued that this guideline is stricter than in the days before Photoshop, when multiple exposures could be used to perfect the presentation of the data. Perhaps it is, but this is just one of the advantages of the digital age to the reviewer and editor, who can now spot these manipulations when in the past an author would have taken the time to do another exposure" (Rossner and Yamada 2004, 13).

This statement points to a difference in the ways digital and analog images can circulate and be scrutinized.[18] When images circulate in digital space, the underlying pixel data remains associated with the image.[19] This availability of numerical data allows digital images to be accessed and scrutinized more closely than analog images for signs of enhancement or adjustment, letting viewers go beneath the surface or "face value" of the image (Coopmans 2011). Thus, as well as affording new opportunities for researchers to adjust images and control their composition, digital technologies also allow editors (and arguably journal readers) increased opportunities for scrutiny and forensic analysis.[20] Digital imaging technologies have been used in laboratories for over 25 years, but it is perhaps no coincidence that the current crisis of trust in images identified by journal editors has paralleled the introduction of electronic manuscript submission systems over the past decade: this crisis may owe in part to improved access to data through technologies of surveillance, not simply to changing standards of ethics or practice on the part of authors (Rossner 2002).[21]

In summary, one key focus of the journal guidelines is on the composition of images submitted for publication, acknowledging the control that digital image-processing technologies afford the author in preparing clean and convincing images for reader scrutiny. Whether any such adjustments enhance or improve the data presented to readers (Lynch 1991a) is not the issue here; what is at stake is the virtue of the scientist who performs these manipulations. In the guidelines and commentaries, the virtuous—and trustworthy—scientist is depicted as one who exhibits self-control and does not interfere with the relationship between data and image after the main experimental work has been done. This has clear parallels with the epistemic ideal of "mechanical objectivity" as discussed by Daston and Galison (1992, 2007). As a reference point, the image-processing guidelines relating to the composition of figures effectively restrict the manipulation of digital data to what has traditionally been possible—and largely accepted—with analog images.[22]

3 Transparency and Traceability

Early commentaries about the growing use of digital cameras and processing software in laboratories voiced concerns that digital imaging technology "gives researchers the ability to edit scientific images without leaving a trace" (Anderson 1994, 317). Promoting transparency and traceability in image-making practices is another key focus for journal guidelines on image manipulation. This concern is not so much about limiting the nature and scope of image adjustments performed, but about encouraging full disclosure—"even drastic changes are sometimes considered tolerable if scientists spell out exactly what they did" (Pearson 2005, 953).The guidelines stipulate that authors should provide details of the equipment used to generate and capture image data, should list all the transformations or adjustments applied to each submitted image, and should provide journal editors with the original images if requested.[23] In disciplines such as astronomy there is a longer tradition of making this type of information available, but for cellular and molecular biology accepted norms are clearly still being negotiated.[24]

In effect, such guidelines amount to providing a "methods section" specifically concerning the images in a journal article, and thus a written context according to which an image might be interpreted. Notably, journal editors do not explicitly discuss the provision of imaging methods for purposes of experimental replication. Rather, the guidelines and associated commentaries reveal a concern with establishing the provenance and history of a digital image, and of providing a means to anchor the published visual or "surface" representation to the experimental data collected. Editors stipulate the need to have "a clear record of what's been done to an image, from editing to data compression," for, it is claimed, "without such a record, the image's *scientific value* becomes questionable" (Anderson 1994, 318, emphasis added). This concern resonates with findings from a number of STS studies on practices of image making. Ethnographic observations have shown how sequential series of processes such as marking, indexing, labeling, cleaning, and normalizing observations help to transform phenomena into something knowable, visible, and "docile" (Lynch 1985b). The series of representations arising from such processes are not exact copies of one another, but they must be "aligned" in such a way that it is possible to follow the representations from one stage to the next (Latour 1995, 175). Crucially, keeping careful account of these transformations becomes necessary for moving between data and representation, and for building a sound and convincing argument that allows a researcher to justify conclusions on the basis of collected data. As Latour suggests, maintaining comparability and traceability across different stages of the scientific process is necessary to uphold the *meaning* of visual representations. Furthermore, "traceability of the stages must allow for travel in both directions. If interrupted at any point, they cease to transport truth—cease, that is, to produce, to construct, and to conduct it" (Latour 1995, 180).

Photographs provide an interesting case with respect to traceability. As they are traditionally understood to be one of the more "mechanically" produced forms of representation (in contrast to more "manual" representations such as drawings and diagrams), it is tempting to assume a close correspondence between data and resulting image.[25] In particular, the "hyper-realistic" quality of photographs (Lynch 1991a, 214) can promote viewer/reader judgment that there is a direct link between experimental data and image, irrespective of how much image processing has taken place. Readers might not be able to discern (or even be aware of) the degree to which photographs have been adjusted unless explicitly told. Digital imaging technologies compound the potential for misunderstanding in this regard. Cambrosio and Keating (2000) discuss how digital imaging processes can collapse the epistemic distinction between representing and intervening; in a similar vein Lynch suggests that "digital imaging and image processing break down the distinction between manual and mechanical reproductions" (Lynch 1991a, 219). Mechanically produced digital images can be adjusted "manually" at the computer screen using image-processing software. With such software, digital photographs can effectively be transformed into simplified "diagrams" while retaining the appearance of a photograph that presents a coherent and "true" snapshot of the phenomenon in question. Mitchell notes that "potentially, a digital 'photograph' stands at any point along the spectrum from algorithmic to intentional" (Mitchell 1992, 31).

In short, in the digital realm images are "interchangeable displays of numerical data" (Lynch 1991a, 221), and it can become difficult to uphold clear distinctions among forms of representation such as diagrams, photographs, charts, tables, and models. Representations that look to readers like photographs might have undergone transformations or abstractions that distance the published image from the "raw data" without it being apparent on the "surface" of the image. This distance or gap can remain imperceptible to the reader;[26] furthermore, the nature of the relationship between data and image can be difficult for readers to surmise, and a misreading could in some cases lead them to draw inappropriate conclusions. The journal guidelines regarding transparency in image processing might be seen as trying to redress a perceived imbalance between the author and reader of a journal article (one that stands to be exacerbated by digital processing capacities), and asserting the role of readers in judging for themselves the content of an image when given an accurate description of how it has been produced. In practice, it is not clear whether or exactly how the availability of image acquisition details and processing steps may help readers to better evaluate the content of an image. However, the presence of these details might provide guidance regarding the proposed role of the image within the publication, and may also engender confidence and trust in the author's practices. Furthermore, by providing details about how an image was acquired, transformed, and prepared for publication, the authors of the article acknowledge in writing their responsibility for its production.

4 Trust and Objectivity in the Digital Age

Let us return to the matter raised by Lynch (and quoted at the beginning of the chapter) about whether digitally processed images might come to be seen as *more* authentic and trustworthy than "raw" image data. Here Lynch associates trustworthiness with transformations performed by software (not by human hands), provided there is general acceptance of the appropriateness of the (quantitative) transformations performed (Lynch 1991a, 221). This resonates with the notion of "digital objectivity" advanced by Beaulieu, for which, "along with the mechanical objectivity of scanning and imaging technologies," we see "the mobilization of computer-supported statistical and quantitative apparatus, which provide a further mechanism for validation and for guaranteeing objectivity" (Beaulieu 2001, 664–665). The suggestion in both these instances is that digital image processing can contribute to the epistemic ideal of objectivity through the deployment of automated processes that have been mathematically validated. There is an associated implication that this reduces the need for manual intervention and data processing on the part of the author. In contrast, the concerns highlighted by journal editors focus on somewhat different stages and practices of image making. Their guidelines propose limits on digital image adjustment made for seemingly aesthetic purposes,[27] as opposed to for explicitly quantitative or statistical processing—adjustments intended for publication purposes, not for data analysis. The view perpetuated through their writings and guidelines is that such image processing threatens the authenticity and objectivity of visual representations in journal articles: "Let's celebrate real data—wrinkles, warts and all. We want to publish gritty documentary movies, not digitally beautified yarns!" (*Nature* 2006b, 203).

The question of whether digital image processing enhances or detracts from objectivity thus needs some disaggregation and refinement. Objectivity is an epistemic virtue that is not simply manifest on the surface of a published image, but is produced and performed through the methods, morals, and metaphysics associated with that representation (Daston and Galison 1992, 84)—the virtues of the scientist making an image, the tools and practices he or she uses, and community understandings of the nature of representations and what they should convey. The relationship between objectivity and digital image processing must be understood in relation to the processes by which digital images are made, used, interpreted, and valued within the scientific community. In this context some of the concerns of journal editors might be better understood, and indeed resonate more closely with STS analyses than might initially be perceived. Compared with issues raised by Lynch (1991a), Taylor and Blum (1991), and Beaulieu (2001), they speak to a somewhat different set of concerns regarding where and how digital processing tools might be used for image making; however, they still invest trust in an ideal of mechanical objectivity (Daston and Galison 2007).

High-quality, clear, and "clean" images have traditionally been taken to indicate careful, patient, and replicable scientific work—interpreted by readers as reflecting skill and expertise on the part of the author (*Nature* 2006a, 892). The quality of images in publications helps to establish the authority of the author and the credibility of the argument being made. The ease of digital image manipulation using software such as Photoshop can effectively reduce the value of scientific practice; it is almost invariably easier to "beautify" an image on the computer than to replicate a particular experiment in preparation for publication. Consequently, readers run the risk of investing trust in images that have been "inappropriately" altered at the computer desktop, believing them to reflect experimental skill on the part of the author. Regardless of whether such images accurately represent the phenomena under investigation, if an author's practices are challenged, the trustworthiness of the findings being reported is also called into question.

Furthermore, STS studies of astronomy have highlighted the difficulty in maintaining a clear and unambiguous distinction between "aesthetic" and "scientific" practices in image making (see Lynch and Edgerton 1988; Kessler 2007). Aesthetic considerations are found to be deeply embedded in practices of image production, but are not usually discussed or deployed by scientists in the name of creativity and beauty. Rather, they are typically directed toward achieving a certain "representational realism" (Lynch and Edgerton 1988, 200), of "composing visible coherences, discriminating differences, consolidating entities, and establishing evident relations" (Lynch and Edgerton 1988, 212). Aesthetic judgments can be useful—indeed, necessary—when using complex configurations of visualization technologies, where experience and expertise in knowing what to look for become key to guiding image processing and interpretation (e.g., Cambrosio and Keating 2000; Alač 2008; Myers 2008). Here again, the journal editors focus on an arguably more "superficial" notion of aesthetic judgment in digital image processing, one that guides presentation but not data analysis. Their guidelines do not focus on issues of author expertise in quantitative image analysis techniques for data processing, but are concerned more with curbing author creativity at later stages of the scientific process. The mention of any image-processing software other than Photoshop is sparse, despite the proliferation of (more or less validated) specialist software tools for visualization and analysis of natural phenomena. Across their guidelines and editorials, journal editors do not treat Photoshop as an "expert" technology for the scientific community.[28] Nor does Adobe market its Photoshop products as expert technology; as a brand, Photoshop celebrates creativity, imagination, and artistic flair in image making, as indicated by the one-sentence introduction to the "Photoshop family" of products on the Adobe website: "The Adobe® Photoshop® family of products is the ultimate *playground* for bringing out the best in your digital images, transforming them into anything you can *imagine*, and showcasing them in extraordinary ways."[29]

This description is virtually orthogonal to the discipline and careful judgment in image processing advocated by journal editors. Their commentaries express concern that the underlying mathematics or physics of transformations in image-processing software is often "black-boxed" (Greene 2005, 143). Not only can the procedures or algorithms used in image-processing software be inaccessible (sometimes deliberately so, for commercial or proprietary reasons; see, e.g., Coopmans 2011), but the interface of software like Photoshop can be deceptively simple, making it possible to manipulate images with little understanding of the mathematical transformations being applied.

This problem is often described as particularly acute in the biological sciences, owing to a perceived lack of training for young researchers in mathematics as well as in good practice in image preparation: "graduate school curricula typically do not offer systematic instruction in microscopy or image formation, with the result that most biology graduate students rely on ad hoc training by more senior students or postdocs" (Peterson 2005, 881). Formal training was perhaps not seen as necessary for the bricolage-like activities once performed on analog images to prepare them for publication, but for digital bricolage it seems the expectations are higher.

In sum, the objectivity of digital images is inextricably associated with trust in the care and skill of those performing the imaging work. Carusi discusses the need for researchers to develop a shared framework, or "a common way of seeing," in order to engender trust in each other's practices (Carusi 2008, 244); the recent journal guidelines can be seen as a step toward establishing community norms for image processing, with a particular focus on cellular and molecular biosciences. For trust to be maintained, journal editors and readers alike must be confident that researchers are appropriately trained to craft images, and are using the right tools for the job at hand. Only if these criteria are satisfied might digital image processing make a positive contribution to the trustworthiness of published images. In this case, the journal guidelines act as a set of prescriptions that restrict scope for personal judgment in practices of representation; this can be interpreted as a strategy for promoting trust—and through this for upholding an ideal of objectivity—in circumstances where training and local practices are not (yet) accepted as sufficiently uniform (Porter 1995).

5 Conclusion

Digital technologies for imaging natural phenomena have grown rapidly in number and scope since publication of the first volume of *Representation in Scientific Practice*. New configurations of material equipment and scientific practices are emerging as researchers push for increased detail and resolution in visualization, or strive to make visible phenomena that have hitherto been out of sight. Alongside this, conventions are being negotiated within the scientific community regarding how to make digital images, how to present them, and how to write about them. Recent attempts on the part of leading journals to articulate boundaries and set guidelines for image processing are a clear and interesting example of this. Shapin describes the development of early journal articles in terms of a "technology of trust" (Shapin 1984, 491), and the same arguably holds true today. Journal articles are highly stylized documents that are structured (and continually being restructured) according to conventions deemed to be appropriate for the communication and evaluation of scientific observations and matters of fact.

Recent attention by journal editors to imaging practices can be taken as an indication that within the scientific community, the digital realm is seen to be associated with new possibilities—and potential problems—for the production, circulation, and interpretation of images. Digital technologies increase the scope for interacting with and crafting images, particularly at the postacquisition stage, and the guidelines being developed by journals can be thought of as an intervention designed to define acceptable limits in this relatively new but rapidly expanding space. For trust in images to be maintained, agreement must be secured regarding acceptable methods and conventions at every stage of image making. The postacquisition stage now seems to be a key locus for articulating and (re)defining norms, and the journal guidelines attempt to render the processes and choices involved in image making more visible, transparent, and consistent across scientific disciplines.

In discussing the advent of digital processing technologies, Taylor and Blum suggested that the "mantle of trustworthiness" might be transferred "from the passive lens and film to the *interactive* program" (1991, 126, emphasis added). What the term "interactive" means seems key to the current discussion: is the interaction under scrutiny an automated one between program and image data (as Lynch and Beaulieu discuss), or is it the interaction between the researcher and software (Photoshop) interface? It is the latter concern that seems to sit at the heart of the current debates about trust. The broad response of journal editors has been to appeal to an ideal of objectivity that is procedural and mechanical, one that constrains the freedom of the author and promotes trust through reliance on more mechanical processing tools and the recording of processing steps. In doing so, their guidelines problematize the relationships between manual and mechanical processes of representation, and between practices of representing and intervening; exactly how these relationships are negotiated in practice as digital technologies evolve is a topic worthy of continued empirical investigation.

Notes

1. The trustworthiness of digital images has been widely debated in relation to journalism, print media, and photography competitions since the late 1980s (see Mitchell 1992, 16–17; Grinter 2005), but has become a more active concern within the life sciences research community since the turn of the century.

2. See for example the guidelines published by *Nature* (available at http://www.nature.com/authors/policies/image.html; accessed 12 September 2011), *Science* (available at http://www.sciencemag.org/site/feature/contribinfo/prep/prep_init.xhtml; accessed 12 September 2011), the *Journal of Cell Biology* (available at http://jcb.rupress.org/site/misc/ifora.xhtml; accessed 12 September 2011), and *PLoS Biology* (available at http://www.plosbiology.org/static/figureGuidelines.action; accessed 12 September 2011).

3. For example, the purification of data from artifacts, and the separation of signal from noise, are understood as central preoccupations in laboratory work (e.g., Latour and Woolgar 1979;

Pinch 1985). Similarly, the transformation of "data" into publishable image involves multiple steps of selection and refinement that have been documented and analyzed in detail in a number of studies (Knorr-Cetina and Amann 1990; Lynch 1985a and 1985b; Lynch and Edgerton 1988).

4. Recent STS scholarship concerning digital image-making practices has explored medical imaging technologies such as brain imaging (e.g., Alač 2008; Beaulieu 2002; Joyce 2006) and mammography (Coopmans 2011), as well as digital imaging in scientific disciplines and subdisciplines including astronomy (Kessler 2007), immunology (Cambrosio and Keating 2000), and protein crystallography (Myers 2008).

5. Elsewhere I analyze in detail how the (guide)lines being defined by journal editors for digital image processing negotiate a number of longstanding and quite practical tensions regarding the role of images in scientific publications (Frow 2012).

6. Indeed, the idea that images in a journal article allow readers to "witness" at a distance the natural phenomena being described can be traced back to the origins of the scientific publication in the mid-seventeenth century (see Shapin 1984).

7. One commentary in *Nature* describes the possibilities for image fabrication thus: "with Photoshop, a few clicks of the mouse can transform a featureless black microscope snap into a starry vista littered with labelled proteins" (Pearson 2005, 952).

8. The journals *Science* and *Nature* are widely considered to be the world's leading interdisciplinary scientific journals; their commentaries and guidelines are thus used as sources to reflect general trends and concerns within the scientific community at large.

9. For more detailed analysis, see Frow (2012).

10. *Nature* online instructions (available at http://www.nature.com/authors/policies/image.html; accessed 12 September 2011).

11. An example of inappropriate image composition can be found in two articles published in *Nature Cell Biology* in 2003 (Sawada et al. 2003a; Sawada et al. 2003b), in which the first author used Photoshop to make composite western blot images by cutting and pasting together bands from several different experiments (Pearson 2005, 953). Formal investigation by the journal editors concluded that although the interpretation of the research findings was not affected by the image manipulation, "the frequency and severity of the manipulations" undertaken necessitated full retraction of the papers (*Nature* 2007, 355; Sawada et al. 2007).

12. *Nature* online instructions (available at http://www.nature.com/authors/policies/image.html; accessed 12 September 2011). Not all of the requested information must be detailed in the main article; it can be included in supplementary information files that accompany the online version of a published manuscript.

13. Some journals do stipulate quite specific guidelines on occasion; for example, the *Nature* guidelines state that when cropping images of gels and blots, "cropped blots in the body of the paper should retain at least six band widths above and below the band" (available at http://www .nature.com/authors/policies/image.html; accessed 12 September 2011). No detailed rationale is provided for this guideline.

14. This applies particularly to phenomena operating at scales or in places not directly accessible to the human eye (e.g., Lynch and Edgerton 1988; Cambrosio and Keating 2000; Alač 2008), or when dealing with very large datasets (e.g., Burri and Dumit 2008, 303).

15. Digital images differ from their analog counterparts in terms of physical composition and informational content (Mitchell 1992). Digital space is represented mathematically as pixels (picture elements) that are uniform in size, evenly spaced, and each able to represent one of a range of discrete values corresponding to degrees of tone or color. Underpinning a digital image is thus a matrix of numbers, with each number corresponding to one pixel (Lynch 1991b, 67–68). This has implications for the composition and adjustment of digital images; as Lynch points out, "pixelated space is manipulatory . . . the 'world' breaks down into arbitrary bits of information that allow its composition and recomposition to be arranged by a *user*" (Lynch 1991b, 64, original emphasis). For published images, these users are the authors who select, frame, and adjust their images in preparation for submission to a journal.

16. Little attention is devoted in the guidelines to norms for image acquisition, or to more complicated image-processing activities such as deconvolution and quantitative analysis (North 2006).

17. "[Digital] functions seem to duplicate work that might otherwise be done in a photography developing lab or a graphics workshop. Keyboard and touch-screen operations displayed on the video monitor replace the classical toolbox of scissor, paper, rule, paintbrush, and bench. A bricoleur's vocabulary is displaced into the digital system, as 'palette,' 'paintbrush,' and 'slice' become electronic operations" (Lynch 1991b, 66–67).

18. The mobility of digital images in the context of "e-science" has been associated with changing sociotechnical configurations, increases in the speed and scale of data circulation, and the ability to compare large datasets (Beaulieu 2001; Coopmans 2006). For journals, a key implication of the mobility of digital images relates to changing relations of data access and image surveillance.

19. This link between image and pixel data is effectively severed when a digital image is printed in hard-copy form. A reader presented with a printed digital image has to treat it as an analog representation, with no access to the underlying numerical data.

20. Ironically, journal editors themselves often make use of Photoshop for forensic purposes. The former managing editor of *JCB* notes: "We examine all digital images in all accepted manuscripts for evidence of manipulation. For black and white images this involves simple adjustments of brightness and contrast in Photoshop. . . . For color images, we use the 'Levels' adjustment in Photoshop to compress the tonal range and visualize dim pixels" (Rossner 2006, 24).

21. Rossner observes that compared with recent improvements in the ability to detect image manipulation, "it is clearly more difficult to determine whether numerical data have been misrepresented, fabricated, or falsified" (Rossner 2006, 24).

22. This is not to suggest that analog images are free from problems relating to their preparation and interpretation. Indeed, the journal editors' guidelines identify a number of tensions regarding

images in scientific publications that may apply to both digital and analog representations (see Frow 2012).

23. *JCB* editorial policies state that an article can be rejected if authors are unable to supply original images upon request, regardless of any evidence of inappropriate image alteration (available at http://jcb.rupress.org/site/misc/ifora.xhtml; accessed 12 September 2011).

24. For example, as Lynch noted when studying digital image processing in astronomy research 20 years ago, "one astronomer said in an interview that virtually anything goes, as long as the 'look up table' (correspondence between palette and intensities) is published" (Lynch 1991b, 71). A similar convention for microscopy is being encouraged in the recent *Nature* guidelines: "The display lookup table (LUT) and the quantitative map between the LUT and the bitmap should be provided" (available at http://www.nature.com/authors/policies/image.html; accessed 12 September 2011).

25. The issue here has less to do with whether this is an appropriate understanding of photographs than with how photographs are understood or interpreted in practice.

26. Latour discusses this difficulty in recognizing representational slips or gaps: "We never detect the rupture between things and signs, and we never find ourselves faced with the imposition of arbitrary and discrete signs upon shapeless and continuous matter. We only see an unbroken series of well-nested elements, each of which plays the role of sign for the previous and of thing for the succeeding" (Latour 1995, 169).

27. The journals considered here make firm statements about the unacceptability of aesthetic criteria as guides for image processing. For example, "In *Nature's* view, beautification is a form of misrepresentation. Slightly dirty images reflect the real world" (*Nature* 2006a, 892).

28. Photoshop was originally developed as customized image-processing software by PhD student Thomas Knoll at the University of Michigan, but over the last 20 years has become a household name for digital image processing—so much so that "to photoshop" is now recognized as a verb. According to the *Oxford English Dictionary* online (accessed 18 January 2011), "to photoshop" means "To edit, manipulate, or alter (a photographic image) digitally using computer image-editing software."

29. Available at http://www.adobe.com/products/photoshop/family/ (emphasis added; accessed 12 September 2011).

References

Alač, Morana. 2008. Working with brain scans: Digital images and gestural interaction in fMRI laboratory. *Social Studies of Science* 38:483–508.

Anderson, Christopher. 1994. Easy-to-alter digital images raise fears of tampering. *Science* 263:317–318.

Beaulieu, Anne. 2001. Voxels in the brain: Neuroscience, informatics and changing notions of objectivity. *Social Studies of Science* 31:635–680.

Beaulieu, Anne. 2002. Images are not the (only) truth: Brain mapping, visual knowledge, and iconoclasm. *Science, Technology and Human Values* 27:53–86.

Burri, Regula Valérie, and Joseph Dumit. 2008. Social studies of scientific imaging and visualization. In *The Handbook of Science and Technology Studies*, 3rd ed., ed. Edward J. Hackett, Olga Amsterdamska, Michael Lynch, and Judy Wajcman, 297–317. Cambridge, MA: MIT Press.

Cambrosio, Alberto, and Peter Keating. 2000. Of lymphocytes and pixels: The techno-visual production of cell populations. *Studies in History and Philosophy of Biological and Biomedical Sciences* 31 (2):233–270.

Carusi, Annamaria. 2008. Scientific visualisations and aesthetic grounds for trust. *Ethics and Information Technology* 10:243–254.

Coopmans, Catelijne. 2006. Making mammograms mobile: Suggestions for a sociology of data mobility. *Information Communication and Society* 9 (1):1–19.

Coopmans, Catelijne. 2011. "Face value": New medical imaging software in commercial view. *Social Studies of Science* 41:155–176.

Council of Science Editors. 2009. White Paper on Promoting Integrity in Scientific Journal Publications, 2009 Update. New York: Council of Science Editors. Available at: http://www .councilscienceeditors.org/i4a/pages/index.cfm?pageid=3331 (accessed 17 November 2012).

Couzin, Jennifer. 2006. Don't pretty up that picture just yet. *Science* 314:1866–1868.

Daston, Lorraine, and Peter Galison. 1992. The image of objectivity. *Representations* (Berkeley, CA) 40:81–128.

Daston, Lorraine, and Peter Galison. 2007. *Objectivity*. Cambridge, MA: Zone Books.

Frankel, Felice. 2002. *Envisioning Science: The Design and Craft of the Science Image*. Cambridge, MA: MIT Press.

Franzen, Martha, Simone Rödder, and Peter Weingart. 2007. Fraud: Cases and culprits as perceived by science and the media. *EMBO Reports* 8 (1):3–7.

Frow, Emma K. 2012. Drawing a line: Setting guidelines for digital image processing in scientific journal articles. *Social Studies of Science* 42 (3):369–392.

Greene, Mott T. 2005. Seeing clearly is not necessarily believing. *Nature* 435:143.

Grinter, Rebecca E. 2005. Words about images: Coordinating community in amateur photography. *Computer Supported Cooperative Work* 14.161–188.

Joyce, Kelly A. 2006. From numbers to pictures: The development of magnetic resonance imaging and the visual turn in medicine. *Science as Culture* 15 (1):1–22.

Kessler, Elizabeth A. 2007. Resolving the nebulae: The science and art of representing M51. *Studies in History and Philosophy of Science* 38:477–491.

Knorr-Cetina, Karin, and Klaus Amann. 1990. Image dissection in natural scientific inquiry. *Science, Technology and Human Values* 15 (3):259–283.

Latour, Bruno. 1995. The "pédofil" of Boa Vista: A photo-philosophical montage. *Common Knowledge* 4 (1):145–187.

Latour, Bruno, and Steve Woolgar. 1979. *Laboratory Life: The Social Construction of Scientific Facts.* London: Sage Publications.

Lister, Martin. 2004. Photography in the age of electronic imaging. In *Photography: A Critical Introduction*, 3rd ed., ed. Liz Wells, 295–336. London: Routledge.

Lynch, Michael. 1985a. *Art and Artifact in Laboratory Science: A Study of Shop Work and Shop Talk in a Research Laboratory.* London: Routledge.

Lynch, Michael. 1985b. Discipline and the material form of images: An analysis of scientific visibility. *Social Studies of Science* 15 (1):37–66.

Lynch, Michael. 1991a. Science in the age of mechanical reproduction: Moral and epistemic relations between diagrams and photographs. *Biology and Philosophy* 6:205–226.

Lynch, Michael. 1991b. Laboratory space and the technological complex: An investigation of topological conjectures. *Science in Context* 4 (1):51–78.

Lynch, Michael, and Samuel Y. Edgerton. 1988. Aesthetics and digital image processing: Representational craft in contemporary astronomy. In *Picturing Power: Visual Depiction and Social Relations*, ed. Gordon Fyfe and John Law, 184–220. London: Routledge.

Mitchell, William J. 1992. *The Reconfigured Eye: Visual Truth in the Post-Photographic Era.* Cambridge, MA: MIT Press.

Myers, Natasha. 2008. Molecular embodiments and the body-work of modeling in protein crystallography. *Social Studies of Science* 38:163–199.

Nature. 2000. Electronic submissions now welcome. Editorial. *Nature* 403: 467.

Nature. 2006a. Not picture-perfect. Editorial. *Nature* 439: 891–892.

Nature. 2006b. Appreciating data: Warts, wrinkles and all. Editorial. *Nature Cell Biology* 8(3): 203.

Nature. 2006c. Beautification and fraud. Editorial. *Nature Cell Biology* 8(2): 101–102.

Nature. 2007. Imagine . . . Spot-checking figures for manipulation. Editorial. *Nature Cell Biology* 9(4): 355.

North, Alison J. 2006. Seeing is believing? A beginners' guide to practical pitfalls in image acquisition. *Journal of Cell Biology* 172 (1):9–18.

Pearson, Helen. 2005. CSI: Cell biology. *Nature* 434:952–953.

Peterson, Daniel A. 2005. Images: Keep a distinction between beauty and truth. *Nature* 435:881.

Pinch, Trevor. 1985. Towards an analysis of scientific observation: The externality and evidential significance of observation reports in physics. *Social Studies of Science* 15:167–187.

Porter, Theodore M. 1995. *Trust in Numbers: The Pursuit of Objectivity in Science and Public Life.* Princeton: Princeton University Press.

Rossner, Mike. 2002. Figure manipulation: Assessing what is acceptable. *Journal of Cell Biology* 158:1151.

Rossner, Mike. 2006. How to guard against image fraud. *Scientist* 20 (3):24.

Rossner, Mike, and Kenneth M. Yamada. 2004. What's in a picture? The temptation of image manipulation. *Journal of Cell Biology* 166:11–15.

Ruivenkamp, Martin, and Arie Rip. 2010. Visualizing the invisible nanoscale study: Visualization practices in nanotechnology community of practice. *Science Studies* 23:3–36.

Sawada, M., P. Hayes, and S. Matsuyama. 2003a. Cytoprotective membrane-permeable peptides designed from the Bax-binding domain of Ku70. *Nature Cell Biology* 5:352–357.

Sawada, M., W. Sun, P. Hayes, K. Leskov, D. A. Boothman, and S. Matsuyama. 2003b. Ku70 suppresses the apoptotic translocation of Bax to mitochondria. *Nature Cell Biology* 5:320–329.

Sawada, M., W. Sun, P. Hayes, K. Leskov, D. A. Boothman, and S. Matsuyama. 2007. Retractions. *Nature Cell Biology* 9:480.

Shapin, Steven. 1984. Pump and circumstance: Robert Boyle's literary technology. *Social Studies of Science* 14 (4):481–520.

Taylor, Peter J., and Ann S. Blum. 1991. Pictorial representation in biology. *Biology and Philosophy* 6:125–134.

13 Legitimizing Napkin Drawing: The Curious Dispersion of Laffer Curves, 1978–2008

Yann Giraud

1 Introduction

The Laffer curve is not truly a curve but an insight.

—Roger Starr, *New York Times*, July 26, 1981

In its textbook version, the tale of the Laffer curve is a straightforward narrative of how professional economists respond to political propaganda. For instance, in the sixth edition of John Sloman's *Economics*, this two-dimensional diagram is displayed as a symmetrical, bell-like curve (figure 13.1) and located in a box bearing the ironic title "Having your cake and eating it." The accompanying text begins with a few words about the political origins of the figure, attributed to President Ronald Reagan's advisor Arthur Laffer. The functioning of the curve is explained without further analytics: suppose that, at a 0 percent tax rate, tax revenues will be equal to 0 and that at a 100 percent tax rate, because there is absolutely no incentive to produce at all, tax revenues will be null as well. Then the curve assumes that there exists a tax rate t_i, at some point between 0 and 100 percent, which will yield maximum revenues. When the tax rate is higher than t_i, tax revenues will not be maximal, so that everyone—taxpayers as well as the government—would be better off with a reduction in tax rates. In spite of the symmetrical shape of the figure, the author states that the curve "may peak at a 40 per cent, 50 per cent, 60 per cent or even 90 per cent rate." The climax of the story is reached when the author attempts to appraise the validity of this visual representation. It is said that while "Laffer and others on the political right argued that tax rates were above t_i," "most evidence suggests that tax rates were well below t_i in the 1980s and certainly are now" (Sloman 2006, 283). In Sloman's treatment, the curve itself does not provide a representation of a measurable magnitude; rather, it holds out the promise that such measurement could be undertaken in the future, using statistical data. As a curve, it gives the reader a sense that there might be a mathematical relationship out there, but the analytical support for this relation is not formally elaborated in the text.[1] In addition, the visual representation itself seems to be associated with the community

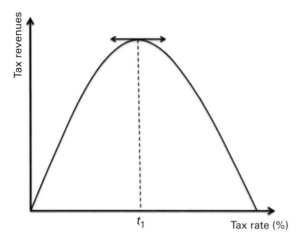

Figure 13.1

The textbook version of the Laffer curve shows that there is a tax rate t_1 that maximizes tax revenues. When the tax rate is higher than t_1, it is possible for the government to cut tax rates while simultaneously increasing its tax revenues.

of political propagandists rather than with professional economists. Whereas the former use the Laffer curve to *argue*, the latter supply *evidence* to show that the political discourse is well-founded or not.

Constructivist sociologists of science will recognize in the preceding depiction of the strict demarcation between the communicational status of the Laffer curve as a visual representation, on the one hand, and the scientific virtues of the mathematical groundwork one can apply to it, on the other hand, a curious variant of the "diffusion model" they have criticized over the past 25 years or so.[2] In this model, concepts and tools are initially invented by geniuses whose reputations go far beyond their modest status as working scientists. These innovations then diffuse to larger audiences—other scientists, students, and the public at large—through a process that appears as though automatic: the actions of many individuals and groups in accepting and adopting such innovations are taken for granted, and the many controversies that the discoveries had engendered in the first place are usually toned down and depicted as inevitable though unwelcome noise in the process of diffusion. In light of the diffusion model, the Laffer curve is a very strange case because, at least in the textbook account of it, its uptake seems to have worked in reverse. Rather than a genius like the Pasteurs and Diesels of the past, the inventor of the curve, Arthur Laffer, is frequently not even granted the title of economist (though he certainly is one). The real heroes of the story are the anonymous researchers working "downstream," who have provided the evidence that the curve is not empirically validated. But, strangely, in spite of the textbook authors' heroic assault against the obvious falsity of the Laffer curve, the curve is still present in the textbook, displayed in its canonical, symmetrical version. We are left with a question that the diffusion model is unable to address in a satisfying way: how did it stick?[3]

Perhaps one of the alternative models sociologists have elaborated in response to the diffusion model will provide a better framework for thinking about the Laffer curve. In the "dispersion"—or "translation"—model, facts and tools travel within various communities through space and time. The attitudes that these different communities have toward the traveling objects—acceptance, defiance, ignorance, or interest—greatly differ depending on various contingencies. To be able to travel, the objects have to be transformed to suit new constituencies. As a result, their shape, meaning, and functioning will be affected, and so will their epistemological status. In the process, the attitudes of the people who interact with them will likely evolve as well (see Latour 1987, 139–140). In recent years, a number of historians of science have followed this framework to build richer narratives of the dissemination of visual representations in various disciplines. David Kaiser's (2005) account of Feynman diagrams in physics relies explicitly on the dispersion model, which in his opinion integrates two important dimensions of the objects it studies: how they circulate and how they persist. In addition, historians have argued that visual representations must be studied at the intersection of well-entrenched dichotomies—laboratories versus museums, arts versus sciences, geometry versus algebra—as it is often in the dialogue between these that visualization is discussed (Wise 2006).

The purpose of this chapter is to provide a historical account of the Laffer curve as a case study of dispersion. It seems to be an ideal case for doing so because, as I hinted earlier, the formal triviality of the curve contrasts sharply with the complexity of its circulation between and among different communities: economists, policy advisors, propagandists, and journalists.[4] For each of these communities—which, in addition, were not necessarily homogeneous within themselves—the Laffer curve had different embodiments and meanings; hence it makes more sense to use the plural *curves* rather than the singular. Nonetheless, we shall see that the dispersion of Laffer curves presented two peculiarities: first, unlike many other diagrams used in economics, popular instantiations of the Laffer curve preceded its "academization" by professional economists; second, in spite of the numerous transformations the curve has undergone in the process of its circulation, its canonical presentation—the symmetric bullet like curve in figure 13.1—was reinforced over time. The circulation of Laffer curves must be explained by the internal dynamics of each of the communities that interacted with them, and by what happens at the intersection of these communities, most importantly as the consequence of the ambiguous position that the economics profession has maintained toward its role in policy advising. Though the discipline has always been eager to offer more legitimate knowledge—from a scientific standpoint—than what is circulated in newspapers and pamphlets, it also has had to find legitimacy as a policy-oriented science. I will begin by depicting the political origins of the curve (section 2); then I will turn to its subsequent "academization" (section 3) and generalization (section 4). As the following discussion will be mostly historical in its content, I will discuss in the conclusion a few implications for the sociology of knowledge and more specifically for the study of representational practices in economics.

2   The Political Meaning of the Original Laffer Curve

The Laffer curve did not originate in the academic literature, and Arthur B. Laffer himself was only partly responsible for its circulation.[5] It was in fact introduced by Jude Wanniski in his 1978 book *The Way the World Works*. Born in 1936, Wanniski had been hired as a columnist at the *Wall Street Journal* in 1972, after years of covering energy policy for various newspapers. When his book came out, he had just been forced to resign from this position after it was discovered that he was distributing leaflets for a Republican senatorial candidate. He decided to begin a career as an advisor to various Republican politicians and created a business for doing so, Polyeconomics. Wanniski was known as an ardent advocate of "supply-side economics," a term he had coined in 1976 in contrast to the demand-side emphasis in Keynesian economics on governmental intervention to stimulate private consumption and investment and reduce unemployment. Supply-side economics was defended by two economics professors, Robert Mundell and Arthur Laffer, who argued that state intervention had detrimental effects on the supply of goods and labor and was the main cause for reduced GNP and increased unemployment. Though Mundell, a former PhD student at MIT who had taught at the University of Chicago until 1971, was respected among economists for his international macroeconomic models, he was becoming marginalized in the profession in the 1970s after his scientific production had declined.[6] Laffer, on the other hand, was considered among fellow economists as a promising young scholar who had mostly ceased academic work to pursue a more politically driven career.[7] After studying at Yale and Stanford and teaching at the Chicago school of business, he had served as an economic advisor for the Nixon and Ford administrations. According to Wanniski, Laffer had drawn the curve that was named after him on a cocktail napkin during a meeting with a few presidential aides in December 1974.[8]

In its original rendition, the Laffer curve appeared as a bullet-shaped diagram in which tax revenues were represented on the horizontal axis and the aggregate tax rate on the vertical axis (figure 13.2). In Wanniski's words, the curve illustrates the idea that "there are always two tax rates that yield the same revenues" (Wanniski 1978, 97). While the lower portion of the curve represents an increasing relationship between tax rates and tax revenues, this relationship becomes decreasing beyond point E. The interpretation of this relation is that when tax rates increase from zero to any percentage below point E, economic agents will pay their taxes while keeping unchanged

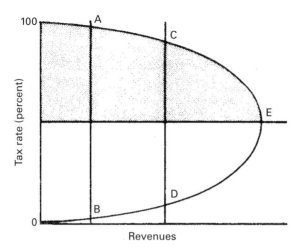

Figure 13.2
The original bullet-shaped Laffer curve (Wanniski 1978, 97) shows that there are two different tax rates—for instance A and B, or C and D—yielding the same amount of tax revenues. The only exception is point E, where tax revenues are maximal.

their supply of labor and goods. In this lower range, higher tax rates will yield higher tax revenues. In "the prohibitive range," however, which is the part beyond point E, people will decide to work less or to participate in a barter (or black) economy, which will lead to a reduction of the tax base and subsequent decrease in tax revenues. Point E thus appears as the point at which tax revenues would be maximal. This point, Wanniski noted, is a "variable number. *It is the point at which the electorate desires to be taxed.* . . . It is the task of the political leader to determine point E and to follow it through its variations as closely as possible" (ibid., 98–99, emphasis in the original).

Wanniski placed the Laffer curve at the center of his 1978 book, the publication of which had been supported by Irving Kristol, a former colleague at the *Wall Street Journal* and one of the leading figures of the then-emerging neoconservative movement.[9] The passage of Proposition 13 in California in June 1978, which resulted in a drastic reduction of property tax rates in that state, was succeeded by tax revolts all over the country. In the run-up to the midterm elections and the next presidential campaign, there was a clear sense among Republicans that more was needed than the traditional conservative discourse, which advocated a cut in government spending and controls over prices and wages to stabilize inflation. For this purpose, the Laffer curve was timely. Unlike the then-orthodox right-wing discourse, which tried to convince the population that sacrifices were needed on the road to economic growth, the Laffer curve suggested that less taxation would result in more wealth to distribute. As a result, supply-side economists did not advocate a cut in public investment: on the contrary, a tax cut contributing to increases in tax revenues would permit *more* government spending in the future. This was, briefly sketched, the "economics of joy" that dominated Ronald Reagan's campaign from 1978 on (Stein 1984).

Wanniski's book and the Laffer curve it promoted provided a stepping-stone for supply-side propaganda, relying on antiexpert populism and the bashing of the economics profession as a whole. Wanniski argued in the beginning of his book that voters knew more than economists, in the sense that their own behavior could contradict the experts' predictions. In this critique, academic economists were chastised alongside economic advisors for providing irrelevant theories and giving unsound advice. In Wanniski's view, the Laffer curve embodied the failure of economists—except for Mundell and Laffer, of course—to realize that there was a thin frontier between a monetary and a barter economy, between work and nonwork. The rest of the book attempted to demonstrate that economic history—from the decline of the Roman empire to past and present economic crises—could be read using the Laffer curve and that past economic theories could all be encapsulated using the curve's framework as well. In chapter 8 the author attacked Keynes and his main opponents, Milton Friedman and the Chicago School monetarists. While Keynes's advocacy of deficit spending without tax cuts would throw the economy in the prohibitive range of the Laffer curve, Friedman's plea for a cut in government spending would result in an increase in interest rates that would in turn counterbalance the expected effects of tax decreases. Wanniski argued that what made these two frameworks equally invalid was that they both modeled the economy in terms of partial equilibrium analysis, referring to Alfred Marshall, who analyzed one market at a time. In contrast, his own model, the Laffer curve, would follow Léon Walras's general equilibrium framework, where all economic effects were simultaneously taken into account. This is why, according to Wanniski, only Laffer and Mundell were right in their approach, while other economists were building "elegant, mathematical edifices atop a foundation of illusion" (ibid., 166). This constituted no less than an insult to most mainstream economists in that postwar period, who claimed they had revived the Walrasian tradition, which, incidentally, was more, not less, mathematical than the Marshallian analysis most had abandoned.[10]

In the face of such disparaging comments, most economists remained silent. Not a single review of *The Way the World Works* was published in academic economics journals, despite the fact that such journals often devoted space to nonscholarly literature.[11] Admittedly, some economists expressed their disapproval of the Laffer curve, but only when writing in the popular press. For instance, Joseph Minarik was quoted in the *New York Times* as stating that he had done the calculation and found that the results were too erratic to certify that the curve existed; "its designers had better iron out some of the wrinkles before offering it to the public" (Rattner 1978). Even Laffer backed off a bit from Wanniski's enthusiasm. Asked by a journalist to explain his curve, he presented it as embodying two conflicting effects that tax rates have on tax revenues: one was the "arithmetic" or "accounting" effect, which implied that lower taxes would lead to a decrease in revenue collected per dollar of tax base; and the other was the "economic effect," which resulted in an expansion of the tax base. However, he did not conclude on which effect would dominate the other (quoted in Jensen 1978). In 1979, three of Laffer's students at the University of Southern California published a short paper in which they tried to build a more elaborate version of the curve, using diagrammatic reasoning only (Canto, Joines, and Webb 1979). In this model, the curve was derived from a once-accepted microeconomic model in which an increase in tax rates creates incentives to move from the market sector (labor) to the household sector (leisure). This model, however, constituted a serious step back from Wanniski's ambitions. It ignored many of the complexities in a generalized Walrasian framework and based its analysis on a single tax rate imposed on market-sector production. In addition, the students had undertaken rather disappointing empirical tests: using the tax cuts of 1962 and 1964 as an experiment, they had to conclude that it was "almost equally likely that the Kennedy tax cuts increased revenues as that they decreased them" (ibid., 38). Two subsequent papers (Canto, Joines, and Laffer 1981; Laffer 1981) tried to build more elaborate models of taxation, but none of them included an actual Laffer curve. These contributions, which did not appear in mainstream economics journals, failed to reach academic economists. The latter, indeed, considered the Laffer curve debate as an unwelcome diversion from serious economic thinking.

3 The "Academization" of the Curve

It took another three years or so before the Laffer curve penetrated the scholarly literature. In the meantime, Ronald Reagan had become president and his administration already had undertaken measures to reduce taxes. Supply-side economists were quite influential in the tax reforms. Acting as an advisor to New York Congressman Jack Kemp, Wanniski helped draft the Kemp-Roth tax cut proposal of 1978, which served as the basis for the Economic Recovery Tax Act (ERTA) of 1981. The bill that was passed by Congress, however, represented a slight retreat from Kemp's initial plan: whereas Kemp had advised a 33 percent tax cut, uniformly applied to all tax brackets, the ERTA reduced top rates from 70 to 50 percent and bottom rates from 14 to 11 percent. Though the tax reform did not scrupulously follow the prescriptions of the Laffer curve, the latter was still seen as one of its main ideological components in the popular press. The rise in tax revenues that supply-side economists had predicted did not occur, and the tax cuts resulted in budget deficits and subsequent efforts to cut spending. One of these efforts involved a proposal to cut social and behavioral science funding by 75 percent. Written by David Stockman, head of the Office of Management and Budget, it boded significant harm to economics research, which depended heavily on grants from the National Science Foundation. The American Economic Association executive committee, which usually avoided political matters, expressed concerns over the situation. Its president, William Baumol, wrote a letter to chairpersons of economics departments in the United States, encouraging them to write to members of Congress to alert them to the detrimental consequences such cuts would have on the pursuit of economic knowledge.[12] In 1981, then, the gospel of supply-siders was materially threatening the economics community. As if this weren't bad enough, the saga of the "failed" tax cuts was also turning the whole profession into a laughingstock in the eyes of other scientists and intellectuals, who considered the Laffer curve as not only a particularly weak example of economic thinking but also a symptom that economics itself was mere ideology backed by mediocre mathematics. Characteristic of this position was the publication in *Scientific American* of an article by Martin Gardner, a well-known specialist in recreative mathematics, titled "The Laffer Curve and Other Laughs in Current Economics." Mocking Wanniski's argument that point E could be located anywhere along the curve, Gardner introduced the neo-Laffer curve (figure 13.3). "Like the old Laffer curve," he wrote, "the new one is also metaphorical, although it is clearly a better model of the real world. Since it is a statistical reflection of human behavior, its shape constantly changes, like the Phillips curve, in unpredictable ways" (Gardner 1981, 27). Not only did Gardner satirize the original Laffer curve, but he also made fun of further attempts to draw estimates from the curve: "Because it takes so long to gather data and even longer to analyze all the shift parameters, by the time an NL curve is drawn it is out of date and not very useful" (ibid.).[13]

Meanwhile, within the Reagan administration there was a growing discontent with supply-side economics. Among advisors, monetarists such as Arthur Burns and Alan

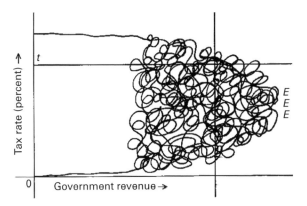

Figure 13.3
On Gardner's (1981, 27) "neo-Laffer curve," which the author intended as an acerbic parody of simplistic models in economics, the inflection point of the curve could be located anywhere and there could be as many of these points as one can possibly imagine.

Greenspan still believed that the control of money supply by increasing interest rates should prevail over tax cuts as a way to ensure long-term economic growth. Monetarism, indeed, was the main principle that had guided the actions of the Federal Reserve since 1979, when Paul Volcker was appointed chairman. In contrast, true believers in supply-side economics estimated that current monetary policy had prevented tax cuts from yielding the expected revenues. As tensions rose, the political debate turned nasty. Wanniski, for instance, wrote in July 1981: "Milton Friedman is not a big man, but he is very heavy. His monetarist economic ideas . . . are crippling the Nobel Laureate's old friend, Ronald Reagan, the United States economy and indirectly, all of our trading partners. Professor Friedman is barely five feet tall, but his shadow falls across the last decade of global inflation" (Wanniski, 1981). In reaction to such acrimony, many economists felt that the time was ripe for a more serious academic inquiry into the Laffer curve, which would not only lead the debate in a more rigorous direction but would also restore the economics profession's credibility. The economists' response, quite naturally, implied increasing the mathematization of the competing models.[14] Two contributions, published in 1982, proved particularly influential in bringing the Laffer curve into the scholarly debate.

First, Don Fullerton (1982) offered a complete analytical model of the curve that would render it a testable hypothesis. He noted that while the typical Laffer curve represents how tax revenues vary with changes in tax rates, the key factor for determining whether we are in the normal or the prohibitive range is the elasticity of the factor supply: the degree to which a variation in tax will affect the supply of the factor—labor or capital—being taxed. As a result, an infinity of Laffer curves can be imagined—with one curve for each value of a given elastic supply factor. Fullerton thus postulated that the Laffer curve was not a "hill," as the canonical representation showed, but a "ridge," since it had to be represented in a three-dimensional diagram with factor supply elasticity as the third dimension. Fullerton only drew the "crest" of the ridge, which showed a downward-sloping relation between tax rates and factor supply elasticity, representing each point along the curve where tax revenues would be maximal (see figure 13.4). The normal range is located below the curve, while the area above the curve corresponds to the prohibitive range. Using a general-equilibrium econometric model of the United States, Fullerton tried to estimate the new curve and eventually the shape of the Laffer curves for various tax rates and supply factor elasticities. He concluded that, in theory, the US could be operating in the prohibitive range in the case where wages are taxed at a very high rate. Yet, taking into account previous econometric estimations of the labor supply elasticities for various groups of the population, he showed that under realistic assumptions the prohibitive tax rate would have to be well above the current rate in the US. In other words, the Laffer curve would begin to decrease at a much higher tax rate than what previous representations had suggested.

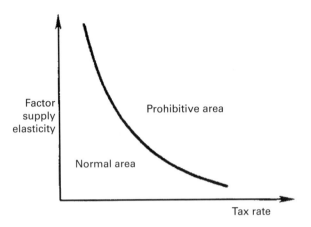

Figure 13.4
Economists rarely work with three-dimensional diagrams. Though Fullerton (1982) added a third dimension to the original diagram, showing that the shape of the curve is also determined by the price elasticity of supply, namely the extent to which changes in tax rates affect labor, capital goods, and other "supply factors," he did not provide the resulting three-dimensional figure. Instead, he displayed a two-dimensional curve (Fullerton 1982, 9) showing a downward-sloping relation between tax rates and factor supply elasticity. Each point on this curve yields maximum tax revenues on a different two-dimensional Laffer curve. Tax rates located on the right side of the curve are considered prohibitive because they discourage taxpayers from offering more factors. This is the relationship Fullerton subsequently uses to provide an econometric estimation of the Laffer curve.

James Buchanan and Dwight Lee's (1982) short paper constituted a completely different approach to the Laffer curve. The authors did not try to question its foundations but rather its consequences within the economic framework of public choice theory. Unlike most economists, who assumed that the State was an outside observer operating in favor of socially desired outcomes, public choice theorists argued that government consisted of utility-maximizing economic agents who acted for their own benefit. From this starting point, Buchanan and Lee sought to explain how a reputedly rational government would end up choosing a less than optimal tax rate. Their answer consisted in arguing that the Laffer curve others had tried to assess so far was in fact the representation of a relationship that would only occur in the long run, while the government was more likely to act according to short-run objectives.[15] In their framework, the difference between short- and long-run effects is the time it takes for taxpayers to adjust to changes in taxation. Visually, their representation (figure 13.5) consists of one long-run Laffer curve (LRLC) and several short-run Laffer curves (SRLC). For each point located on the LRLC there exists a different SRLC that is more favorable to the government. This means that because taxpayers do not fully adjust

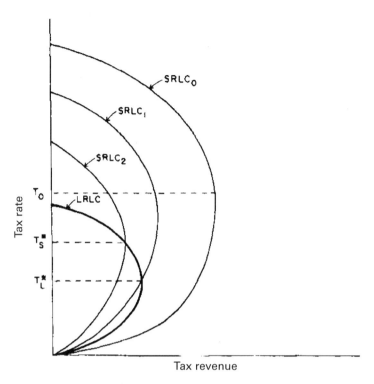

Figure 13.5

Buchanan and Lee's (1982, 817) recreation of the Laffer curve attempts to explain why a profit-seeking government would set a tax rate that does not yield maximal tax revenues. This, they explain, happens because, whereas the definitive Laffer curve (LRLC) represents a relationship that occurs in the long run, the government acts according to a short-run Laffer curve (SRLC). In the short run, taxpayers are not going to react immediately to tax increases, which is a more favorable situation for the government. Setting a very high tax rate, such as T_0, would completely eliminate the tax base, as it would be outside of the LRLC. The equilibrium, therefore, has to be located on the LRLC. However, it will not be located at T_L, the tax rate that would maximize tax revenues in the long run, but at T_S, which is the point on the LRLC that will maximize tax revenues in the short run and is therefore located at the top of a different SRLC.

their behavior to the existing tax rate in the short run, it is always better for the government to increase its tax rate above the level that would yield maximum revenues in the long run. However, the government has to choose a tax rate located on the LRLC because it is the only way it can secure a tax base. If the tax rate were located beyond the curve, there would be no income to tax at all. Eventually, the chosen tax rate will be one that yields maximum revenues on an SRLC and that is simultaneously located somewhere on the decreasing portion of the LRLC. This situation, Buchanan and Lee argue, is stable because there is no incentive for the government to lower its taxes in the short run, which is the only planning horizon it is concerned with. Therefore, the nonoptimal situation is going to persist over time. Buchanan and Lee's contribution was particularly striking: unlike previous attempts to deepen the analytical implications of the Laffer curve, their version of the latter involved no other mathematical formalism than its diagrammatical representation. The reason for this is that the authors, unlike Fullerton (1982), were not interested in questioning the shape of the curve in accordance with real-world data. Instead, they focused on the mere possibility that the curve exists. By doing so, they could address a question of political theory: how do rational governments end up acting against their own interests? In this sense, the Laffer curve appeared not as a model in need of refinement, but as an engine of discovery.[16]

Even though they took the Laffer curve in two different directions, Fullerton's and Buchanan and Lee's respective contributions participated in its legitimation as a scholarly tool by providing a respectable starting point for further research.[17]

4   The Generalization of Laffer Curves

It is striking that precisely at the moment that the Laffer curve was assaulted as a political project, it began to attract increasing attention from academic economists. By 1986, though new tax cuts were in preparation, the Laffer curve had mostly disappeared from public debate.[18] In 1984, Herbert Stein, Nixon's former economic advisor, had summarized the general feeling: "Between 1981 and 1983, the country moved from a flush of enthusiasm for tax reduction to a sad recognition that taxes were too low—that we are . . . an undertaxed society" (Stein 1984, 356). In Stein's book, the Laffer curve was represented as a sharply asymmetrical bell-like curve with the shift from the normal into the prohibitive range occurring approximately at an 80 percent tax rate. Stein noted: "The conventional picture in which the curve is symmetrical and reaches its high point at 50 percent has only an aesthetic justification" (ibid., 247). Yet the same symmetrical Laffer curve was subsequently dispersed through the scientific literature along three different lines of development: first, as a basis for further econometric hypotheses; second, for deeper explorations of its underlying analytics; and finally, for applications beyond public finance.

In the first line, for developing testable hypotheses, the Laffer curve did not assume much significance. In fact, the difficulty in making empirical estimations from it was that there were too many variables that had to be simplified or ignored: changes in the income distribution, the complexity of tax structures, the way tax revenues were redistributed among taxpayers, etc. Consequently, empirical estimations of the peak varied from one study to another, covering a range between 40 and 90 percent. Moreover, the Laffer curve model provided no clear basis for comparing such estimates with one another. One solution was to move from testing the Laffer curve to treating the real-life consequences of particular tax cuts as experimental findings. This is what Austan Goolsbee, an economist at the University of Chicago, did in 1999. Studying various tax changes in US history, Goolsbee argued that no particular effect on revenues could be identified that would inform policy. He concluded that "[t]he notion that governments could raise more money by cutting rates is, indeed, a glorious idea. . . . Unfortunately for all of us, the data from the historical record suggest that it is unlikely to be true at anything like today's marginal tax rates" (Goolsbee 1999, 44). What is interesting here for our purpose is that Goolsbee, though he referred throughout his article to the Laffer curve, did not really test the *curve*, but only the *idea* that higher marginal tax rates may reduce government revenues. Goolsbee's target in fact was Martin Feldstein, an economist with more solid academic credentials than Laffer. Feldstein had provided theoretical justifications for tax reduction as early as the mid-1970s and had been involved in tax reform during Reagan's second term, precisely when the supply-siders had lost their influence in the government. Goolsbee (1999, 9) distinguished Feldstein's academic work from the "conventional Laffer curve," which he said "does not exist." He cited no article by Laffer, did not reproduce the famous curve anywhere in his paper, and made no reference to the, by then, almost universally derided version that Wanniski and other supply-siders had offered in the late 1970s. In a discussion and commentary section published with the article, Robert E. Hall began by noting that Laffer should not have been mentioned in the title at all (in Goolsbee 1999, 48). What Goolsbee's paper signaled, however, was that by the late 1990s the term "Laffer curve" had become somewhat decoupled from its origins to serve as a trademark for the academically respectable argument that tax cuts, by increasing labor supply, would help increase both national and governmental revenues.

The second type of academic study tried to explore the deeper analytical complexities of the Laffer curve. Instead of trying to estimate where the peak of the curve would occur, these studies showed an increasing interest in the question of whether a switching point *ever* occurs, and therefore questioned the shape of the curve itself. Malcomson (1986), for instance, studied the possibility that the curve would slope upward for all average tax rates. This discussion was pursued in the pages of the *Journal of Public Economics* between 1986 and 1991. Gahvari (1989) proved that in Malcomson's framework, a sufficient condition for the curve to have a downward-sloping portion is for the tax revenue to be redistributed to taxpayers in the form of a lump sum. Guesnerie and Jerison (1991) attempted to generalize the Laffer curve by suggesting that it could be replaced by an infinity of curves corresponding to various tax revenue functions in a general-equilibrium framework. While their paper said that it was impossible to draw policy conclusions, because it was impossible to predict whether a decrease in tax rates would yield higher tax revenues, their major contribution was to broaden the relevance of the Laffer curve. While in the original Laffer model the focus had been exclusively on tax revenues, Guesnerie and Jerison argued that a general-equilibrium framework should concentrate on social welfare as well. Because tax revenues are used to finance public goods, higher tax rates will result in two conflicting effects: first, higher rates will increase social welfare, because people will benefit from the additional public goods; but, second, the resultant rise in the prices of private goods will have a detrimental effect on social welfare. This leads to multiple possible equilibria which would yield maximum tax revenues, but it will be virtually impossible to know which rates would actually maximize social welfare. By seeking generalization and analytical rigor, these models rely on increasingly complex formalisms. While they undoubtedly demonstrated that the textbook version of the Laffer curve was a special case of a more general relationship with very little chance to occur in real life, they also helped substantiate the claim that the curve could exist in theory. In consequence, by the late 1990s one was very likely to come across standard Laffer curves in the mainstream literature.

A third aspect of the use of the Laffer curve in recent economics literature was derived from the Buchanan-Lee model of 1982, which offered the most easily transposable framework for studying practical cases. One significant example is an article by Clark and Lee (1996), which invokes the Laffer curve in a study of criminal sentencing policy. Their article plotted the "sentencing Laffer curve" for each average sentence length that required prison space (figure 13.6). According to their model, a reduction in average sentence length, which should normally reduce the need for prison space, could lead to the opposite effect, as crime rates rise in response to greater incentives to violate the law. Using an analysis quite similar to Buchanan and Lee's (1982), Clark and Lee showed that the government would likely choose an inefficiently low average sentence length that would not minimize the need for prison space. As in the Buchanan-Lee model, this suboptimal situation might persist, because in the short term the government would have no interest in extending sentence lengths.

Shmanske (2002) provided another application of the Buchanan-Lee framework, this time in a study of the economics of education. Shmanske's starting point was the notion that some schools might lighten the load of their curriculum in the expectation that this would achieve a higher level of enrollments. Beyond a certain point, however, they would end up with insufficient enrollments because of the degradation of the school's reputation. Moreover, such degradation would prevent the school from

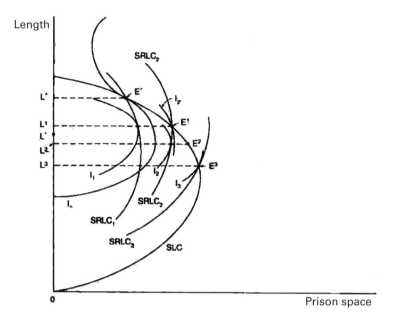

Figure 13.6

Clark and Lee's (1996, 252) "sentencing curve" is an application of the analytics of the Laffer curve to the economic analysis of crime. The resulting diagram resembles that of Buchanan and Lee (1982, figure 6), but it reads differently. The sentencing Laffer curve (SLC) depicts a long-run relationship between the average sentence length and the required prison space. Along the decreasing (upper) portion of the SLC, higher sentence terms deter criminals and save prison space. Along the increasing (lower) part of the SLC, the decrease in prison space is due solely to the government's indulgence while crime rates are high. Social preferences are represented by the indifference curves I: citizens tolerate long sentence terms only to the extent that they effectively reduce the crime rate, hence the optimal outcome E^*. Yet, as in Buchanan and Lee, the government acts according to the short-run Laffer curve (SRLC), in which criminals do not fully adjust their behavior to the existing average prison length. This pushes the government to shorten the average sentence length, which will decrease the need for prison space in the short run but will encourage crimes in the long run. The final equilibrium on the SLC will thus move to the less favorable outcome E_3.

toughening its curriculum in the future, because it would not be credible enough to do so. Such metaphorical applications of the Laffer curve beyond the domain of public finance began to appear in academic publications in the mid-1990s. In the popular press, too, there had been indications of such metaphorical use since the late 1980s. In the *New York Times*, Democratic politician Ed Markey (1988) referred to a "thermo-nuclear version of the Laffer curve," which he explained as the paradoxical notion that "[u]nder Reagatomics, we now build more nuclear weapons and improve our first-strike capabilities in order to have fewer nuclear weapons and reduce the threat of a disarming nuclear first strike." The use of the curve outside the traditional boundaries of economics seemed to meet a larger social demand for simple economic models with practical implications for real-life situations.[19] In this context, the Laffer curve became a widespread metaphor to explain the wrongdoings of a supposedly rational government.

5   Conclusion

John Maynard Keynes once remarked that diagrams in economics make up part of "that elegant apparatus which generally exercises a powerful attraction on clever beginners, which all of us use as an inspirer of, and a check on our intuitions and as a shorthand record of our results, but which generally falls into the background as we penetrate further into the recesses of the subject" (Keynes 1924, 332–333). Recalling the Laffer curve story, we can partly agree with Keynes's remark: the Laffer curve exerted fascination upon "clever beginners"—in this case, mostly noneconomists—and served as an "inspirer" for more elaborate economic models. It is, however, more difficult to state that the original curve fell into the background as models became increasingly technical, because its canonical representation seems to have been reinforced in the process. While the preceding narrative has explained *how* this has happened over the past 30 years, it is time to reflect on the reasons for the curve's persistent circulation. Two lines of interpretation can be put forward, one relying on the status of the Laffer curve as a visual representation located at the intersection between research and politics, and the other involving the specific institutional context surrounding its circulation.

This brief history of the Laffer curve shows that it was initially conceived as a simple representation of a reality that was supposed to be complex: a reality that involved a complicated tax structure and as many different behaviors as individual taxpayers could adopt in response to changes in tax rates, not to mention the cumulative effects that such behaviors could generate at the macroeconomic level. Yet economists were in no position to criticize the pretension of a simple curve to depict such intricate relationships, because in the recent past they had used several equally simple curves to depict the functioning of the economy as a whole. Why are such simple visual representations so persistent? The answer is that while the subsequent mathematization of such simple curves may increase their credibility in the eyes of more theoretically inclined researchers, it also makes them increasingly less appealing for public policy. In the case of the Laffer curve, its simplicity helps explain why the most mathematically sophisticated versions of the curve have not circulated widely, even among economists.[20] Therefore, one general conclusion that can be drawn from this narrative is that economic representations that claim strong policy relevance and accordingly travel across the boundary between academic research and politics are likely to persist over time despite, and perhaps even because of, their failure to hold up in the process of technical refinement.

Still, the academization of the Laffer curve must also be understood in light of the broader sociological and institutional context over the past century that shaped the role of economists in policymaking in the United States. Fourcade (2009) has argued that what was specific about American economics was its academic entrenchment, in contrast to the more bureaucratic kind of economic expertise that prevailed in continental Europe. From the 1930s onward, in the absence of a preexisting technocratic elite, American governments increasingly relied on academic economists to perform various technical tasks. Economists' legitimacy as policy experts turned on expectations of their solid academic standards, but also on their ability to penetrate the marketplace and sell their skills through public and private nonprofit organizations that had no definite political agendas, such as the Ford Foundation or the Brookings Institution. Hence the relation of economists to politics was often indirect and ambiguous. The dispersion of the Laffer curve(s) took place in the latter part of the twentieth century, when this model of economic expertise faced a crisis, as a new kind of crudely ideological think tank tried to compete with academic economists in the market for technical expertise. When some of these think tanks, like the Cato Institute with which Laffer and his allies were involved, began to influence politics, economists' relationship to politics became even more ambiguous. While some academically respectable economists were interested in this new opportunity to spread their ideas, others estimated that they should further elevate the scientific standards of their discipline to provide a bulwark against what they considered a threat to their legitimacy. These conflicting attitudes may explain the disdain that prevailed among academic economists in the early reception of the Laffer curve, as well as its later appropriation as an acceptable research object. While, over the past two decades or so, sociologists of science have sought to show that the study of representations in scientific practice must pay attention to the social contexts in which these representations are produced, the story of the Laffer curve shows that, for particular representations, the notion of context would have to be extended to include the larger cultural, societal, and institutional aspects that frame and legitimize scientific discourse.

Acknowledgments

I would like to thank Béatrice Cherrier and José Edwards for helping me locate some of the early papers on the Laffer curve, Roger Middleton for sharing unpublished drafts, and Fabian Gouret for helpful comments. Also, I benefited from insightful remarks, corrections, and additional references provided by three anonymous referees and by the editors of this volume. The usual caveats apply.

Notes

1. This, apparently, is not peculiar to the Laffer curve or to economic representations in general, as Lynch (1991) has argued that the same visual rhetoric is often used in social theory.

2. See Latour and Woolgar (1979), Knorr-Cetina (1981), Lynch and Woolgar (1990), and Pickering (1995). More precisely, the critique of the "diffusion model" is elaborated in Latour (1987, 132–144).

3. The persistence of the Laffer curve is shown in the fact that it is mentioned in most economics textbooks but also in many dictionaries and encyclopedic volumes for professional economists. See for instance Fullerton (2008) and Middleton (2010). In addition, it is still used or referred to in works published in top-ranking journals (e.g., Laroque 2005).

4. It must be emphasized, as will be further developed in this chapter, that several of these individuals belonged to more than one community, with some journalists acting as political propagandists and some economists serving as policy advisors.

5. The claim has been made that the idea behind the Laffer curve can be located in many past traditions in the economic literature, ranging from preclassical economists to John Maynard Keynes (Fullerton 1982; Laffer 2004). This paper, however, will focus solely on the contributions that followed from Laffer's visual representation and not on those that may have served as an inspiration for it.

6. A *New York Times* article, published in 1986, described him as an "eccentric" who "[did] not write much in scholarly journals, although he frequently did so in the past" (Bennett 1986).

7. On Laffer's reputation as an economist, see Jensen (1978) and Kristof (1986).

8. While it is sometimes claimed that Donald Rumsfeld and Dick Cheney were the other participants, Laffer (2004) admits that he has no memory of this meeting.

9. See Goodman (1981). Kristol also happened to own *The Public Interest*, a public policy journal in which the chapter of *The Way the World Works* that introduced the Laffer curve had already been published.

10. Friedman was one exception, as he explicitly expressed his devotion to Marshall's framework (Friedman 1949).

11. There was one review in the *Journal of Economic History* and it was only modestly critical of Wanniski's position, even stating that "it is hard to be harsh on a first effort which appears to develop a case not too different from the results of cumulative scholarship" (Gunderson 1978).

12. On the role of economists in securing NSF funding, see Mata and Scheiding (2012).

13. The same visual representation was used years later by Daniel Dennett (1991, 109).

14. For an account of the mathematization of economics, see Weintraub (2002) and Blaug (2003).

15. Interestingly, Buchanan and Lee do not mention the previous literature anywhere in their paper. The latter is written in such a way that it remains ambiguous whether the authors are addressing the academic or the larger political debate over the curve.

16. The idea of a visual representation as an "engine of discovery" is reminiscent of economics in late nineteenth-century Britain, where, under the influence of William Whewell's philosophy of science, Marshall and his disciples at Cambridge considered geometrical analysis to be a suitable tool for building economic theory (Klein 1995). In their conception, supply and demand curves were not pedagogical simplifications used to teach the mathematically illiterate, but should serve as serious means of demonstration using projective geometry. This view was challenged in the middle of the twentieth century as the discipline moved further toward mathematization, but the method has remained influential in the domain of economic education (Giraud 2010).

17. A search in the Social Science Citation Index reveals that Fullerton (1982) has been cited 61 times from 1982 to 2009, while Buchanan and Lee (1982) has been cited 33 times. By contrast, Laffer (1981) has been cited only 12 times. Other contributions such as Canto, Joines, and Webb (1979) and Canto, Joines, and Laffer (1981) are not indexed.

18. Ironically, in the same year Laffer ran unsuccessfully for the Republican nomination for a US Senate seat in California.

19. On the question of "relevance" in recent economics, see Fleury (2012).

20. It is meaningful that one recent instantiation of the Laffer curve in a top economics journal (Laroque 2005) references Buchanan and Lee (1982) but does not mention the mathematically refined models and numerous econometric estimations that were subsequently published.

References

Bennett, Robert A. 1986. "Eccentric economist: Robert A. Mundell; Supply-side's economic intellectual guru." *New York Times* (12 January).

Blaug, Mark. 2003. The formalist revolution of the 1950s. In *A Companion to the History of Economic Thought*, ed. Warren J. Samuels, Jeff E. Biddle, and John B. Davies, 395–410. Oxford: Blackwell.

Buchanan, James M., and Dwight R. Lee. 1982. Politics, time and the Laffer curve. *Journal of Political Economy* 90 (4):816–819.

Canto, Victor A., Douglas H. Joines, and Arthur B. Laffer. 1981. Tax rates, factor employment, and market production. In *The Supply-Side Effects of Economic Policy*, ed. L. H. Meyer, 3–32. St. Louis and Washington: Center for the Study of American Business.

Canto, Victor A., Douglas H. Joines, and Robert L. Webb. 1979. Empirical evidence on the effects of tax rates on economic activity. *Proceedings of the Business and Economic Statistics Section*, 30–40. Washington: American Statistical Association.

Clark, J. R. and Dwight R. Lee. 1996. Sentencing Laffer curves, political myopia and prison space. *Social Science Quarterly* 77 (2):245–255.

Dennett, Daniel. 1991. *Consciousness Explained*. Boston: Little, Brown.

Fleury, Jean-Baptiste. 2012. The evolving notion of relevance: An historical perspective to the "economics made fun" movement. *Journal of Economic Methodology* 19 (3):303–319.

Fourcade, Marion. 2009. *Economists and Societies: Discipline and Profession in the United States, Britain and France, 1890s to 1990s*. Princeton: Princeton University Press.

Friedman, Milton. 1949. The Marshallian demand curve. *Journal of Political Economy* 57 (6):463–495.

Fullerton, Don. 1982. On the possibility of an inverse relationship between tax rates and government revenues. *Journal of Public Economics* 19: 3–22.

Fullerton, Don. 2008. Laffer curve. In *The New Palgrave Dictionary of Economics*, 2nd ed., ed. Stephen N. Durlauf and Lawrence E. Blume IV, 839–841. London: Palgrave.

Gahvari, Firouz. 1989. The nature of government expenditure and the shape of the Laffer curve. *Journal of Public Economics* 40:251–260.

Gardner, M. 1981. The Laffer curve and other laughs in current economics. *Scientific American* 245:18–31.

Giraud, Yann. 2010. The changing place of visual representation in economics: Paul Samuelson between principle and strategy, 1941–1955. *Journal of the History of Economic Thought* 32 (2):175–197.

Goodman, Walter. 1981. Irving Kristol: Patron saint of the new right. *New York Times* (6 December).

Goolsbee, Austan. 1999. Evidence on the high-income Laffer curve from six decades of tax reform. *Brookings Papers on Economic Activity* 30 (2):1–64.

Guesnerie, Roger, and Michael Jerison. 1991. Taxation as a social choice problem: The scope of the Laffer argument. *Journal of Public Economics* 44 (1):37–63.

Gunderson, Gerald. 1978. Review of *The Way The World Works*. *Journal of Economic History* 38 (4):1059–1060.

Jensen, Michael C. 1978. The principles of a tax-cut guru. *New York Times* (17 June).

Kaiser, David. 2005. *Drawing Theories Apart: The Dispersion of Feynman Diagrams in Postwar Physics*. Chicago: University of Chicago Press.

Keynes, John Maynard. 1924. Alfred Marshall, 1842–1924. *Economic Journal* 34 (135): 310–372.

Klein, Judy L. 1995. The method of diagrams and the black arts of inductive economics. In *Measurement, Quantification and Economic Analysis*, ed. Rima Ingrid, 98–137. London: Routledge.

Knorr-Cetina, Karin. 1981. *The Manufacture of Knowledge*. Oxford: Pergamon.

Kristof, Nicholas D. 1986. Stalking a Senate seat: Arthur B. Laffer; professor takes a new course, again. *New York Times* (2 February).

Laffer, Arthur B. 1981. Government exactions and revenue deficiencies. *Cato Journal* 1 (1): 1–21.

Laffer, Arthur B. 2004. The Laffer curve: Past, present and future. *Backgrounder* (Washington, D.C.) 1765:1–16. http://www.heritage.org/research/taxes/upload/64214_1.pdf, accessed 3 January 2011.

Laroque, Guy. 2005. Income maintenance and labor force participation. *Econometrica* 73 (2):341–376.

Latour, Bruno. 1987. *Science in Action: How to Follow Scientists and Engineers through Society*. Cambridge, MA: Harvard University Press.

Latour, Bruno, and Steve Woolgar. 1979. *Laboratory Life: The Social Construction of Scientific Facts*. London: Sage.

Lynch, Michael. 1991. Pictures of nothing? Visual construals in social theory. *Sociological Theory* 9 (1):1–21.

Lynch, Michael, and Steve Woolgar, eds. 1990. *Representation in Scientific Practice*. Cambridge, MA: MIT Press.

Malcomson, James M. 1986. Some analytics of the Laffer curve. *Journal of Public Economics* 29:263–279.

Markey, Edward J. 1988. "Reagatomics." *New York Times* (7 November).

Mata, Tiago, and Tom Scheiding. 2012. National Science Foundation patronage of social science, 1970s and 1980s: Congressional scrutiny, advocacy network, and the prestige of economics. *Minerva* 50 (4):423–449.

Middleton, Roger. 2010. The Laffer curve. In *Famous Figures and Diagrams in Economics*, ed. Mark Blaug and Peter Lloyd, 412–418. Cheltenham, UK: Edward Elgar.

Pickering, Andrew. 1995. *The Mangle of Practice: Time, Agency and Science*. Chicago: University of Chicago Press.

Rattner, Steven. 1978. Washington watch: Trimming cost of regulation. *New York Times* (27 November).

Shmanske, Stephen. 2002. Enrollment and curriculum: A Laffer curve analysis. *Journal of Economic Education* 33 (1):73–82.

Sloman, John. 2006. *Economics*. 6th ed. Harlow, UK: Prentice Hall.

Stein, Herbert. 1984. *Presidential Economics: The Making of Economic Policy from Roosevelt to Reagan and Beyond*. New York: Simon and Schuster.

Wanniski, Jude. 1978. *The Way the World Works*. New York: Simon and Schuster.

Wanniski, Jude. 1981. The burden of Friedman's monetarism. *New York Times* (26 July).

Weintraub, E. Roy. 2002. *How Economics Became a Mathematical Science*. Durham: Duke University Press.

Wise, M. Norton. 2006. Making visible. *Isis* 97 (1):75–82.

14   How (Not) to Do Things with Brain Images

Joseph Dumit

An article in the 30 July 2004 issue of *Science* (Beckman 2004) discussed a case that was scheduled to come before the US Supreme Court, which concerned whether a 17-year-old convicted murderer named Christopher Simmons should be eligible to receive the death penalty (*Roper v. Simmons*, 543 U.S. 551, 2005). Specifically, the question was whether adolescents are protected from capital punishment under the Eighth Amendment prohibition of cruel and unusual punishment, due to their *relative* incapacity for decision making compared with adults.[1] The argument put forward by the defense was that adolescence was a distinct period of human life characterized by risky, immature behavior. Drawing on neuroscience findings, among other sources of evidence, the defendants sought to have a "bright line" drawn by the court declaring that no one under 18 could be determined beyond a reasonable doubt to be mature enough be fully culpable. Citing precedent from a previous Supreme Court decision in which the mentally retarded were determined to be not fully culpable, the case was made for similar standing.

The article appeared in *Science* because a number of *amicus curiae* briefs (solicited and unsolicited "friend of the court" briefs providing the court with information) were submitted in this case from professional groups including the American Psychiatric Association, American Psychological Association, and American Bar Association, which argued that recent neuroscience findings demonstrated that teens have immature brains and were therefore less culpable than adults. One self-described "science brief" explicitly argues that neuroscience and the visual power of brain imaging are decisive:

Adolescents' behavioral immaturity mirrors the anatomical immaturity of their brains. To a degree never before understood, scientists can now demonstrate that adolescents are immature not only to the observer's naked eye, but in the very fibers of their brains. (AMA et al. 2004, 10)[2]

Adolescents are strongly asserted here to be a category of people who are socially recognizable and recognizably immature. The brief also asserts that the category has biological meaning since its members' immaturity can be objectively visualized and measured. The flourish of "in the very fibers of their brains" rather than "in their brains," where "fibers" has no precise meaning, suggests a deeply polemical stance. The implicit claim being made is that what is obvious to the eye about adolescents is insufficient to determine that they are truly immature: neuroscience is needed to document it. Concealed in this appeal to biological explanations is a categorical slippage from a stereotypical claim about the immaturity of adolescents in general to a neurological claim about the immaturity of all adolescents.

Are adolescents immature? Is this even a proper question? On one hand, this age grouping is socially defined. Societies like the US draw lines that define persons under 21 as unable to drink legally, those under 16 as unable to drive cars legally, those under 18 as unable to vote, and so on. Each of these lines is justified through a combination of tradition, ideas of social immaturity, and a conventional notion that we have to draw a line somewhere. Each of these lines in fact defines immaturity socially by specifying a legal age limit for various acts. On the other hand, asking whether all adolescents are immature makes immaturity into an essential and biological claim, and places the problem of defining immaturity beyond social convention and legal stipulation. Many of the neuroscience studies brought to bear in the *amicus* briefs were used to claim both that adolescents are behaviorally immature and poor risk-takers, and that there is brain evidence to confirm this.

But the claim that adolescents are immature because they are prone to take behavioral risks is both a social truism and not clear at all. One could easily point to any number of adults who take terrible and stupid risks, and to teenagers who are exceptionally careful, cautious, and "mature." Indeed, the recent mortgage and banking crises, the very existence of Las Vegas, and the strategy of multilevel marketing could be used to suggest that adults take more risks than teenagers. This suggests that trying to combine age, social truisms, and neuroscience is a very tricky proposition that needs careful evaluation of the ways in which evidence is generated and promoted.

Most fascinating about the *Roper* case, and the ongoing discussion of whether neuroscience evidence can demonstrate adolescents' immaturity, is the seemingly obvious equation of neuro-riskiness with lesser culpability. Less than three decades ago, there was a focus on a very different type of adolescent, and a very different type of brain: dangerous ones. Legal analysts point to a trend in legislation and punishments in which youthful violence was seen as a symptom of intractable dangerousness in those individuals. This resulted in harsh sentences against youth in federal and many state courts, "adult time for adult crime" (Maroney 2009; Scott and Steinberg 2008). At that time, and continuing today, research into the neuroscientific signs and roots of dangerousness produced a large body of evidence claiming that someday dangerousness would be visualized and predicted through brain imaging (e.g., Raine and Yang 2006).

The current legal trend (post-2004) toward both lesser penalties and the use of neuroscience evidence is a response to the previous era. For two neuroscientists leading this change, "new scientific knowledge . . . provides the building blocks for a new legal regime. . . . They put adolescent offenders into an intermediate legal category—neither children, as they were seen in the early juvenile court era, nor adults, as they often are seen today" (Scott and Steinberg 2008, 16). They argue, in other words, for a new category of person, the neuro-adolescent, specifically in response to an overly harsh legal regime that was eliminating adolescence as a legal category altogether. Both in courts and in neuroscience, then, there is a fascinating and very important problem to unravel: a conflict between two ways of seeing adolescent persons and brains, one defined in terms of pre-maturity (adolescent immaturity), and the other in terms of an existing trait (dangerousness). As we will see, this is an ongoing struggle, and during the *Roper* trial the prosecutor sought to make the defendant Simmons more culpable and more dangerous *because* he was an adolescent.

The long history of how behavioral labels, such as mental retardation or adolescence or dangerousness, come to be equated with biological differences—hereditary, genetic, neurological—has been charted by historians of science and medicine. Scientific findings "have become a popular resource precisely because they conform to and complement existing cultural beliefs" about identity, age, gender, and family (Nelkin and Lindee 2004, 195). Michael Hagner has described the episteme of the modern brain as emerging in the 1800s, encompassing both a localizationist paradigm (the assumption that different mental functions and behavioral traits correspond to different regions of the brain) and brain-person types (the female brain, the criminal brain, brain of genius, pathological brain), what he calls "narratives of the brain." As he notes, "this is quite crucial for an understanding of modern brain research as a human science" (Hagner 2010, at 2 hr. 38 min.). While calling brain research a human science may seem obvious or trivial, it suggests that social and neurological categories depend on each other *in practice*, within both legal and scientific productions of knowledge.

At stake in these arguments in courtrooms, scientific journal articles, and popular sources are the statuses of categories such as adolescents and dangerous persons. Is the status the same across these categories, and should it be? Can neuroscience draw a bright line that shows us what adolescence means and what to do with adolescents? And if so, does this mean that neuroscience can provide a test for immaturity and culpability, and a test for dangerousness? In this chapter, I will interrogate these questions by examining how neuroscience claims are deployed in battles over adolescent immaturity in courtrooms, in popular media, and in some of the neuroscience studies themselves.

On one hand, the studies work, in the sense that they *do things*: they travel and produce effects in the different arenas in the world; they report findings about adolescent immaturity and provide clear pictures of it. They seem to conform precisely to the dream of drawing bright categorical lines between categories of people who are not fully culpable and others who are. Even when some of the neuroscientists explicitly warn that the studies are not yet ready for such application, lawyers and the media may ignore these warnings in using their texts and images. On the other hand, a close reading of the evidence in some of these brain studies of adolescence suggests that they are not able to answer the questions that law and society put to them. If anything, the studies seem to suggest that the brain is not a good location for resolving these questions. My argument is that these types of studies cannot be neutral arbiters of efforts to assign classes of people to either immaturity or dangerousness, because they are embedded within the very categories they are being asked to resolve.

1 Brains' Bright Line

BRIGHT LINE: A judicial rule that helps resolve ambiguous issues by setting a basic standard that clarifies the ambiguity and establishes a simple response.[3]

In court, a bright line is a shared and clear guide for everyone, one that does not admit of nuance (e.g., the age for statutory rape). Biological criteria have been invoked, with varying degrees of success and authority, in cases attempting to define personhood in embryos and stem cells, and the sex of a transsexual.[4] In death penalty cases, Aronson and Cole (2009) have argued that science, including DNA profiling, cannot provide the epistemic certainty that reformers want it to (see also Lynch et al. 2008). One of the questions before the Supreme Court in *Roper* was whether neuroscience could have a say in setting such a line around legal immaturity. On the one hand, there was an implicit assumption that immature people *must* have immature brains, and that determination of brain immaturity might have a bearing on our understanding of adolescent immaturity. On the other hand, people of various ages can be said to be immature, a different kind of thing from classifying them as immature because they are members of a specific age category. The notion of a bright line requires a logic, a form of explanation, that compels judgment. A law review article by historian of science Jay Aronson laid out the stakes for the court:

The use of scientific evidence in *Roper* was interesting because brain images were not used to make gross pathologies of the brain visible, which is how neuroscience has traditionally been invoked in the criminal justice system over the past two decades (for example, in the case of John Hinkley). Rather, the Simmons defense team sought to narrow the legal category of culpability by constructing a model of a normal, mature adult brain that was capable of supporting the functions of a reasonable man and contrasting that model with the developmental chaos of a teenager's brain. They sought to have both anatomical and cognitive normalcy and pathology defined by age rather than by some diagnosable medical condition or mental state. In other words, Simmons's legal team argued that, as a population, adolescents' brain structure and function have not yet matured to the level found in a normal population of adults. (Aronson 2009, 920)

The attempt was to argue that adolescents, defined by age, were specifically immature compared with adults, and therefore that invoking neuroscience was simply a way of helping to assign individuals to the correct legal category. This role is an ethical one for neuroscience: the question of whether neuroscience *should* draw the line for the courts misses the fact that we need to know whether or not neuroscience can speak to classifications of people, and whether it can hold up for each of the categories in a particular classification. When the Supreme Court made its decision in *Roper*, it decided in favor of the bright line for defining juveniles as being under 18. However, it neither accepted nor rejected neuroscience evidence (the court essentially argued that society draws arbitrary lines all the time and that this was a good case for such a line).[5]

The press coverage of *Roper*, however, focused on the *amicus curiae* brief and its strong "bright line" claims that drew on neuroscience studies of samples of people from different populations. In both the AMA *amicus* and the popular press, brain imaging in particular was invoked via what Kelly Joyce (2008, 50) has called a cultural narrative of transparency: "[t]he idea that nature can be known without human mediation and that scientists witness and thus reveal the natural world." The rhetorical force of the causal connection being made, however, depended on sliding between claims about individuals, groups, populations, and types of people.

Summary of Argument: *The adolescent's mind* works differently from ours. Parents know it. This Court has said it. Legislatures have presumed it for decades or more. *And now, new scientific evidence sheds light on the differences*. . . . Scientists have documented the differences along several dimensions. *Adolescents as a group*, even at the age of 16 or 17, are more impulsive than adults. They underestimate risks and overvalue short-term benefits. They are more susceptible to stress, more emotionally volatile, and less capable of controlling their emotions than adults. In short, *the average adolescent* cannot be expected to act with the same control or foresight as a mature adult. (AMA et al. 2004, 2, emphasis added)

What must be unpacked in these *amicus* statements are the constantly changing meanings or referents of "adolescents" and the signatory status of "new scientific evidence." The language shifts from "the adolescent's mind" to "adolescents as a group" to "the average adolescent." "Average" here slides between "most common" or "typical" and a statistical mean. If "the average adolescent" is all that "the adolescent mind" refers to (even ignoring the slide between mind and brain), then claims about such an adolescent have much less rhetorical force, since, viewed in terms of statistical means, any two groups that differ in their averages on almost any measure can have wide if not complete overlap in the range of that measure (e.g., Duster 2003; Marks 1995). Research in the field of science and technology studies has demonstrated this for the problem of categories in race, sex, genetics, and clinical trials (Duster 2003; Epstein 2007; Beaulieu 2000; Star 1989; Shim 2000). Steven Epstein notes that "[w]hen race is used as a variable in research, there is a tendency to assume that the results obtained are a manifestation of the biology of racial differences . . . this presupposition is seldom warranted" (2007, 207). Individual differences in these cases are stripped out, and the often massive overlap among individuals across groups is treated as noise to be reduced rather than as evidence of the weakness of the initial use of the social category as a variable, or even as a proxy.

Comparing the use of race or gender to the use of adolescence in the context of *Roper* emphasizes the danger of reifying a neuroscientific difference on the basis of categories. In an excellent review of the status of neuroscience in courtrooms, Terry Maroney found that, in general, the courts agree with this critique by recognizing that "imaging studies show group trends . . . [and] do not show that all individuals in the group perfectly reflect the trend" (Maroney 2009, 225). Together with the fact that there is no specific measure (no biomarker) for a mature brain, nor reliable predictions for future maturity, "[c]ourts thus have a strong basis for deeming brain science irrelevant to many highly individualized claims, such as whether a defendant was able to form specific intent" (228).

There are two components of this analysis. The first is that at present there is no clear neuroscientific criterion with which to replace or support judicial determination of maturity or decision-making capacity. The second is that there might someday be such a measure. If we imagine we have such a measure, however, it would call into question the precise association of immaturity with age. A brain measure would instead *replace* the age measure, and perhaps even a behavioral assessment, and we might discover that some middle-aged men are immature and some preteens are mature, *regardless of their actual behavior*. It is important to pay attention to this issue in the design of studies of adolescents, because it shows the conundrum that the neurosciences are in when they address social categories of persons and attempt to put forward scientific claims about a category.

Among other problems, Maroney stated that if one were to follow such logic further and ask about adolescent maturation by gender, there would have to be two different age lines, one for each gender. "If structural brain maturity were the correct metric, it would counsel that boys and girls become subject to juvenile court jurisdiction, and age out of it, at different times; indeed, one testifying expert has conceded as much" (Maroney 2009, 69). This problem is recursive. If brain maturity by group is materially relevant, then any subgroup would be subject to the same examination, including groups defined by race, geographic birthplace, religion, sexuality, etc. There is no a priori reason why any group could not be tested for their brain maturity.

At this point we should realize that if brain maturity can be measured and shown to correlate with age, however strongly, and if the age group should be treated differently because of such brain immaturity, then in fact brain maturity and not age should be the reason to treat members of that group differently. The relevant group in this case would not be those whose ages fall into the range correlated with brain immaturity, but those whose individual brains exhibit the measured brain immaturity itself, regardless of age. Adjacent to dangerousness, expert medico-legal opinion creates here the "risky adolescent" as a different "third term" between delinquency and illness.

> Expert medico-legal opinion offers in fact a third term, that is to say, I want to show that probably it does not derive from a power that is either judicial or medical, but from a different type of power that for the moment I will provisionally call the power of normalization. (Foucault 2003, 42)

Brain imaging has in fact been proposed as a test of maturity, to "'prove' whether a child is mature enough to be tried for a crime" (Firth 2010).

2 Neurorealism: Neuroscience to the Rescue

In the passage quoted above (*And now, new scientific evidence sheds light on the differences*), "new scientific evidence" serves to document what we know to be true—socially, legally, and scientifically.[6] Maroney, in his review, describes the problem as one of "common sense" claims that "purport to rest on a different empirical basis—that of neuroscience—and to result in more unshakeable conclusions, as a biological basis for immaturity ostensibly shows immaturity to be more deeply rooted and involuntary than does a psychological basis" (Maroney 2009, 117). Similarly, science reporter Sharon Begley claims that reporters love "when science can confirm what your grandmother told you" (Begley 2010).

One reading of this commonsense and yet bizarre desire to know that the brain is responsible is through what neuroscientists have analyzed as the "neuro-fallacy" of "neuro-realism," a cultural Cartesianism in which behavioral—social, cultural, or psychological—claims are seen as proven only when there is evidence of a neurological activity corresponding to them:

> neuro-realism reflects the uncritical way in which an fMRI investigation can be taken as validation or invalidation of our ordinary view of the world. Neuro-realism is, therefore, grounded in the belief that fMRI enables us to capture a "visual proof" of brain activity, despite the enormous complexities of data acquisition and image processing. (Racine et al. 2005, 160)

Brain evidence, that is, can "prove the behavior is real," effectively taking over the authority of the claims of the behavioral observation.[7] The critique of neurorealism tends to locate it at the level of the lay public belief: "how coverage of fMRI investigations can make a phenomenon uncritically real, objective or effective in the eyes of the public" (ibid.). The critics attribute it to improper reading of science taking place in the context of a "Cartesian culture," which is reinforced through runaway journalism (mostly the normal way in which science journalism covers stories).

But media-promoted neurorealism is not a full explanation for the appeal of using neuroimaging in legal contexts. Despite a general biological determinism present in the common belief that adolescents are different from adults, the demand for a more "deeply rooted" explanation reveals a gap within biology. Neither genes nor hormones are enough to account for cognitive behaviors. The demand goes beyond a neurorealist claim that something is more real when there are neural correlates. In the briefs and testimony to the court, neuroscience appears to become an obligatory passage point: without brain verification, a phenomenon is less real, despite being obvious. With brain-based evidence, political contests over social and psychological processes are claimed to be settled by science (cf. Nelkin and Tancredi 1989).

Begley (2010) describes an extreme example of this with another brain-person type, the addict:

For years and years, Alan Leshner [as head of the National Institute of Drug Abuse] would go before congressional committees and explain that addiction was biological disease, and they wouldn't believe him—[they insisted] that it was will, a moral failing, lack of self-control, etc. It [addiction] lacked the reality that cardiovascular disease had. So he brought with him fMRIs. He was sitting at witness table . . . this time he showed the pretty pictures we have all seen and he said: Here you see the extra activity when the coke addict walks past the corner where he got his fix . . . Here in the reward circuitry of the brain you can see that subsequent hits of coke . . . do not produce the same amount of reward . . . Here you can see this part of the brain that drives us to goal seeking behavior. So he was really just saying that addiction leads to tolerance and is cued.

Suddenly the scales fell off the congressmen's eyes, now they saw addiction is a brain disease. It is real, it is biological, and it is not the addicts' fault. It is the addicts' brain center's fault. Again a very Cartesian dualistic way of looking at it. But if you believe in doing research to help people who are addicted, that is a good thing. And NIDA should get its fair share of taxpayer money, this is a good thing from the use of fMRIs.

Begley is right that many people today see neuroscience as telling them something more about a social truth, especially when it verifies an intuition they already have. But if we think back to the analogy between criminal brains and adolescent brains, we might notice another aspect of Begley's description. Neuroscience is positioned carefully to reinforce one type of brain narrative—that addiction is biological and therefore not the addicts' fault. At other times and places, attributing addiction to an addict's biology reinforces a different narrative—that the addict cannot be helped and needs to be incarcerated.

In *Picturing Personhood* (Dumit 2004), I followed the use by the National Institute of Drug Abuse (NIDA) and Leshner of a PET brain imaging study of MDMA (ecstasy) users. The narrative changed step by step. The original *Lancet* study by McCann et al. showed PET scans of two individuals (a user and nonuser) as representative of a statistical correlation of a neural marker in their brains. A NIDA newsletter reprinted the images and changed this correlative claim to one in which "you can see the damage done to the brain of an ecstasy user." Finally Leshner, speaking before Congress, stated of the same images, "you can see that the brain on the top belongs to someone who has never used ecstasy." No study in fact has ever implied that one can diagnose from brain scans either ecstasy use or damage or a drug-free life. Nonetheless, brain images juxtaposed with captions made them exemplary of categories, enabling such clear but false readings to go unremarked.

Leshner and NIDA, in other words, used neuroscience in order to produce a *particular kind of neurorealist narrative*, and they were quite willing to exaggerate or even make up claims about the brain and brain-person types in order to persuade congressional funding to flow in one direction rather than another. Begley concludes that if you believe in the work they are doing, then "this is a good thing from the use of fMRIs." She concedes, in other words, that brain imaging *can* and *should* be used for polemical purposes when those purposes are good. This polemical use of expertise recalls similar use of psychiatric evidence by psychiatrists in courts when making predictions of dangerousness (cf. Szasz 1988; Burns 2008). We will discuss this notion of the *ethical misuse of neuroscience* in the conclusion.

Returning to the AMA et al. *amicus curiae* brief, we can see a similar polemical use of evidence that reads the neuroscience literature as confirming claims that accord with a particular brain narrative—claims that the scientific papers themselves do not make and often argue against.

Contrary to what some laypersons may believe . . . the evidence reveals that these older adolescents do not have adult levels of judgment, impulse control, or ability to assess risks. Recent advances in brain-imaging technology confirm that the very regions of their brains involved in governing these behavior-control capacities are anatomically immature. (AMA et al. 2004, 4)

The evidence summarized to this point, based upon studies of normal adolescents, leads to the conclusion that normal adolescents cannot be expected to operate with the level of maturity, judgment, risk aversion, or impulse control of an adult. Adolescents cannot be expected to transcend their own psychological or biological capacities. (ibid., 20)

In these arguments claiming neuroscientific authority there is a continued slippage from "average adolescent" to "normal adolescents" to "the adolescent brain" to apparently "all adolescents": "Adolescents cannot be expected to transcend their own psychological or biological capacities."[8] The brief appears to be operating on three levels: (1) overtly—a particular brain narrative is being privileged over other ones; (2) culturally—neurorealism is being reinforced; (3) legally—neuroscience is being positioned as necessarily involved in adjudicating social issues as an obligatory passage point.

3   Scientific Overlap and Interpretation: Immature Brains, Immature People

One could safely assume that the visualizing trend in medicine was promoted by mass media eager to exploit the power of fascinating, authoritative images. But the opposite is equally true: doctors and hospitals, keen on public relations, recognized the enormous publicity value of intriguing bodily images. (van Dijck 2005, 10)

The papers being cited in the court cases and *amicus* briefs, however, reveal a more complicated relationship. Neuroscience publications are themselves dependent on social categories and are drawn into social adjudication. They therefore often participate in the polemical use of their own findings. They make use of the brain as a switch-point between social and technical categories, and adjudicate brain narratives without subjecting those narratives to scientific standards of rigor. Thus they also sometimes commit the fallacy of neurorealism.

One way we can understand how this happens is by analyzing how neuroscience journal articles, and neuroimaging articles in particular, begin from or build upon social categories of personhood like adolescence or criminality (or female or smart or gay, etc.). Visual images often erase or delete the technical and social work practices and assumptions needed to generate them (Star 1989; Joyce 2008). In so doing, they often show a strong tension between scatterplots, images, and claims. Scatterplot data often make it clear that there is very large variation within each of two groups being contrasted, and much overlap in variation between the groups, illustrating that, given a particular brain measure, there is no way to know which group a subject is in. At the same time, there are statistical trends in the data that suggest that there may be something interesting to investigate further. This type of work is best described as hypothesis-generating. Yet when images are made, they almost always reduce the scatterplot data to contrasting categories, rendered as absolutely opposed to each other visually. I have argued that this type of "extreme image" makes visual claims often at odds not only with the data but also with the textual claims made in the articles (Dumit 2004). In this section, I want to examine articles cited in the *amicus* briefs to see how the images together with the textual claims become polemical. My aim is not to single out these particular cases as problems; indeed it is a common practice within neuroscience to use images in this way, and state-of-the-art in terms of methods. Rather I want to use these examples to reflect on the role that neuroscience has come to play within courtrooms, and perhaps vice versa. One example of how categories are confused in the most careful of articles can be seen by returning to the striking passage in the AMA et al. *amicus curiae* brief:

Adolescents' behavioral immaturity mirrors the anatomical immaturity of their brains. To a degree never before understood, scientists can now demonstrate that adolescents are immature not only to the observer's naked eye, but in the very fibers of their brains. (AMA et al. 2004, 10)

The citation for this claim is Gogtay et al. (2004), and the pictures referred to are illustrations of brain maturation. The images from this paper appear in the *Science* coverage of the Roper case (Beckman 2004), visually reinforcing the point.

The design of the Gogtay visuals is elegant and careful. Rather than trying to generalize development from average groups at different ages (as other labs have), they tracked individuals longitudinally. "Thirteen healthy children for whom anatomic

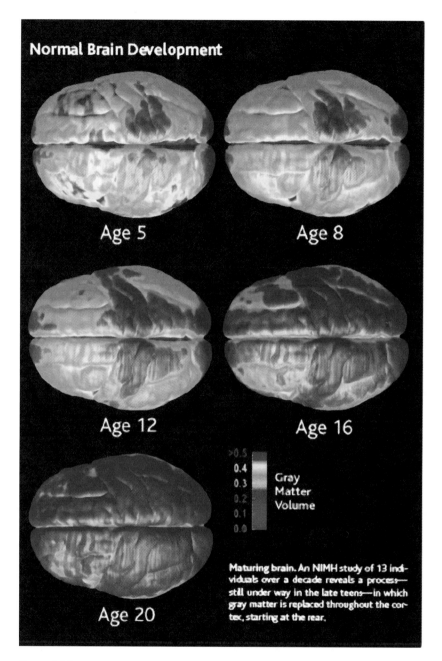

Normal Brain Development

Age 5

Age 8

Age 12

Age 16

>0.5
0.4
0.3
0.2
0.1
0.0

Gray
Matter
Volume

Age 20

Maturing brain. An NIMH study of 13 individuals over a decade reveals a process—
still under way in the late teens—in which
gray matter is replaced throughout the cortex, starting at the rear.

Figure 14.1

Normal brain development, as illustrated in Beckman (2004, 597; using images from the study described in Gogtay et al. 2004). The use of the biological process of brain tissue "maturation" in this composite illustration from an article in *Science* suggests that it should be read as an indication of social or cognitive maturity, though no measures of any kind of maturity other than actual age were made. The arbitrary color scale also implies a cessation of neural growth around age 20, though the brain continues to change throughout one's life. The colors in each brain represent average neural maturity at various points, a claim that figures 14.2 and 14.3 call into question.

brain MRI scans were obtained every 2 years, for 8–10 years, were studied." At the same time, the group was averaged together. The paper is carefully put together and makes clear that it is hypothesis-generating, given that the researchers were only able to study 13 children. Together, the MRI scans were analyzed for changes in gray matter density and thickness over time. The result is a map of "maturation." The term "maturation" is used technically to refer to "brain maturation" or "gray matter maturation," bio-mathematically defined as the cessation of further cortical growth during aging in an area. All 61 uses of "mature" and "maturation" in the paper are used in this technical way. Those areas that did not grow during the age range studied were declared to have "matured early."

Never does the paper use "mature" to refer to socially defined immaturity or maturity, nor does it use "immature" at all. Of course the connotations of the words "mature" and "maturation," and their fluid cycling between social and biological uses, is hard to ignore, illustrating "the enormous heuristic value that universalist claims have in guiding scientific research" (Hornstein and Star 1990, 431). Furthermore, brain and neural connections continue to change throughout life, so the notion of matu-rity as a realized state is deeply misleading. So where does the *amicus* brief's summary arise from, its claim that "[a]dolescents' behavioral immaturity mirrors the anatomical immaturity of their brains"? Here is the relevant passage:

Thus, the sequence in which the cortex matured agrees with regionally relevant milestones in cognitive and functional development. Parts of the brain associated with more basic functions matured early: motor and sensory brain areas matured first, followed by areas involved in spatial orientation, speech and language development, and attention (upper and lower parietal lobes). Later to mature were areas involved in executive function, attention, and motor coordination (frontal lobes) (Gogtay et al. 2004, 8177).

The connecting statement is very tentative: the sequence of maturation *agrees* with milestones of cognitive and functional development. There is no more said about this very weak relationship.

In the scatterplots paired with the images, each data point represents one of the 13 subjects at a particular age (figures 14.2 and 14.3). Looking at the scatterplots makes it clear that while the overall sequence moves, the range of gray matter volumes for any area varies greatly for ages 10–20, so greatly that there is no way of looking at any brain region to assess maturity within this age range. Group trends are turned into what look like individual findings. The quite variable scan numbers become a single gray mat-ter (GM) change number ("units of GM volume"); and then the range of gray matter numbers is clumped into windows of color.

Coloring brain images can drastically change the visual impact of the data (Alač 2011; Beaulieu 2000). The continuous variations in numbers are turned into quali-tative fields—a striped rainbow of colors—in order to visually highlight differences (from dark red to bright yellow to green to bright light blue to dark blue to purple). Put

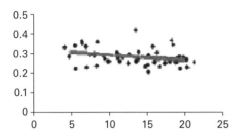 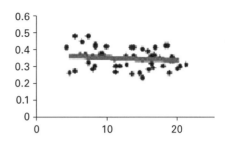

Figure 14.2
Scatterplots, from Gogtay et al. (2004).

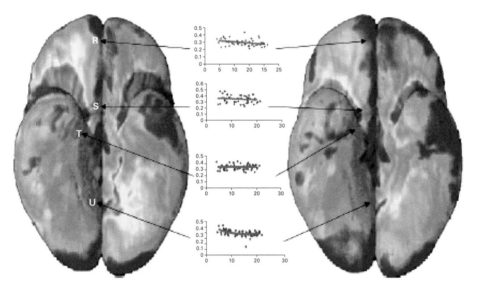

Figure 14.3
Brains with scatterplots, from Gogtay et al. (2004). Bottom view of the brain showing early and late time-lapse images. Points correspond to anterior and posterior ends of the olfactory sulcus (R and S) and collateral sulcus (T and U); mixed-model graphs corresponding to the regions of interest on the right hemisphere are shown in the middle. *X*-axis values show ages in years, *y*-axis values show GM volumes. These images demonstrate the difficulty and complexity of trying to obtain meaningful data through brain mapping. Each colored point in a brain corresponds to the average change in that area across thirteen individuals. The scatterplots of the individuals shows the tremendous variation among those individuals, variation that is erased in reducing them to average changes.

another way, no matter how small the actual difference in average change, the visual differences will appear stark. In this manner the highly variable data are converted into distinct, homogeneous "territories," creating a maximally different brain appearance (Dumit 2004). The resulting colored image of maturation (figure 14.1) thus appears visually dramatic and easy to interpret, belying the underlying variability, especially for smaller age ranges.

Taken at face value (the paper states a number of caveats and cautions), the Gogtay article makes an excellent case for differential gray matter volume and density increase and cessation of increase in brains between the ages of 4 and 21. Its recruitment to underpin social "immaturity," however, is solely based on a neurorealist notion that social immaturity, no matter how obvious, is not "real" unless there are accompanying brain changes. One may be tempted to call this reductionism, but as Beaulieu notes, "[t]o accuse brain mapping of reductionism is to miss the ways in which it powerfully redefines concepts like behaviour, nurture, culture and environment. The relational role of the map is therefore to link context, mind and brain" (2003, 563).

The Gogtay article does not attempt to connect these changes with behavioral changes. Reading it as the *amicus* does, as "explaining" these changes, is an example of a social recruitment: "To a degree never before understood, scientists can now demonstrate that adolescents are immature not only to the observer's naked eye, but in the very fibers of their brains." The *amicus* acts as if the images are the argument of the Gogtay paper, and that they directly demonstrate the strongest possible case linking anatomical and social "maturities." The question of adolescent immaturity raised in the court case, the neuroscience claims addressed to the question, and the images themselves traveled quite widely, appearing not only in *Science* but in daily newspapers, on National Public Radio, and so on, often with the images completely disconnected from the experimental context and carefully guarded wording of the study. When some of the authors of the articles were questioned, however, they tried to mitigate the force of such runaway claims:

Neuroscientist Elizabeth Sowell of UCLA says that *too little data exist* to connect behavior to brain structure and imaging is far from being diagnostic. "We couldn't do a scan on a kid and decide if they should be tried as an adult," she says. (Beckman 2004, 599, emphasis added)

There's *still a long way to go* in untangling how brain development influences what teens do and why they do it, remarks Jay N. Giedd of the National Institute of Mental Health in Bethesda. Courts and legislatures grappling with the juvenile death penalty *nonetheless need to consider* the brain's unfinished status during adolescence, especially in the frontal lobes, according to Giedd, a pioneer in research on brain development. (Bower 2004, 300, emphasis added)

Both of these responses state that the meaning of brain changes is not understood, even if the data are suggestive (and hypothesis-generating). The authors hedge the future, indicating that there is no doubt that neuroscience, with more data, eventually will make the connections, but only after a long time. These scientists' confirmations of the potential of the technology, while distancing themselves from currently untenable statements, are characteristic of promissory science: expressing confidence that neuroscience is posing the question properly even if the results are not yet in.[9] By not explicitly explaining how conceptual slippage and individual variability matter, and by leaving the timing of the future indeterminate, these responses allow other neuroscientists, like Ruben Gur in court and Daniel Amen in popular media, to claim that the technology can *now* enable them make such claims. Gur and Amen do look at individual scans and claim to make diagnoses of mature brains and dangerous ones, and claim to reliably predict an individual's behavior from their brain scans.

4 Explanatory Wars: Adolescence versus Dangerousness

Modern expert medico-legal opinion has a very precise function: it makes possible an exchange between juridical categories . . . and medical notions. . . . You can see how notions like those of perversity make it possible to stitch together the series of categories. . . . This set of notions functions, then, as a switch point, and the weaker it is epistemologically, the better it functions. (Foucault 2003, 33)

What the *amicus curiae* briefs conclude about the studies they cite is that neuroscience explains why teens are less culpable than adults, and therefore not deserving of the death penalty. But in the case of *Roper v. Simmons*, the prosecutor argued during the trial that the defendant's being a 17-year-old was actually more reason to recommend that he receive the death penalty.

In rebuttal, the prosecutor gave the following response: "Age, he says. Think about age. Seventeen years old. Isn't that scary? Doesn't that scare you? Mitigating? Quite the contrary I submit. Quite the contrary." (*Roper v. Simmons*, 543 U.S. 551, 2005, 558).

The prosecutor implies that Simmons is truly dangerous, or evil, and that his young age implies a deeper streak of it rather than something he would grow out of. At issue here are two different brain narratives, one concerning the "adolescent brain," the other the "dangerous" or "criminal brain." If the adolescent brain narrative is one of developmental maturity, predicting that immature brains will tend to become more mature, the criminal brain narrative is one of character predestination, predicting that some brains are inherently more dangerous or evil than others (Reznek 1997). Both of these narratives are still alive and well in our culture and scientific literature.

The criminal brain has been the object of neurological research for over a century. The long history of assuming that criminality is biologically detectable and visible (Galton 1883; Beaulieu 2000), especially that there must be a brain trace that can account for criminality, has fought with the criminal justice system's dependence on free intentionality as a criterion for punishment. For over one hundred years, psychologists and psychiatrists have played the role of translators, recruiting "the brain" and

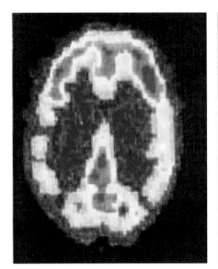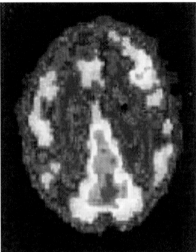

Figure 14.4
Murderer versus normal, from Raine (1999). Brain scan (PET) of a normal control (left) and a murderer (right), illustrating the lack of activation in the prefrontal cortex in the murderer. The figures are a transverse (horizontal) slice through the brain, so you are looking down on the brain. The prefrontal region is at the top of the figure, and the occipital cortex (the back part of the brain controlling vision) is at the bottom. Warm colors (e.g., red and yellow) indicate areas of high brain activation; cold colors (e.g., blue and green) indicate low activation. Based on a study comparing the brains of 22 murderers to 22 normal individuals, these two images were selected to starkly show the differences between the two groups (though many brains of murderers were not different from those of normal individuals). The implication of Raine's papers is that murderers and other psychopaths differ throughout their life in their brains and other biomarkers.

heredity to authorize assessments of dangerousness and legal insanity, even when their sciences did not address these categories. Similarly, neuroscience has produced study after study attempting to locate and show the neuro-differences between persons who are dangerous or potentially dangerous and those who are not (who are normal). Some researchers have published images of a "murderer" brain compared with a "normal" one (Raine et al. 1997).

The results of neuroscience studies of dangerousness have the same problems as those that study adolescence. The researchers must first decide how to select representatives for each comparison group, and then they must assume that identifiable differences between the samples will stand in not just for the groups, but also for the socially relevant characteristic of that group. Or rather, the results of their studies must show the differences between the two groups of brains (represented as averages, extremes, or some other group function) as if they are brain-person types.

Thus we have a set of studies that assume and show that dangerousness is the result of brains that are abnormal from early childhood on, and we have another set of studies that assume and show that "normal" risky adolescence is the effect of having brains that are different from adult brains. The AMA et al. *amicus* brief intervened within this contested field by positing the adolescent studies as the only relevant ones and not citing the dangerousness research at all. The team of lawyers who wrote the brief stated that this selection process was intentional and politically motivated. They found the science that supported their social and legal advocacy aims.

We did have a political position regarding the juvenile death penalty—we were against it. . . . We harnessed the available research to the existing law and argued that the death penalty's goals of retribution and deterrence were not served by its application to juveniles. Thus, the science brief was born. (Haider 2006, 379)

When researching scientific data, we looked for articles that addressed adolescents' inability to engage in logical reasoning, the inability to control impulses, and to be followers rather than leaders. (ibid., 382)

These studies were presented to the Court as evidence that adolescents are biologically different. (ibid., 381)

The lawyers were guided in their efforts by an advisory panel of scientists whom they chose and who in turn helped the team select and analyze the studies chosen. "In short, we did not push the science beyond its limits" (ibid., 379). The role of the brief was thus to shut down one way of using neuroscience—to ground claims of biological dangerousness—and to reinforce instead the association of neuroscience with the verification of adolescent immaturity. Given the history of the use of biological notions of dangerousness, this polemical use of neuroscience is not surprising, and may even be laudable. Both legal analyst Maroney and science journalist Begley agree with the brief and many neuroscientists:

Ongoing research on the links between brain maturation and psychological development in adolescence has begun to shed light on why adolescents are not as planful, thoughtful, or self-controlled as adults, and, more importantly, it clarifies that these "deficiencies" may be physiological as well as psychological in nature. (Steinberg and Scott 2003, 1016)

Accepting this polemical use of neuroscience analysis comes with a price, however: neuroscience today, like psychiatry, finds itself always speaking with courts and with politics in mind, perhaps to the detriment of its ability to focus on properly neuroscientific questions. For instance, neuroscience research into development seems to reveal greater variability in brain maturation than stereotypical views of adolescent riskiness have suggested, and should at least be investigated for how such research might refigure our commonsense notions of age limits. Instead neuroscience findings are almost only used to reinforce the stereotypes, along with the idea that more research will make the connection clearer. Given that the same problem arises with

research into the criminal or dangerous brain, this suggests that neuroscience continues to be a subordinate field to the more powerful fields of popular culture and the law.
Its presence in courtrooms and mass media is one of reinforcement and thus similar to
that of the brain at the turn of the last century: "the weaker it is epistemologically, the
better it functions" (Foucault 2003, 33).

5 Conclusions: Polemical Brains

Each standard and each category valorizes some point of view and silences another. This is not
inherently a bad thing—indeed it is inescapable. But it is an ethical choice, and as such it is dangerous—not bad, but dangerous. (Bowker and Star 1999, 7)

Verifying what your grandmother said, verifying stereotypes about groups, is immensely
satisfying on many levels. There is therefore a definite tendency to stop asking questions
when scientific research reaches a point where it seems to confirm what we already know.
Just as "dangerousness" is a legally actionable category that psychiatrists are called upon
to "predict," even though it isn't one of their categories, so immaturity is a moral category
and not a neuroscientific category, though neuroscientists might have something to say
about brain development.[10] There is nothing in the papers discussed here that could tell us
how long a person with an "immature brain" would take to "mature." Seen therefore as a
"real but unknown risk," the "immature brain" turns out to be a variant of the "dangerous
brain," polemically positioned so as to look like a mitigating factor.

   The same research can demonstrate both immaturity as a stage and dangerousness
as a trait. Each statement about the research that relates it to the courtroom reinforces
one meaning in preference to the other. In essence, the statements about correlation
that correspond to a brain-person type shift the burden of proof. The choice among
brain narratives is therefore a social choice prior to the experimental design. Seen this
way, the experiments are polemical in their inception.

   Neuroscientists Elizabeth Phelps and Laura A. Thomas presciently called attention
to the problem in a strongly worded 2003 paper:

Recent advances in brain imaging techniques have allowed us to explore the neural basis of
complex human behaviors with more precision than was previously possible. As we begin to
uncover the neural systems of behaviors that are socially and culturally important, we need to be
clear about how to integrate this new approach with our psychological understanding of these
behaviors. . . . [I]t is inappropriate to assume that the results of neuroimaging studies of a given
behavior are more informative than the results of psychological studies of that behavior. (Phelps
and Thomas 2003, 747)

Refusing the neurorealistic assumption that the brain and therefore neuroscientific
results are somehow logically prior to or explanatory of human behavior or social categories, Phelps and Thomas instead try to place neuroscience and psychology in conversation. Thinking with truly realistic assumptions means thinking with the brain's

plasticity: "Showing a behavior 'in the brain' does not indicate that it is innate, 'hard-wired,' or unchangeable. Every experience leads to an alteration in the brain" (Phelps and Thomas 2003, 754).

They point out that the fact that there are correlations between brains of selected groups should not be surprising. This is the assumption and the basic empirical starting point of the neuroscience experiment. Correlations alone explain nothing, especially when they are weaker than our selection criteria! They may be as useful for proposing social or experiential causes as they are for genetic or biological ones. Phelps and Thomas are trying to help us recognize the epistemological weakness that subtends the dangers in our social habit of being excited by neurological correlations.

In conclusion, by proposing to address the social categories of "adolescence," "riskiness," and "dangerousness," rather than calling them into question, neuroscience risks losing its analytic focus on actual behavior and the meaning of variability (scientifically and socially). The fact is that neuroscience has come to have explanatory power far in excess of its confirmatory ability. This neurorealism is incredibly tempting, not just to journalists and lawyers but to neuroscientists themselves. The tendency to think that neuroscience results, and images especially, can and should be used in the service of social categories, no matter how well intentioned, is a mistake. It does produce social effects, but it is scientifically wrong and I think dangerous to the discipline. I have tried to show that in the case of adolescence, (1) it is unwarranted, (2) it reinforces wrong views about the brain, about neuroscience, and about the integrity of science and scientists, and (3) it fails even on their own polemical grounds, because then opposing polemical uses are equally justified, and leaves in its wake no criteria for distinguishing the neuroscientific wheat from the chaff.[11]

Acknowledgments

Thanks to Michael Lynch, James Griesemer, Natasha Myers, Orit Halpern, Jake Kosek, Lochlann Jain, and two anonymous reviewers for helping improve this paper.

Notes

1. It must be noted that the arguments in this case are not about excusing adolescents or saying that they are not responsible. "The public debate about criminal punishment of juveniles is often heated and ill-informed, in part because the focus is typically on excuse when it should be on mitigation" (Steinberg and Scott 2003, 1010). "Mitigation" refers to factors that lessen without removing responsibility and therefore can be brought to bear on how severely someone can be punished without affecting their guilt. James Boyle has complained that this entire framework is bizarre, since the very idea of measured culpable capacity cannot be bracketed to just the penalty phase of trials. If accepted, it goes to the heart of the criminal justice system and its jury adjudication of intent (James Boyle, personal communication, 5 February 2011).

2. The signers included the American Medical Association, the American Psychiatric Association, the American Society for Adolescent Psychiatry, the American Academy of Child and Adolescent Psychiatry, the American Academy of Psychiatry and the Law, the National Association of Social Workers, and the National Mental Health Association.

3. See "Bright Line Rule," from West's Encyclopedia of American Law (2005). Available at http://www.encyclopedia.com/doc/1G2-3437700625.html (accessed 10 November 2012).

4. See, e.g., Mulkay (1997) and McConvill and Mills (2003).

5. Some commentaries on the *Roper* decision claim that neuroscience did play a strong role (e.g., Gruber and Yurgelun-Todd 2006; Ross 2009), but most see a more general reliance by the Court on a wide variety of sociological and other studies (Denno 2006).

6. "Behavioral scientists have observed these differences [between adolescents and adults] for some time. Only recently, however, have studies yielded evidence of concrete differences that are anatomically based. Cutting-edge brain imaging technology reveals that regions of the adolescent brain do not reach a fully mature state until after the age of 18" (AMA et al. 2004, 2).

7. While the studies that explicitly tried to study neurorealism have their own problems, the overall effect seems surprisingly robust and obvious: claims like this are made quite often.

8. There is also a misuse of experimental tasks (specific cognitive tasks performed repeatedly during an experiment) that are morphed into social categories like maturity and judgment. See Fitzpatrick (2012): "But are these topics actually well suited for study by functional brain imaging? I hope this essay will convince readers that the answer to this question is 'no.' Despite their seeming accessibility, functional brain imaging studies are difficult to design, execute, and interpret. In my experience, functional imaging studies posing questions about aspects of human cognition that are poorly understood at the psychological and neurobiological level contribute to general misconceptions about brain and mind, and even more generally, the links between biology and behavior."

9. Such caveats are repeated in the critiques of neuroscience in courtrooms as well (cf. Aronson 2009; Maroney 2009). According to Maroney (2009, 58), "Neither structural nor functional imaging can determine whether any given individual has a 'mature brain' in any respect, though imaging might reveal gross pathology. Researchers therefore agree that developmental neuroscience cannot at present generate reliable predictions or findings about an individual's behavioral maturity."

10. I'm indebted to Michael Lynch for this formulation.

11. I take this phrase from the James S. McDonnell Foundations's website, Neuro-Journalism Mill. Available at: http://www.jsmf.org/neuromill/about.htm (accessed 10 November 2012).

References

Alač, Morana. 2011. *Handling Digital Brains: A Laboratory Study of Multimodal Semiotic Interaction in the Age of Computers*. Cambridge, MA: MIT Press.

AMA et al. 2004. Brief for the American Medical Association et al. as amici curiae in support of Respondent. *Roper v. Simmons*, 543 U.S. 551 (2005) (No. 03-633).

Aronson, Jay D. 2009. Neuroscience and juvenile justice. *Akron Law Review* 42:917–930.

Aronson, Jay D., and Simon A. Cole. 2009. Science and the death penalty: DNA, innocence, and the debate over capital punishment in the United States. *Law and Social Inquiry* 34 (3):603–633.

Beaulieu, Anne. 2000. The space inside the skull: Digital representations, brain mapping and cognitive neuroscience in the decade of the brain. PhD diss., University of Amsterdam.

Beaulieu, Anne. 2003. Brains, maps and the new territory of psychology. *Theory and Psychology* 13 (4):561–568.

Beckman, M. 2004. Neuroscience, crime, culpability, and the adolescent brain. *Science* 305 (5684):596–599.

Begley, Sharon. 2010. Why the media never met an fMRI they didn't like: The press and neuroscience. Colloquium presentation, UC Davis (10 November).

Bjork, J. M., A. R. Smith, C. L. Danube, and D. W. Hommer. 2007. Developmental differences in posterior mesofrontal cortex recruitment by risky rewards. *Journal of Neuroscience* 27 (18):4839–4849.

Bower, Bruce. 2004. Teen brains on trial. *Science News* 165 (19):299–301.

Bowker, Geoffrey C., and Susan Leigh Star. 1999. *Sorting Things Out: Classification and Its Consequences*. Cambridge, MA: MIT Press.

Burns, Stacey. 2008. Demonstrating "reasonable fear" at trial: Is it science or junk science? *Human Studies* 31 (2):107–131.

Denno, Deborah. 2006. The scientific shortcomings of Roper v. Simmons. *Ohio State Journal of Criminal Law* 3: 379–396.

Dumit, Joseph. 2004. *Picturing Personhood: Brain Scans and Biomedical Identity*. Princeton: Princeton University Press.

Duster, Troy. 2003. *Backdoor to Eugenics*. 2nd ed. New York: Routledge.

Epstein, Steven. 2007. *Inclusion: The Politics of Difference in Medical Research*. Chicago: University of Chicago Press.

Firth, Niall. 2010. Brain scans could be used to "prove" whether a child is mature enough to be tried for a crime. *Daily Mail* (7 October).

Fitzpatrick, Susan M. 2012. Functional brain imaging: Neuro-turn or wrong turn? In *The Neuroscientific Turn: Transdisciplinarity in the Age of the Brain*, ed. Melissa M. Littlefield and Jenell M. Johnson, 180–199. Ann Arbor: University of Michigan Press.

Foucault, Michel. 2003. *Abnormal: Lectures at the Collège de France, 1974–1975*. New York: Picador.

Galton, Francis. 1883. *Inquiries into Human Faculty and Its Development*. London: J. M. Dent.

Gogtay, N., J. N. Giedd, L. Lusk, K. M. Hayashi, D. Greenstein, A. C. Vaituzis, T. F. Nugent 3rd, et al. 2004. Dynamic mapping of human cortical development during childhood through early adulthood. *Proceedings of the National Academy of Sciences of the United States of America* 101 (21):8174–8179.

Gruber, Staci A., and Deborah A. Yurgelun-Todd. 2006. The mind of a child: The relationship between brain development, cognitive functioning, and accountability under the law: Neurobiology and the law: A role in juvenile justice? *Ohio State Journal of Criminal Law* 3: 321–543.

Hagner, Michael. 2010. A brief history of "homo cerebralis." Podcast. Brain Self and Society Symposium, "Personhood in a NeurobiologicalAge," London School of Economics, BIOS Centre.

Haider, Aliya. 2006. *Roper v. Simmons*: The role of the science brief. *Ohio State Journal of Criminal Law* 3: 369–377.

Hornstein, Gail A., and Susan Leigh Star. 1990. Universality biases: How theories about human nature succeed. *Philosophy of the Social Sciences* 20 (4):421–436.

Joyce, Kelly A. 2008. *Magnetic Appeal: MRI and the Myth of Transparency*. Ithaca: Cornell University Press.

Lo, Yian. 2010. Are juvenile offenders incorrigible. *NeuroLaw Blog* (posted 23 November). Available at: http://blog.neulaw.org/?p=1569 (accessed 5 February 2012).

Lynch, Michael, Simon Cole, Ruth McNally, and Kathleen Jordan. 2008. *Truth Machine: The Contentious History of DNA Fingerprinting*. Chicago: University of Chicago Press.

Malabou, Catherine. 2008. *What Should We Do with Our Brain?* New York: Fordham University Press.

Marks, Jonathan. 1995. *Human Biodiversity: Genes, Race, and History*. New York: Aldine de Gruyter.

Maroney, Terry A. 2009. The false promise of adolescent brain science in juvenile justice. *Notre Dame Law Review* 85:89–176.

McConvill, James, and Eithne Mills. 2003. Re Kevin and the right of transsexual persons to marry in Australia. *International Journal of Law, Policy and the Family* 17 (3):251–274.

Mulkay, Michael J. 1997. *The Embryo Research Debate: Science and the Politics of Reproduction*. New York: Cambridge University Press.

Nelkin, Dorothy, and M. Susan Lindee. 2004. *The DNA Mystique: The Gene as a Cultural Icon*. Ann Arbor: University of Michigan Press.

Nelkin, Dorothy, and Laurence Tancredi. 1989. *Dangerous Diagnostics: The Social Power of Biological Information*. New York: Basic Books.

Phelps, Elizabeth A., and Laura A. Thomas. 2003. Race, behavior and the brain: The role of neuroimaging in understanding complex social behaviors. *Political Psychology* 24 (4):747–758.

Racine, Eric, Ofek Bar-Ilan, and Judy Illes. 2005. fMRI in the public eye. *Nature Reviews. Neuroscience* 6 (2):159–164.

Raine, Adrian. 1999. Murderous minds: Can we see the mark of Cain? Dana Foundation. Available at http://www.dana.org/news/cerebrum/detail.aspx?id=3066.

Raine, Adrian, Monte Buchsbaum, and Lori Lacasse. 1997. Brain abnormalities in murderers indicated by positron emission tomography. *Biological Psychiatry* 42 (6):495–508.

Raine, A., and Y. Yang. 2006. The neuroanatomical bases of psychopathy: A review of brain imaging findings. In *Handbook of Psychopathy*, ed. C. J. Patrick, 278–295. London: Guilford.

Reznek, Lawrie. 1997. *Evil or Ill? Justifying the Insanity Defence*. London: Routledge.

Ross, Catherine J. 2009. A stable paradigm: Revisiting capacity, vulnerability and the rights claims of adolescence after *Roper v. Simmons*. In *Law, Mind and Brain*, ed. Michael Freeman and Oliver R. Goodenough, 183–198. Burlington, VT: Ashgate.

Scott, Elizabeth S., and Laurence Steinberg. 2008. Adolescent development and the regulation of youth crime. *Future of Children* 18 (2):15–33.

Shim, Janet K. 2000. Bio-power and racial, class and gender formations in biomedical knowledge production. *Research in the Sociology of Health Care* 17:173–195.

Star, Susan Leigh. 1989. *Regions of the Mind: Brain Research and the Quest for Scientific Certainty*. Stanford: Stanford University Press.

Steinberg, Laurence, and Elizabeth S. Scott. 2003. Less guilty by reason of adolescence: Developmental immaturity, diminished responsibility, and the juvenile death penalty. *American Psychologist* 58 (12):1009–1018.

Szasz, Thomas. 1988. *Insanity: The Idea and Its Consequences*. New York: Wiley.

Van Dijck, José. 2005. *The Transparent Body: A Cultural Analysis of Medical Imaging*. Seattle: University of Washington Press.

West's Encyclopedia of American Law. 2005. Bright line rule. Available at: http://www.encyclopedia.com/doc/1G2-3437700625.html (accessed 5 February 2012)

Reflections

15 Preface

Steve Woolgar

Our final section comprises brief commentaries by seven authors who have had long-standing and even formative roles in studies of representational practice. Each of their commentaries offers broad reflections on the field, and each calls attention to both persistent and changing issues in studying representation in scientific practice.

A first issue is, perhaps inevitably, the very notion of representation. Lorraine **Daston** speaks of intractable conceptual problems associated with the very idea of representation. Studies of representation, whether "in practice" or otherwise, can inherit those problems by reproducing the conceptual limitations with the very terms they use to describe detailed practices of "doing" representation or "interpreting" representations. Daston argues that the time is ripe to go "beyond" representation, and calls for a shift from epistemological to ontological treatments of images. Michael **Lynch** reflects on the ways in which philosophical pictures hold us captive, in earlier times with respect to reference, and more recently with respect to information. A focus on representational activities as practice allows us to understand information as a contingent product, just as it allows us to appreciate the contingency of reference. Steve **Woolgar** looks back on various attempts to problematize and displace the notion of representation since the publication of the earlier volume, and comments on current concerns to replace a focus on epistemology with a focus on ontology. Are all efforts to exorcise representation doomed to failure?

A second issue is reflexivity. Lucy **Suchman** calls for a measure of reflexivity in our studies of representation: how are we "scientifically representing" representation in scientific practice? She raises questions about the very idea of "practice" and how we "represent" this elusive demon in our own studies. What exclusions do we generate through our own practices of articulating and bounding the phenomena we study? John **Law** asks us to attend to the "collateral realities" that are being "done" at the periphery of what representation in scientific practice is ostensibly about. These collateral realities are "social and material on the one hand, and metaphysical on the other"—indeed, Law insists that acknowledgment of the metaphysical aspect is necessary for developing "the capacity to see the things that aren't being literally described."

A third issue is trust. Martin **Kemp** discusses the idea that visual images and graphics are used, whether effectively or in a suspect way, to evoke "reality," as though directly on the page or screen. He observes that when such images of novel, previously unknown realities are mediated by scientific instruments, "the rhetoric of reality is complemented by the 'rhetoric of instrumental objectivity.'" He reemphasizes how scientific imagery puts us in the position of "witness," and the rhetorical devices employed to engender trust in science as an enterprise that provides direct access to reality. By contrast, Bruno **Latour** suggests, as he did in his seminal contribution to *RiSP*, that the supposed "gap" between previously unknown realities and visual images is densely populated by "long cascades of successive traces," traces that themselves trace back to instrumental achievements, and create further gaps for novel instruments to address. He argues that what's important about scientific visual imagery is not "the visual," and that a focus on "the visual" actually distracts from what such imagery can really teach us.

These seven short pieces comment on the nature and prospects of studies of representation in general. In this vein, they reference the "big" themes of representation—for example, epistemology, ontology, visualization, and trust. The commentaries thus provide an interesting complement to the empirical case studies. Whereas the latter deliver a crucial deflationary effect—"science" is brought down to earth, made commonplace and subject to epistemic leveling; the "elevator words" in philosophy of science are unloaded at the ground floor (Hacking 1999, 21ff)—these final commentaries remind us about the traps, troubles, and taken-for-granted assumptions that continue to characterize even the very best empirical studies of representational work.

Reference

Hacking, Ian. 1999. *The Social Construction of What?* Cambridge, MA: Harvard University Press.

16   Beyond Representation

Lorraine Daston

After some twenty years of remarkable work on visualization in science,[1] it is now astonishing to recall how blind historians of science once were to anything but words: scientific texts were purely textual; when we came to an image (a drawing, a graph, a table, a diagram, a photograph, it was all one), we just flipped the page. Illustrations in history of science monographs, insofar as there were any, consisted almost exclusively of portraits of past scientific luminaries. Pick up almost any recent book or article in the field now and it is likely to be peppered with images, many of which are as essential as the well-chosen quotation is to making the author's point. Images have come into their own as a source for the history of science, even if we are still learning how to interpret them and to emancipate ourselves from text-centered analogies such as "reading images" and "visual literacy."

Perhaps the most tenacious of these word-obsessed metaphors is the idea of representation itself: an image is said to "represent" something in the world, as a word "refers" to it. The same processes of duplication are implied: ideally, the representation mirrors the world, as the word corresponds to it. The perfect dictionary contains words for everything that is in the world and only those things (forget "unicorn" and "phlogiston"); the perfect atlas is its pictorial complement. Correspondence theories of scientific truth echo correspondence theories of linguistic reference. Jorge Luis Borges (1998, 325) parodied this ideal of perfect and perfectly useless representation with his parable of the map made in one-to-one scale, indistinguishable from the empire it surveys. To re-present in the literal sense is just as futile as the Borgesian map.

Yet the ideal of perfect representation has proved particularly tenacious in the realm of the image, and nowhere more so than for the scientific image. Although the history of art was forced at last by the advent of photography to abandon its long fascination with mimesis (dating back to Pliny's tales of Zeuxis and Parrhasius deceiving each other with *trompe l'oeil* grapes and curtains, endlessly repeated for centuries, in *Historia naturalis*, 35.36.65), the same specter of the perfect copy still haunts the history of science, despite the best efforts of scientists and artists to explain that the most faithful rendering is often not one that could be mistaken for the original. Paradoxically, photography seemed to liberate art from mimesis while at the very same time apparently enslaving science—at least in the eyes of influential romantic critics like Charles Baudelaire (1962, 319), who exhorted artists to paint what they dreamed, not what they saw, but in the next breath commended photography to scientists in the interests of "absolute material exactitude." On this view, still widely held despite the overwhelming evidence submitted against it by historians of photography,[2] subjective art could flaunt representation while objective science was obliged to knuckle under to it.

Perhaps this is the reason why so much of the literature on scientific visualization in the 1990s was focused on the processes of *transforming* object into image ("series," "mediation," and "work" were favorite antidotes to any presumption of transparency or self-evidence), in order to exorcise the demon of representation and its duplicate worlds. Knowing what they were up against, editors Michael Lynch and Steve Woolgar recommended that readers of their seminal collection on *Representation in Scientific Practice* "consider serial, 'directional,' relations between representations, and differences in the abstracted or naturalistic form of representations, to be relations between technical products in a work process. The 'direction' is not a movement away from or toward an originary reality, but a movement on an assembly line" (Lynch and Woolgar 1990, 8 [1988, 108]). When these lines were published in 1988, their tenor was deliberately counterintuitive. Read now from a distance of some twenty-five years, they seem almost obvious—a measure of the success of the analysis, certainly, but also of the way in which hands-on experience with computer programs like Photoshop is changing intuitions about the plasticity of even mechanically produced images, including those of scientists.[3] The time is ripe to think about images beyond representation.

"Representation" is an intrinsically epistemological notion. Whether understood as direct mirroring or serial manipulation, whether the chain between object and image is long or short, straight or curvy, representation assumes a prior "presentation." (Presentations need not be some bedrock reality—the word itself has a Kantian ring, of a world already packaged for our sensibility and understanding, decidedly not the *Ding an sich*.) To *re*present conjures up associations with spectacle and the stage, more gripping and intelligible than workaday life but also at several removes from it. Re-presentation lends itself to metaphors of both refinement (as in the extraction of metal from ore) and falsification (as in the fictionalization of facts). From the standpoint of naive realism, representation is suspect, a vision glimpsed through a glass darkly; from that of neo-Kantianism, it is essential, the precondition for experience and *a fortiori* for knowledge. Whatever the philosophical standpoint, representation is always derivative from some presentation, and therefore directs attention toward how rather than what we know, epistemology rather than ontology.

What would an ontological account of scientific images look like? For starters, it would collapse the distance between presentation and representation: the image *is* the presentation, the working object of science. Computer simulations, the greatest revolution in scientific empiricism since the canonization of observation and experiment in the late seventeenth century, provide the most spectacular example of this collapse, especially in fields like cosmology or meteorology.[4] Other imaging techniques, such as the atomic force microscope, simultaneously make and make visible, present and represent all at once (Daston and Galison 2007, 397–402). These new practices sometimes ignite controversy just because they conflate presentation and representation. Do data generated by algorithms programmed into a computer really count as empirical? Where should the line between cleaning up and massaging an image be drawn? Editorials in leading scientific journals are furiously debating these questions and, more importantly, laying out guidelines for best practice that both reflect and shape new intuitions about what is and is not real: ontology in the making.[5]

Dramatic as these developments are, their novelty should not be exaggerated. Long before computer simulations and Photoshop, humble scientific observation has been fruitful in crystallizing the objects of scientific inquiry out of the buzzing swarm of ordinary experience. Since science became extravagantly visual in the sixteenth century, images have supplied many disciplines with their working objects. But anxieties about how subjective presuppositions, both personal and theoretical, might distort objective facts fostered a view of observation as nothing more than simple perception, in order to guarantee that it was "theory-free"—and therefore an unadulterated presentation, not a representation of the world (Daston 2008). These anxieties are still rampant in editorials in science journals that attempt to draw a bright line between "beautification and fraud" in the age of image-processing programs. The epistemological opposition between subjective and objective is as robust as ever. Nonetheless, the practices of what some have called e-science, most importantly the rise of computer simulations as a new kind of scientific empiricism, presage an era that may take the discussion about images beyond representation.

Notes

1. The literature on scientific visualization is by now large and rich, but trailblazing studies include, for the history of science, Rudwick (1976); Latour (1986); Lynch and Woolgar (1990); and for the history of art, Alpers (1983); Crary (1990); Stafford (1994); and bringing together both disciplinary perspectives, Jones and Galison (1998).

2. For scientific photography, see Tucker (2005).

3. See for example *Nature Cell Biology* (2006).

4. For a discussion of the philosophical implications of simulations, see Humphreys (2007).

5. See Emma Frow's chapter in this volume, and also Frow (2012).

References

Alpers, Svetlana. 1983. *The Art of Describing: Dutch Art in the Seventeenth Century*. Chicago: University of Chicago Press.

Baudelaire, Charles. 1962. Salon de 1859. In *Curiosités esthétiques. L'Art romantique et autres oeuvres critiques*, ed. Henri Lemaître. Paris: Editions Garnier.

Borges, Jorge Luis. 1998. On exactitude in science. In *Collected Fictions*, trans. Andrew Hurley, 325. New York: Penguin.

Crary, Jonathan. 1990. *Techniques of the Observer: On Vision and Modernity in the Nineteenth Century*. Cambridge, MA: MIT Press.

Daston, Lorraine. 2008. On scientific observation. *Isis* 99:97–110.

Daston, Lorraine, and Peter Galison. 2007. *Objectivity*. New York: Zone Books.

Frow, Emma K. 2012. Drawing a line: Setting guidelines for digital image processing in scientific journal articles. *Social Studies of Science* 42: 369–392.

Humphreys, Paul. 2007. *Extending Ourselves: Computational Science, Empiricism, and Scientific Method*. New York: Oxford University Press.

Jones, Caroline A., and Peter Galison, eds. 1998. *Picturing Science, Producing Art*. New York: Routledge.

Latour, Bruno. 1986. Visualization and cognition: Thinking with eyes and hands. *Knowledge in Society* 6:1–40.

Lynch, Michael, and Steve Woolgar. 1988. Sociological orientations to representational practice in science. *Human Studies* 11(2/3): 99–116. Reprinted in Lynch and Woolgar 1990.

Lynch, Michael, and Steve Woolgar, eds. 1990. *Representation in Scientific Practice*. Cambridge, MA: MIT Press.

Nature Cell Biology. 2006. Appreciating data: Wrinkles, warts and all. Editorial. *Nature Cell Biology* 8: 203.

Rudwick, Martin J. S. 1976. The emergence of a visual language for geological science, 1760–1840. *History of Science* 4:149–195.

Stafford, Barbara Maria. 1994. *Artful Science: Enlightenment Entertainment and the Eclipse of Visual Education*. Cambridge, MA: MIT Press.

Tucker, Jennifer. 2005. *Nature Exposed: Photography as Eyewitness in Victorian Science*. Baltimore: Johns Hopkins University Press.

17   Representation in Formation

Michael Lynch

One of the most often-quoted aphorisms from Wittgenstein's (1958) *Philosophical Investigations* is the line "a picture held us captive, and we could not get outside it" (§115). Of course, Wittgenstein is not concerned with pictures as such, but with the picture theory of language: the ancient and modern preoccupation with *reference* as the primary linguistic function of interest for philosophy. Specifically, it is a preoccupation with words standing proxy for objects, signs for meanings, pictures for things depicted, and signifiers for things or ideas signified. Moreover, Wittgenstein is not simply making an observation about the picture theory's hold on the philosophical imagination. To use another metaphor, he is suggesting that it is a box that philosophers have been unable to think outside of. But, in addition to proposing to show philosophers a way out of their conceptual entrapment (with yet another metaphor of a fly trapped in a bottle), he also suggests that our ordinary language already positions us outside the confines of what philosophy has constructed with such rigorous analytical care.

As Lorraine Daston points out in her brief but incisive comment in this volume, reference and its close cousin *representation* also have had a tenacious hold on research in the history and social study of science, and for an obvious reason. If we figure that scientific research is a matter of discovering and inventing novel ways of seeing, identifying, naming, organizing, mapping, and manipulating worldly things and their relationships, then what could be more important than reference? The surprising answer that Wittgenstein gives to this question is *just about everything*! That is, most of what we *do* with language is *not* a matter of referring to, or representing, some thing or idea that resides outside of language. I take it that he is not declaring that reference is impossible, or asserting that there is no "real world" that stands outside of language; rather, he is observing that referring to things, ideas, "imaginaries," or whatever is far from the only, or even most important, thing we do with language and other communicative media. Further, I understand him to be saying that when we do refer to things, whether in science or daily life, what we say or depict cannot be comprehended by working within the narrow parameters of the picture theory's schema of words standing for things or ideas and pictures resembling objects.

As Daston points out, several of the contributions in the earlier volume of *Representation in Scientific Practice* (Lynch and Woolgar 1990) show how the concept of representation has temporal and practical dimensions—it is not limited to a static epistemological confrontation between picture and object that frames age-old questions about referential certainty. In the laboratory and field sciences, things and data are presented and re-presented in sequential chains of practice. Accordingly, re-presentation becomes a matter of presenting an initial *something* again and again; transforming, transposing, and translating the material/semiotic forms of that *something*; and serially disclosing and detailing what that initial, inchoate *something* was all along. The companion term *visualization* signals an orientation to the temporal and practical formation of *what* becomes observable, measurable, comparable, and accountable. Understood in this way, visualization is far from limited to practices that allow something to be seen, observed, or depicted. It also includes the full array of practices that *make* and *render accountable* the concrete material forms of the things and fields that scientists investigate.

Steve Woolgar and I settled for *Representation in Scientific Practice* as the title for the 1990 volume, though with some trepidation. With Catelijne Coopmans and Janet Vertesi joining the editorial conversation for the present volume, once again we retained the core terms of that title, and again with some trepidation. In both instances, we had no problem with the term "practice," though I suppose we could have had qualms in light of Stephen Turner's (1994) critique of theories that deploy the term *practice*—with its concrete, material, and down-to-earth sense—while smuggling in conceptions of practice that are abstract, indistinct, and recondite. However, as the contributions to the volume exemplified, we located practices in concrete situations. Our main problem was, and is, with the term *representation*. The term carries lots of baggage, but it also has its advantages. One advantage, as Hannah Pitkin (1972) elucidated and Bruno Latour (1993, 27) later elaborated in his own way, is that the term bridges the world of political representation (representing people) and that of the sciences (presenting and re-presenting things, kinds, constants, and causal relationships with texts, equations, graphs, diagrams, and images). In both domains, as conventionally understood, a representative or representational device stands proxy for what (or who) it represents. Actions that would be impossible to perform in the domain represented (among the masses; in the world-out-there) become manageable and effective in the restricted domains of the parliament and the laboratory. There also is a potential for rupture, leading the masses or the world-out-there to become alienated and disenfranchised in relation to a particular representative or an entire mode of representation. Another advantage of the term is its temporal implication—re-presentation, presenting again (and again and again, indefinitely). One theme developed in several of the chapters in the earlier volume was that of chains, series, or "cascades" (Latour, 1990) of traces, images, or "renderings" (Lynch, 1990) in laboratory work. The downside is that the concept of representation carries the heavy baggage of the correspondence "theory" of reference: the classic idea that a representation stands in a mimetic or symbolic relation to an independent reality. I put "theory" in quotes, because the baggage-bearing conception of meaning is more diffuse and pervasive than is the case for the more familiar varieties of theory.

Avoiding the word "representation" would not unload the baggage that concerns us. Moreover, the word remains in common use, where it is not bound to a particular philosophy or even a classical metaphysical picture. As Wittgenstein (1958, 169) reminds us, the concept of a representation is "very elastic . . . [and] intimately connected" with that of "what is seen." The emphasis on practice both in the 1990 edition of *Representation in Scientific Practice* and in this new volume is meant to draw attention to the many ways in which words do more and other than refer, pictures do more and other than depict, and representations do more and other than correspond to objects and/or ideas. The point of emphasizing practice is not to suggest that words, pictures, and other renderings that are commonly called "representations" *fail* to refer, *fail* to depict, or *fail* to represent anything real. Instead, the point is that their practical uses establish *what* they do, *how* they mean, and what is done *with* them. They may very well refer, depict, or represent, in terms of relevant standards of adequacy and accuracy, but they may also be found to do other jobs as well. They are performative, but not because "performativity" is an inherent property of the particular images or signifiers that are deployed. Instead, they are performative as a function and consequence of the actions and circumstances in which they are used.

When settling for the term "representation," in the title of both this and the earlier volume, we chose *not* to qualify it with the word "visual." In part, this is because not all of the chapters in either volume are about visual images, though I suppose it would be possible to say that written words, equations, and other inscribed traces and signs are no less "visual" than graphic and pictorial images. In addition, *visualization* is not specific to a particular sensory modality. Graphic and pictorial images are used to visualize and analyze sounds, inaudible vibrations, fluctuations in infrared radiation, variations in tunneling current, and many other sensory and nonsensory objects and qualities. To put this another way, visualization includes the arrangements of materials, instruments and their outputs, and the embodied practices that *produce* visual displays. The literary end products may be visual and graphic, but the technologies through which "raw data" are processed often operate quite differently from the mind's eye (or the eye's mind). Visualization is as much the work of hands—often many hands—as it is of the so-called "gaze."

Mention of "data" can remind us that we are often said to be living in the "information age." If a philosophical picture of reference held Wittgenstein's contemporaries captive, the picture that holds our contemporaries captive is that of *information*. Digital image processing has been in widespread use in numerous fields for decades, but it has a much more prominent place in the present volume than in the 1990 volume. Conceptions of information are built into the devices, software, observation language, and data management strategies and technologies in the sciences today. At the same time, the notion of information is profoundly ambiguous. This ambiguity runs through the engineering sciences that build and program information-processing technologies, the sciences that use them, and the analytical disciplines (including those in and around science and technology studies) that investigate their construction and use. The ambiguity is between a commonplace conception of information as novel, informative communication and a theoretical conception of information as a ubiquitous stuff that travels through wires, is encoded in bits, bytes, and pixels, and shows up on screens and in print. At times, it can be easy to appreciate the difference between the commonplace sense of information and the (increasingly popularized) engineering sense. However, the two often are conflated, as when we read in popular or social science sources that a rapid increase in information today threatens to overwhelm us. Though it is undeniable that we live in noisy environments, it is not as though a unitary ambient substance increasingly fills the spaces in which we live. The sheer volume of digital communication does indeed create information management problems for organizations that record, archive, and analyze vast numbers of electronic communications, but that problem differs from the problem of finding helpful and reliable information in the midst of all the uninformative clutter. The engineering sense of information fails to distinguish between what is and what is not informative.

Both the 1990 and the present volume place practices at the origin of representations. Although information technology and digital image processing are topics that have become far more prominent in the intervening decades, this does not mean that information provides the raw material for making and interpreting representations. Instead, close studies of how information technologies are used to constitute "raw" data as well as to interpretively construct processed images allow us to understand that *what counts as information* is itself a contingent product of representational activities. Consequently, not only does a focus on practices call attention to representation in formation, it does so for information as well (see Prentice 2013 for a related use of "in formation").

References

Latour, Bruno. 1990. Drawing things together. In Lynch and Woolgar 1990, 19–68.

Latour, Bruno. 1993. *We Have Never Been Modern*. Cambridge, MA: Harvard University Press.

Lynch, Michael. 1990. The externalized retina: Selection and mathematization in the visual documentation of objects in the life sciences. In Lynch and Woolgar 1990, 153–186.

Lynch, Michael, and Steve Woolgar, eds. 1990. *Representation in Scientific Practice*. Cambridge, MA: MIT Press.

Pitkin, Hannah Fenichel. 1972. *The Concept of Representation*. Berkeley: University of California Press.

Prentice, Rachel. 2013. *Bodies in Formation: An Ethnography of Anatomy and Surgery Education*. Durham: Duke University Press.

Turner, Stephen. 1994. *The Social Theory of Practices: Tradition, Tacit Knowledge and Presuppositions*. Cambridge, UK: Polity Press.

Wittgenstein, Ludwig. 1958. *Philosophical Investigations*. Oxford: Blackwell.

18 Struggles with Representation: Could It Be Otherwise?

Steve Woolgar

Empirical studies have characterized representation in scientific practice as involving lengthy struggles with research materials to reconstruct them in a way that facilitates scientific analysis. Rather less attention has been paid to the parallel struggles that characterize our own analytical efforts to come to terms with representation. What are the consequences of our own struggles with research materials? What is the status of "representation" in our analytic accounts? These questions continue to permeate studies of scientific practice. Numerous scholars both in and beyond science and technology studies (STS) have pursued various kinds of skeptical stance, and deployed tactics of disturbance, disruption, and denial in relation to representation. If these can be thought of as attempts to trouble representation—to displace, replace, or exorcise it—it is notable that commitments to representation persist in spite of them. In what sense are our studies of representation critical, what can we make of moves to rid ourselves of representation, and how successful have these moves been?

As noted in our introduction, we editors chose to use the term "representation" with some trepidation. We remain uncomfortable with it because the very idea of representation seems unavoidably to connote the antecedent existence of something (some person, object, or entity) that is being represented. Regardless of how we choose to cast our analytic idiom, our use of "representation" retains and projects the possibility of something "out there." This in turn imposes an important limitation on our analysis: it reads our critical analysis of representational practice as though it is about revealing sources of error—of *mis*representation.

STS has tried to resist such a reading through declarations of impartiality. Michael Lynch (in his reflection on researching representational practice in this volume) says that "the point of emphasizing practice is not to suggest that words, pictures, and other renderings that are commonly called 'representations' *fail* to refer, *fail* to depict, or *fail* to represent anything real" (original emphasis). This is in line with a central canon that STS should try to avoid taking a view on the status of the claims made by the scientists under study. And yet it is important to acknowledge that this commitment to impartiality is an aspiration rather than a guaranteed achievement. In other words, declarations of impartiality are insufficient to block interpretations to the effect that analyses of representational practices are critical of the efficacy of those practices. Trying to maintain impartiality, and what exactly that means, is a key feature of our struggle with representation.

Lorraine Daston (in her reflection in this volume) says that the term "representation" should not be retained because it is "an intrinsically epistemological notion." But Lynch says that avoiding the word would not divest us of the philosophical baggage that comes with it. So what to do? It is perhaps helpful to situate the problem in the context of the more general thrust of STS. As a rough caricature of some of its more interesting incarnations, one could do worse than say that STS has pursued, in a wide variety of ways, the slogan "it could be otherwise." That is, a characteristic impulse of STS research has been to demonstrate contingency: to show that any established fact, claim, statement, truth, assumption, etc. could have turned out differently. In its more provocative and critical mode, something like this slogan is an identifying feature of much STS research in recent decades.

Compromise solutions have always seemed unsatisfactory in the effort to challenge the root assumption that there is a world out there available and waiting to be represented. Early social constructivism proposed that representations are partly due to social forces and partly due to the nature of the world. For example, Bijker et al. (1987), in their influential proposal for the social construction of technology, argued that technology and society co-construct new technical solutions. Karen Barad's (2007) more recent use of the phrase "meeting the universe halfway" suggests a similar compromise. Such compromise solutions carry a suspicion of incoherence and tend to replay the very divide they are intended to overcome. They beg the question: How do we decide on any occasion how much of each pole is involved? In proposing a way of bridging the gap between world and representation they reassert this dualism.

We can think of the history of STS as a history of analytic "infills," a sequence of the invocation of different analytic resources to show how some fact or event could have turned out differently. The sequence has included the use of terms such as "interests," "performativity," and "materiality." In this volume and its predecessor the focus has been on "practice." Each term offers a slightly different way of situating, tying down, and taming the troubles that would otherwise stem from disturbing fundamental assumptions about the world out there.

Consider one such recent move in STS, which advocates a focus on "ontology" as a preferred move beyond epistemology (Woolgar and Lezaun 2013). The key idea here is that, whereas an epistemological focus has treated scientists' activities in terms of their response to a preexisting world of phenomena, an ontological focus treats scientists' activities as bringing (aspects of) the world into existence in the first place. The "turn to ontology" is more "thoroughgoing," on the view that epistemology implies acquiescence to the assumption of a preexisting world. For example, Annemarie Mol (1982) argues that different practices give rise to distinct, multiple ontologies (even when these bear the same label). So practices do not merely represent; they enact. However, it is as yet unclear how much a focus on enactment differs from a strong form of social constructivism; to what extent are epistemology and ontology distinct? Woolgar and Lezaun's (2013) analysis of the moral order built around the apprehension of an ordinary "bin bag" suggests not so much that ontologies are enacted in multiple ways, but that on any specific occasion participants' activities are organized to preserve ontological singularity.

Take the example of visual representation. In many settings the visual seems to have a special allure. What accounts for that allure? What is the basis for the adage that a picture is worth a thousand words? The metaphor is telling. It declares that pictures are far more efficient as representational devices than are mere words. In many circumstances, this claim has a powerful hold: a photofit is reckoned to be a better representation of an assailant's face than a mere verbal description could ever be; a video clip is said to show the assailant's actions more "like it is" than a mere statement from a witness at the scene. STS encourages us to interrogate the key assumptions involved: exactly how and for whom does the visual (or some technological version of it) count as more "efficient"? But we can also think reflexively about our own attraction to the visual. How does our own engagement with the visual affect our analysis? How easy or difficult is it to sustain impartiality with respect to the purported representational capacities of the visual materials we deal with? What can our experiences of trying to maintain ethnographic distance tell us?

In sum, the key is to acknowledge that revisiting representation is a struggle, and that this struggle can itself be a fruitful topic. To paraphrase Woolgar (1989, xix): "The problem is in conceiving of representation as a problem in the first place; as if it was at all profitable to seek an escape from realist ontology once we have committed to conventions of representation which buttress just that particular ontology. The better strategy is to sustain and explore the paradoxes which arise when we attempt to escape the inescapable, not to attempt their resolution."

In our editorial introduction we cast the issue as "the concept formerly known as representation." This formulation foregrounds the struggle involved in attempting to establish a distance from the term. As with the rock star trying to reinvent himself, something of his former identity stubbornly remains—it will not simply go away, it cannot just be entirely abandoned. Indeed, there may be advantages in reflecting upon, rather than trying to sever, the link with representation. If we accept that there is no escape from representation, that trying to do so is a misplaced ambition, and that much may be gained by reflecting on our struggles, we can at least look forward to further fruitful revisiting.

References

Barad, Karen. 2007. *Meeting the Universe Halfway: Quantum Physics and the Entanglement of Matter and Meaning*. Durham: Duke University Press.

Bijker, Wiebe E., Thomas P. Hughes, and Trevor J. Pinch, eds. 1987. *The Social Construction of Technological Systems: New Directions in the Sociology and History of Technology*. Cambridge, MA: MIT Press.

Mol, Annemarie. 1982. *The Body Multiple: Ontology in Medical Practice*. Durham: Duke University Press.

Woolgar, Steve. 1989. Foreword. In Malcolm Ashmore, *The Reflexive Thesis: Wrighting Sociology of Scientific Knowledge*, xvii–xx. Chicago: University of Chicago Press.

Woolgar, Steve, and Javier Lezaun. 2013. The wrong bin bag: A turn to ontology in science and technology studies. *Social Studies of Science* 43 (3):321–340.

19   Reconfiguring Practices

Lucy Suchman

As science and technology researchers, how do we make the objects of our research? One way to address this question is through the figure of "practice"—both in the sense of research methods as practice, and in the sense of "practice" as itself an object of research. So my opening question can be rephrased as this one: What are the implications of the fact that we are an integral part of the practices through which our research objects are made? This is of course a longstanding question for science and technology studies (for example, Ashmore 1989; Woolgar 1988), but it seems that our thinking about it has recently taken a more radical turn. Feminist science studies scholar Karen Barad (2007), for example, has elaborated the sense of *the apparatus* in ways that extend it beyond the by now well accepted premise that instruments have material effects in the construction of scientific facts, to more deeply conjoin agencies of observation, including subjects, and their objects. She emphasizes that we are neither outside of the world looking at it nor inside it. Rather we are *of* it. She writes:

> The point is not simply to put the observer or knower back *in* the world (as if the world were a container and we needed merely to acknowledge our situatedness in it) but to understand and take account of the fact that we too are part of the world's differential becoming. And furthermore, the point is not merely that knowledge practices have material consequences, but that *practices of knowing are specific material entanglements that participate in (re)configuring the world.* (Barad 2007, 91, original emphasis)

Knowing subjects and objects known, in other words—the distinction that underwrites the classic Western philosophical differentiation of epistemology from ontology—are mutually constituted, including in their enactment as separate things. And delineating lines around and between things is, as we know, a practice of boundary making. It follows that responsible knowing requires an attentiveness to the reiterative, material-discursive practices through which object boundaries are drawn, and to the constitutive relations—and exclusions—that boundary making enacts.

In an argument that I read as deeply resonant with Barad's construct of the apparatus, John Law (2004, 14) characterizes these practices of knowledge making as a "method assemblage"; that is, enactments of "relations that make some things (representations, objects, apprehensions) present 'in-here,' while making others absent 'out-there.' The 'out-there' comes in two forms: as manifest absence (for instance as that which is represented); or, and more problematically, as a hinterland of indefinite, necessary, but hidden Otherness," meaning Otherness, in this context, as that which is taken for granted, unknowable within particular knowledge systems, or actively repressed. So Law takes us, explicitly, to questions of method, of practices of drawing things together, and of making difference.

These arguments resonate for me as well with the ethnomethodological dictum that method—understood as members' of the society's everyday practices of ordering, of making the social world intelligible—rather than being taken by social science to be *its* distinctive provenance and resource, is rather an integral part of our subject matter (Garfinkel 2002). It is in this sense that social science methods are radically *reflexive*; that is, our own work of making sense of the world relies upon the same basic competencies through which its intelligibility is collectively enacted in the first place. Another of ethnomethodology's insights is that, like method, theory is not the exclusive province of the social scientist: the world is full of mundane theories. One form that these take is that of normative prescriptions of various kinds—plans, policies, procedures, rules, conventions, instructions for how things should be done, maps, and the like. And of course social science methods can be formulated prescriptively as well. Conventionally, these prescriptions are taken as separate from, standing outside of, practice: "In theory," we say, things happen this way, but "in practice" it is different—where usually practice is seen as a flawed approximation of the ideal. But a radically different strategy has emerged over the past two decades within science and technology studies, one that treats such prescriptive formulations as themselves particular kinds of artifacts. Like all artifacts, representations are made in specific times and places to be put into use in others, with all of the problematic relations between locations of design and use that have become familiar to us through the study of technologies (Suchman 2007).

The work of representing practice involves narrating a sequence of events in a way that is aimed at revealing what is arguably the ordering work of practitioners themselves. At the same time our orderings, like theirs, place outside the frame an open-ended horizon of details, contingencies, and so forth that are presupposed but not fully articulated. It is these that constitute what Law (2004) characterizes as "mess." Order and mess have of course colloquially been used as normative, evaluative terms, a classic dualism with the first term privileged over the second. It is these politics that it is Law's project to challenge. Order and mess are mutually constitutive, he argues: order obscures mess; mess obscures the practices of ordering for which it is, in Law's terms, the necessary hinterland.

So where does this leave us with respect to representations of scientific practice? Where are the boundaries of a practice, both spatially and temporally? We delineate practices (our own and those of our research subjects) in images framed in particular ways, in transcripts, and in stories. But we could of course redraw those boundaries, following connections out in various other directions. The objects that we articulate, while arguably relevant to practitioners, are also analytic ones of our making. They are, in short, part of a practice, our practice as researchers, writers, and speakers. So I return to the problem with which I started, and ask of practices of representation in scientific practice: is this order, or mess? Like all object making, the delineation of a practice is always and irremediably part *of* a practice that informs what constitute productive and coherent units of analysis. It is that which makes us responsible and accountable for our research and its inclusions. And it is that which calls on us to be attentive to our own practice's systematic and necessary exclusions, and respectful of its constitutive outsides.

References

Ashmore, Malcolm. 1989. *The Reflexive Thesis: Wrighting Sociology of Scientific Knowledge*. Chicago: University of Chicago Press.

Barad, Karen. 2007. *Meeting the Universe Halfway: Quantum Physics and the Entanglement of Matter and Meaning*. Durham: Duke University Press.

Garfinkel, Harold. 2002. *Ethnomethodology's Program: Working Out Durkeim's Aphorism*. Ed. with an introduction by Anne Rawls. Lanham, MD: Rowman and Littlefield.

Law, John. 2004. *After Method: Mess in Social Science Research*. London: Routledge.

Suchman, Lucy. 2007. *Human-Machine Reconfigurations*. New York: Cambridge University Press.

Woolgar, Steve, ed. 1988. *Knowledge and Reflexivity: New Frontiers in the Sociology of Scientific Knowledge*. London: Sage.

20   Indistinct Perception

John Law

> It is not in the object of their knowledge, but in its modification, that monads are bounded. They all reach confusedly to infinity, to everything; but they are limited and differentiated by their level of distinct perception.
>
> —Gottfried Wilhelm Leibniz, *Monadology* (1998, 276, paragraph 60)

If the practices of science represent and visualize the world, then at the same time they don't. Or at least not directly. Science and technology studies (STS) work on representation attends to what is actually represented and how this is done. That is how it should be. But what happens if we think about the elusive whatever-it-is that lies at or beyond the periphery of vision? What happens if we attend to that which we don't quite see?

Leibniz's thinking has some purchase in STS, and especially within actor-network theory (see, for example, Latour 1988). Monads are Leibniz's elementary particles. He tells us that the universe is made of indivisible monads. People, animals, objects, and God, all are monads. He also tells us that monads mirror the universe. Locations of consciousness, they also know or include everything. "They reach . . . to infinity." I want to hang on to this sentiment. This is because it suggests the importance of attending to the nonvisual—indeed to the nonrepresentational; of attending to that which is known but only "confusedly." The problem with Leibniz, and why he is also less helpful, is his commitment to an omniscient God. What is it that distinguishes God from other monads? For Leibniz the answer is that like other monads God reaches to infinity, but unlike them she or he isn't confused (Leibniz 1998, 276, paragraph 60). And since God is perfect, the whole is consistent or "compossible." The universe might have been some other way, but God ordered it in one particular way. It is, as it were, coherent, even if other monads are not able to appreciate its perfection.

But what happens if we do away with God and God's perfection? This was Whitehead's (1978) monadological question. And the answer is that if we do away with the assumption of compossibility then we are left with the prospect, indeed the likelihood, that the universe isn't coherent. Or, to use Leibniz's own terminology, that the "object" of knowledge as well as its "modification" by those monads that are less than perfect is confused. We're also left to wrestle with versions of knowledge that seek to practice what Donna Haraway (1988, 581) refers to as the "god trick." For the divinity may or may not be dead, but a monotheistic version of God is alive and kicking in the secular successor projects to Judeo-Christianity, and these successor projects work on the assumption that they can know the world as a whole, consistently.

So where should we look for confusion or "indistinct perception"? The contemporary STS response comes in several styles. One, it looks for that which is hidden and is *shaping* representation. That's the original descriptive default response that came from the sociology of scientific knowledge (SSK).[1] Two, sometimes, but mostly not, it staples this to an argument about political *distortion* (on the grounds that good methods free from interference know better). It isn't just that (for instance) the social *shapes* representations: it *misshapes* them as well.[2] Three, sometimes instead it attaches the default response to an argument about expertise (for example, Collins and Evans 2002). Knowledge is shaped by the social, but it is argued that the professionals who actually do the work probably see less confusedly than the rest of us—though what counts as "professional" is up for debate.

These are arguments that understand confusion by looking at what it is that *shapes* scientific practices and their representations. But it's also possible to look *forward* and ask about the *effects* of those practices, about what it is that representations and their practices *do* more or less confusedly. So what do we learn if we attend to the *performativity of science practices*? A response might take the form of a list that moved from distinct to indistinct perception, from lesser to greater confusion.

What does science practice do? First and most obviously, it *represents*. Science practices are endlessly inventive about the ways in which they visualize whatever it is they are describing, and such practices are exactly what this book is about. They're fascinating, but we're not in the realm of Leibniz's indistinct perception.

Second, there is the object, the process, the relation: whatever it is that is being *represented*. There are many beautiful STS studies that illustrate how it is that technoscience work doesn't just generate representations of objects, but simultaneously performs whatever it is that is being represented. The argument is that *realities are enacted in difficult technoscience practices*.[3] Emphasis on the "difficult."

Third, there's another old STS concern, *authority* (Collins 1975; Shapin 1984; Shapin and Schaffer 1985). To represent a reality is also to claim the authority to represent it. So, for instance, the views of the teams at CERN are taken seriously by other cosmologists whereas the views (say) of the STS community are not. Of course there are contested authorities, and whose views count may change. But even so, authority is implied in scientific practice.

And then, fourth, there are what one might think of as *collateral realities*. These are all the other more or less indistinct realities also being done in practice in science to which no one attends (see Law 2011a and 2011b; and for a social science example, Law 2009). So how to see these? How to raise them from indistinction? It seems to me that one way of approaching this is to cultivate the art of allegory, that is, *the capacity to see the things that aren't being literally described*. How to do this? One way is to attend to the *social and material* on the one hand and the *metaphysical* on the other. Let me touch on each of these in turn by thinking about a possible allegorical form: that of archaeology.

What do archaeologists do? A simple answer is that when they uncover an object they ask what it is "saying" even though it doesn't speak.[4] That is what the object *does* once it's dug up: those are its effects. Foucault, of course, conducts a deconstructive discursive archaeology, but it is also possible to do this more materially. What does (for instance) a paper in *Nature* tell us that it isn't *actually* saying in as many words? Materially and socially? What is it *doing*? As Leibniz implies, once we start asking questions the list is endless. Some thoughts: A paper in *Nature* does a professional and a disciplinary structure (which intersects with authority). It does a system of refereeing, editorial and linguistic conventions, and an education system with an economy to support this, the research itself, the readers of the research and its circulation. It also tells us about—it does—industries that invent and manufacture instruments of inscription, computers, sensing devices, printing presses, and distribution networks including telephones, databases, Internet connections, and power supplies. And this is just a beginning (truly Leibniz was right). For (to gesture and then to stop) it also (for instance) embeds and enacts advanced agricultural practices (famine is probably *not* a preoccupation).

Something like that happens if we look materially and archaeologically at a scientific representation. But the same is true if we recruit Foucault's brand of deconstructive archaeology to explore issues of metaphysics. So if we explore the enactment of the conditions of possibility, we discover at least five threads in the weave. One, there is a world *out there* being done beyond the experiment. Two, the world being done is at least partially *independent* of those who describe it. Three, this world *predates* the experiment and wasn't created by it. Four, it is a world with a *definite form*. And then, five, it is indeed a single world. Or, to put it differently, there's only one reality. It does a *singular* universe, a compossible cosmology. Such are some of the metaphysical collateral realities being indistinctly reached out to—indistinctly reproduced and enacted—in most of the representations of technoscience.[5]

To summarize: technoscience practices generate representations. They offer literal representations of realities, or they can be read that way. They generate objects and authorities too. And then, the key point, they confusedly recognize and enact other collateral realities, so to speak by stealth. Collateral realities, I've said, are social and material on the one hand and metaphysical on the other. But to read them we need to move from the literal to the allegorical and sensitize ourselves to Leibniz's confusions and indistinct perceptions. This is a shift that allows us to explore what it is that visualizations do apart from reporting back on whatever it is they are most obviously depicting.

But why? Why do this? The quick answer is that what *isn't* being said or visualized is as interesting as what *is* being said. Indeed, since that which is indistinct frames, conditions, and is reproduced in technoscience, it's arguably more important than what's being made explicit. There's a lot of metaphysical—not to say social and material—work being done in the representations of technoscience, and it's not all agreeable. Then again, it's also important to understand that no particular allegorical reading tells us the truth. Such readings are better understood as alternative practices for making literal. They have their confusions too, it could be no other way, but even so they interfere with technoscience's own understanding of itself. In so doing they render what previously lay at the periphery of technoscientific vision, that which was indistinct or confused, more explicit.

Perhaps, and this is the real hope, they also render patches of it sufficiently literal to make these discursively and politically contestable. For playing with metaphysics is not just a matter of play. Or if it is, then it is play that is also deadly serious. This is because metaphysics are embedded in—and reproduce—powerful versions of reality. So, for instance, Western visions typically work to reproduce a single and compossible world. And this is a compossibility that displaces noncompossible alternatives. Whereas what is at stake—for instance in many postcolonial encounters—is *precisely* the question of noncompossibility. The struggles between mining companies and indigenous people are both political *and* metaphysical. Or, to put it differently, metaphysics is politics by other means.[6]

Notes

1. See Haraway (1989), though her book does much additional work.

2. For an early and sophisticated version of this argument, see Barnes (1977).

3. See, for instance, Hacking (1992), who, however, confines his argument to laboratory science. For other examples see Latour and Woolgar (1986), Latour (1999), and the book by Haraway (1989) cited above. For an exemplary account in health care see Mol (2002).

4. I follow Laura Watts (2008) in asking the question about contemporary materials. See Foucault (1972).

5. I develop this argument at greater length in Law (2004). For the specificity of European metaphysics when compared with those of China see Hall and Ames (1995). See also Law and Lien (2013).

6. See, for instance, de la Cadena (2010), Escobar (2008), and Verran (1998).

References

Barnes, Barry. 1977. *Interests and the Growth of Knowledge*. London: Routledge and Kegan Paul.

Collins, H. M. 1975. The seven sexes: A study in the sociology of a phenomenon, or the replication of experiments in physics. *Sociology* 9:205–224.

Collins, H. M., and Robert Evans. 2002. The third wave of science studies: Studies of expertise and experience. *Social Studies of Science* 32:235–296.

De la Cadena, Marisol. 2010. Indigenous cosmopolitics in the Andes: Conceptual reflections beyond "politics." *Cultural Anthropology* 25 (2):334–370.

Escobar, Arturo. 2008. *Territories of Difference: Place, Movements, Life, Redes*. Durham: Duke University Press.

Foucault, Michel. 1972. *The Archaeology of Knowledge*. London: Tavistock.

Hacking, Ian. 1992. The self-vindication of the laboratory sciences. In *Science as Practice and Culture*, ed. Andrew Pickering, 29–64. Chicago: University of Chicago Press.

Hall, David L., and Roger T. Ames. 1995. *Anticipating China: Thinking through the Narratives of Chinese and Western Culture*. Albany: State University of New York Press.

Haraway, Donna J. 1988. Situated knowledges: The science question in feminism and the privilege of partial perspective. *Feminist Studies* 14 (3):575–599. Available at http://www.staff.amu.edu.pl/~ewa/Haraway,%20Situated%20Knowledges.pdf (accessed 2 January 2013).

Haraway, Donna J. 1989. *Primate Visions: Gender, Race and Nature in the World of Modern Science*. London: Routledge and Chapman Hall.

Latour, Bruno. 1988. Irréductions. In *The Pasteurization of France*, 153–236. Cambridge, MA: Harvard University Press.

Latour, Bruno. 1999. Circulating reference: Sampling the soil in the Amazon forest. In *Pandora's Hope: Essays on the Reality of Science Studies*, 24–79. Cambridge, MA: Harvard University Press.

Latour, Bruno, and Steve Woolgar. 1986. *Laboratory Life: The Construction of Scientific Facts*. 2nd ed. Princeton: Princeton University Press.

Law, John. 2004. *After Method: Mess in Social Science Research*. London: Routledge.

Law, John. 2009. Seeing like a survey. *Cultural Sociology* 3 (2):239–256.

Law, John. 2011a. Collateral realities. In *The Politics of Knowledge*, ed. Fernando Domínguez Rubio and Patrick Baert, 156–178. London: Routledge.

Law, John. 2011b. The explanatory burden: An essay on Hugh Raffles' *Insectopedia*. *Cultural Anthropology* 26 (3):485–510.

Law, John, and Marianne Elisabeth Lien. 2013. Slippery: Field notes on empirical ontology. *Social Studies of Science* 43 (3):363–378.

Leibniz, Gottfried Wilhelm. 1998. Monadology. In *G. W. Leibniz: Philosophical Texts*, ed. R. S. Woodhouse and Richard Franks, 268–281. Oxford: Oxford University Press.

Mol, Annemarie. 2002. *The Body Multiple: Ontology in Medical Practice*. Durham: Duke University Press.

Shapin, Steven. 1984. Pump and circumstance: Robert Boyle's literary technology. *Social Studies of Science* 14:481–520.

Shapin, Steven, and Simon Schaffer. 1985. *Leviathan and the Air Pump: Hobbes, Boyle and the Experimental Life*. Princeton: Princeton University Press.

Verran, Helen. 1998. Re-imagining land ownership in Australia. *Postcolonial Studies* 1 (2): 237–254.

Watts, Laura. 2008. The future is boring: Stories from the landscapes of the mobile telecoms industry. *21st Century Society: Journal of the Academy of Social Sciences* 3 (2):187–198.

Whitehead, Alfred North. 1978. *Process and Reality*. New York: Free Press.

21   A Question of Trust: Old Issues and New Technologies

Martin Kemp

The production and manipulation of images in the digital age has been seen as precipitating a crisis of trust. The basic issues, however, are as old as the earliest uses of images and diagrammatic presentations to convey information. There have over the ages been a series of tensions of which participants were aware at the time: episodes such as the dearth of anatomical pictures in early humanist publications of Galenic medicine, alongside the great anatomical picture books (Roberts and Tomlinson 1992); the use of optical devices, such as telescopes and microscopes, to show things that most people could not witness; Linnaeus's injunctions about the primacy of the specimen over illustration in botany (Linnaeus 1751, aphorism 11); the battle in the nineteenth century (and earlier) between the "warts and all" school of naturalistic representation (led in part by William Hunter) and those who believed in synthesizing the ideal specimen (Kemp 1993); and the ready use of the new technology of photography to make rigged images, most famously the snapshots of the Cottingley fairies which fooled Conan Doyle among others (Kemp 2006).

Any visual presentation of evidence or data is designed to endow the author's enterprise with a special sense of conviction. We are asked, in effect, to "see" the obvious truth. This applies to tables of data, neatly packaged to produce significant patterns, to graphs and pie charts etc., that exude an air of irrefutable precision, and not least to images that work within a naturalistic framework, which claim to put us in the same position as an objective eyewitness as the maker of the image.

Many of us will be able to remember the neat curves that our physics teacher at school encouraged us to draw on graphs of ragged data—collected by incompetent experimenters with poor equipment. I can recall edging my little crosses on the graph just that bit nearer the desired line. Yes, we "proved" whatever law we were seeking to verify. Of course professional science is not like this—at least not so blatantly. But there is always some degree of tidying up involved in the visual reporting of the untidy process of hypothesis-experiment-conclusion. It is the tidy versions of the visual presentations that emerge into the public domain and that underline their makers' claims to irrefutable precision—and that solicit our trust.

Extensive tables of numbers look impressive. It takes a major effort on the part of the reader to work through all of them, a bigger effort to replicate their gathering, and even more again to revise the starting assumptions. This is precisely what Stephen Jay Gould did with the anthropological data about racial brain sizes in his classic *The Mismeasure of Man* (Gould 1981). He apparently demonstrated how both the obtaining of data and its interpretation were skewed by the desired end. Ethnographic investigators' neatly arrayed and convincing pictures of skulls provided an additional level of conviction, letting us "see" that the frontal profile of a "negro" skull was closer to that of a monkey than it was to the *Apollo Belvedere*. However, some of Gould's apparently secure data analysis has now in turn been reworked by Jason Lewis and his coauthors (Lewis et al. 2011), who have questioned at least one of Gould's imputations of bias. In any event skull size is now known not to be a significant factor in comparing human intelligences, and the collecting of such data therefore no longer serves the preconceived purpose.

The naturalistic image, which overtly or implicitly claims significant elements of matching with the observer's direct experience of what is being represented, is particularly potent in evoking our trust. We are set up, perceptually and cognitively, to trust the ordering capacities with which we make functional coherence of the sensory impressions that arrive in our eyes and ears. We automatically pick out those elements in what we see so that we can, say, walk round a table without bumping into it. The naturalistic image, as developed over the years, has come to exploit certain of the key perceptual triggers in such a way that we can form an illusionistic image of a table that we feel we could walk round. The gravitational pull in our reaction to a naturalistic image is toward trust.

Illustrators have often enhanced this automatic sense of trust with accessory devices that convince us of the reality of the representation. These conjoined mechanisms I have called the "rhetoric of reality" (to set alongside the "rhetoric of irrefutable precision"). The supporting devices exploit such things as settings, textures, light effects, and realistic details (like pins in anatomy or clamps in physics) to say to us that we are looking at the "real thing." We are implicitly being set up as surrogate eyewitnesses in specific times and places.

Let me give a historical example of the potency of the rhetoric of reality. The persistence of the classic and highly contrived image of the rhinoceros by Albrecht Dürer in 1515, based on a description and still being used to advertise actual rhino shows in the eighteenth century, is largely explained by all the naturalistic devices of modeling and patterning that he exploited to render his picture convincing. His military rhino ended up looking more like the armored beast that people wanted to see than any actual specimen could ever do. He is showing us the essential *rhinoceros unicornis*, and invites us to trust him (Kemp 2012). When he portrayed a pointedly naturalistic unicorn in his *Abduction of Proserpina* (Poesch 1964), the great printmaker similarly used his prodigious skill to grant the chimerical animal a visual status not less than his rendering of a horse.

We all use our trust every day, as when we watch the news on TV or look at the latest photographs of an atrocity in a newspaper. We trust images of things we have never seen and are unlikely to see. I have shown lecture audiences a slide of a duck-billed platypus. I ask what it is. People know. I ask how many people have seen one in the flesh. Very few have. I ask how many people do not believe in the existence of this animal. No one volunteers. We trust what has been represented to us by those who produce surrogate eyewitness pictures. We have little reason to feel superior to those who believed in the unicorn having seen a good picture of it in a great picture book like Conrad Gessner's (1551–1558) *Historia animalium*, which also directly copied Dürer's rhino. When the pictures of animals actually move, like those on YouTube, they have their own special air of authority, specifically because they involve even more of our ordering faculties than a static presentation.

With images produced with the assistance of optical and digital devices, or even generated without evident human intervention, the rhetoric of reality is complemented by the "rhetoric of instrumental objectivity." An image produced by the latest high-tech gadget, whether a perspective machine in the Renaissance, a *camera lucida* in the nineteenth century, or an fMRI scan in our time, promises levels of disinterested precision beyond that of even the most scrupulous human investigator (Daston and Galison 2007). The starting assumptions, hypotheses, data gathering, and image generation are in modern technologies all secreted more or less invisibly in the black box (or brushed aluminum cabinet). The relationship between trust and scientific transparency is now subject to even thicker filters than before, and they contrive to be as out of reach as possible.

An essential element in the technologically generated image is that it should look high-tech. We know what computer images look like. They have a distinct style, with their own stock ways of rendering surfaces, forms, and space. They have set ways of translating the unseeable into something that can be seen. We can readily recognize an image that uses the latest technologies, compared to one made twenty years ago. Even if the scientist's visual point could be made with a rudimentary low-tech image, he or she is highly likely to translate it into a presentation that brandishes its high-tech quality. The journal or publisher is unlikely to expect less.

In the broadest sense, every act of representation is purposefully selective, just as are our acts of seeing. With even the most basic of photographs, depth of field, focus, point of view, lighting, and exposure all serve advertently or inadvertently to include or emphasize some features and to exclude or downplay others. This is to say nothing of its subsequent printing and transmission. The scientist, even more than the casual maker of family snapshots, is in the business of selective visual pointing. And the scientist makes sure that the look of the image manifests all the signs of authenticity that are current at the time of its making and reception.

The essential nature of the issue of visual trust in representations has not changed over time, even if the scope for convincing deception has increased in ways that increasingly defy ready detection. We have in practice to work with certain levels of visual trust, otherwise we could not realistically proceed. Visually, we are trusting beings. As such, our current vulnerability is not essentially different in kind from our predecessors'; it is just far greater in extent.

References

Daston, Lorraine, and Peter Galison. 2007. *Objectivity*. New York: Zone Books.

Gessner, Conrad. 1551–1558. *Historiae animalium*. 4 vols. Zurich.

Gould, Stephen Jay. 1981. *The Mismeasure of Man*. New York: W. W. Norton.

Kemp, Martin. 1993. "The mark of truth": Looking and learning in some anatomical illustrations from the Renaissance and the eighteenth century. In *Medicine and the Five Senses*, ed. W. F. Bynum and Roy Porter, 85–121. Cambridge: Cambridge University Press.

Kemp, Martin. 2006. *Seen and Unseen: Art, Science, and Intuition from Leonardo to the Hubble Telescope*. Oxford: Oxford University Press.

Kemp, Martin. 2012. "The testimony of my own eyes": The strange case of a mammal with a beak. *Spontaneous Generations* 6(1): 43–49. Available at jps.library.utoronto.ca/index.php/SpontaneousGenerations (accessed 2 January 2013).

Lewis, Jason E., David DeGusta, Marc R. Meyer, Janet M. Monge, Alan E. Mann, et al. 2011. The mismeasure of science: Stephen Jay Gould versus Samuel George Morton on skulls and bias. *PLoS Biology* 9(6), 2011: e1001071.

Linnaeus, Carl. 1751. *Philosophia botanica*. Stockholm, Amsterdam.

Poesch, Jessie J. 1964. Sources for two Dürer enigmas. *Art Bulletin* 46:78–86.

Roberts, K. B., and J. D. W. Tomlinson. 1992. *Fabric of the Body: European Traditions of Anatomical Illustration*. Oxford: Oxford University Press.

22 The More Manipulations, the Better

Bruno Latour

Focusing researchers' attention on the visual aspects of various scientific practices has been of great import because it has brought down to earth many philosophical claims about objectivity. And yet focusing on the visual per se might lead in the end to a blind alley. The reason is that image making in science is very peculiar, so peculiar indeed that following its odd ways offers an excellent way to define what is "scientific," after all, in science.

At first, the temptation is great to treat the visual aspects of so many scientific instruments, papers, posters, and displays in the same ways as art historians have considered visualization in their own fields of practice. But if it is true that paintings, photographs, engravings, installations refer many times to other works of art by practicing a form of overt or hidden citation, allusion, parody, or displacement, in science the connection between visual documents is completely different. Every image refers to another image—or better, an inscription—that comes before it and that is itself transformed, yet again, by another inscription down the road, thus forming long cascades of successive traces. Those traces are separated by *gaps* that the evolution of instruments allied to that of interpretive skills tries to narrow down as much as possible. But this narrowing down—that's what is so odd—is obtained by multiplying yet again the number of steps along those cascades of transformations.

It is fair to say that the degree of objectivity of a scientific discipline may be defined by the width of those gaps and by the number of transforming steps necessary to fill them. The referential quality of a discipline, that is, its ability to reach objects inaccessible otherwise and to transport them into a site where they can be evaluated by peers, is entirely dependent on the quality of those chains. The more steps there are *in between* the objects and those who make judgments about them, the more robust those judgments will be.

Such visualization practices remain very paradoxical when considered from the point of view of art history or iconography, since their degree of "realism" is entirely different from that of works of art or any other type of illustrations. It is different first because it never treats the relation between one copy and one model as if it could be

Figure 22.1
Copyright ScienceCartoonsPlus.com. Cartoon by S. Harris.

limited to only *two* steps: the gap then would be much too wide to allow for a robust judgment about the exact connections between two inscriptions. The idea of science as a "mirror of the world" is a spurious import from the history of figurative paintings into epistemology. In science, it is more as if the mirror is situated at the very end of long series of transformations between traces, none of which is an exact replica of the former. In other words, scientific imagery is never mimetic. If it were, there would be no gain of information between one step and the next. It is the *difference* between each step that allows the reference to move on. As indicated by the etymology of the verb "to refer," this is the only way to *bring back* some state of affairs in order to handle it.

If the extension, complexity, and cost of those referential chains are taken into account, one can easily see why *isolating* one inscription from the flow in which it is taken would make one lose its referential quality. An isolated scientific image, strictly speaking, has no reference. The possibility of referring is given by what an inscription inherits from another one, upstream, and what it transfers—by transforming it—to another trace, downstream. Reference is a movement, a deambulation, a trajectory, not a property of a "realistic" image. This is why the number of transformations undergone along a chain, a number that shocks common sense so much because it could be taken as so many "manipulations," makes a lot of sense for practitioners. (Harris's cartoon is right on target when he has the scientist say: "Makes sense." No irony is indeed intended).

White coats know full well that, without those long series of manipulations, they could not narrow the gaps, and that they would have to rely on too many arbitrary judgments to jump over them. What for common sense is manipulation—"Please state the matters of fact as directly and as naturally as possible"—is exactly what, for practicing scientists, insures the quality of the reference—"I can't obtain objectivity without multiplying the transformations."

Of course, as soon as scientists leave their laboratories, they fall back on the commonsense version of science as the mirror image of the world. They are suddenly more than happy to display *one isolated* image extracted out of the chains as "the definitive proof" of the phenomenon they wish to describe. Then, but only then, and only for pedagogical or public relations purposes, are we asked to see *one* image as the copy of *one* phenomenon. But no matter how convincing this display might seem, other practitioners know full well that in order to judge the quality of such an isolated image, one should not try to compare it to its "model" out there, but should check what it has retained from another inscription, before, and what it can send to another inscription, after.

One could even argue that the "model" for which this displayed image is a "copy" is actually an *afterimage*, a mere replication of the interrupted inscription. The mystery of its "realism" would seem less mysterious if this replication were taken into account: of course it is "realist," since it is twice the same thing. Many a quandary of epistemology would be dissipated if those two different positions of the same image could be documented. But in order to do so, one should not isolate the scientific imagery and shoehorn it into the types of question raised by iconography. There is nothing visual in scientific visual imagery. Literally, there is nothing to be "seen."

Now that there is a vast literature on scientific image making, the next frontier is probably to understand what is *not* visual in those chains that comes from the gap situated between two successive inscriptions, a topic that has been hidden by the flood of disputes about the degree of "resemblance" between an image and its copy. Of course, we know that this gap is made of two contradictory features: what is kept from one trace to the next; what is changed from one to the next *so that something* at least is kept constant. But the study of those "immutable mobiles"—that is, how much mobility you can obtain by regulating those two opposite traits—requires a different take than the one obsessed by the "scopic" regime of so much philosophy of science: a philosophy just as much interested in realism, but where realism is generated by moving along the referential chain—"the more mediations the better"—and not jumping out and interrupting the flow of images—"if only there was no mediation at all, how much more accurate our knowledge would be!"

What is needed to fathom scientific image-making processes is probably the equivalent of what Gibson (1986) started to study for ordinary vision: an ecological interpretation that manages to focus not on vision per se but on the deambulation of active
bodies registering features of the landscape by judging the relative proportion of what
changes and what is transformed. "The extracting and abstracting of invariants are
what happens in both perceiving and knowing. To perceive the environment and
to conceive it are different in degree but not in kind. One is continuous with the
other" (Gibson 1986, 258). The research program Gibson so radically proposed many
years ago has not yet been completed: "The very notion of an image as a flattened-out
object, a sort of pancake of a solid body, is shown to be misleading. It begins to appear
that most of what has been written about pictures and images over the centuries is
misleading, or hopelessly vague. We should forget it all and start fresh. The information for the perception of an object is not its image. The information in light to specify
something does not have to resemble it, or copy it, or be a simulacrum or even an exact
projection. Nothing is copied in the light to the eye of an observer, not the shape of a
thing, not the surface of it, not its substance, not its color, and certainly not its motion.
But all these things are specified in the light" (ibid., 304). Only once the mimetic and
scopic obsession of an image as a copy has been put aside will it be possible to study
scientific imagery. Then, the magnificent body of work done over the years by so many
scholars on so many aspects of the joint history of art, science, and perception in order
to foreground the visual in scientific practice will really have come to fruition.

Reference

Gibson, James J. 1986. *The Ecological Approach to Visual Perception*. London: Lawrence Erlbaum
Associates.

Contributors

Morana Alač is Associate Professor in Communication and Science Studies at the University of California, San Diego. Alač's research deals with ordinary, interactional, and practical aspects of science. Her recent work has focused on how scientists study cognition in environments heavily sustained by advanced technologies—brain imaging and machine learning laboratories. She works with video to focus on the dynamics of embodied social interaction. In 2011, the MIT Press published her *Handling Digital Brains*.

Michael J. Barany studies the history, sociology, and material culture of modern mathematics as a PhD candidate in the Princeton University Program in History of Science. His dissertation examines the practices and contexts of postwar mathematical analysis by following the course of the theory of distributions, or generalized functions, as it was developed and deployed in elite mathematical research and pedagogy. His recent research has covered topics ranging from early modern treatments of the Euclidean point to Victorian views on the prehistories of numbers and measures to the phenomenon of blog mathematics.

Anne Beaulieu is Project Manager of the Groningen Energy and Sustainability Programme at the University of Groningen. Previously she spent several years as senior research fellow at the Royal Netherlands Academy of Arts and Sciences (KNAW), notably at the Virtual Knowledge Studio for the Humanities and Social Sciences. Before that, she investigated the development of neuroimaging and cognitive neuroscience, leading to her PhD dissertation (University of Amsterdam, 2000). A dominant theme in Beaulieu's work is the importance of interfaces for the creation and circulation of knowledge. Past research projects focused on data sharing, on knowledge networks, and on visualization and visual knowledge. She also has done extensive work in the field of digital humanities, on new (ethnographic) research methods, and on ethics in e-research.

Annamaria Carusi is Associate Professor in Philosophy of Medical Science and Technology at the University of Copenhagen. Her recent research has focused on three main areas: modeling and simulation in biomedical sciences, particularly systems biology; the role of images and visualizations in science; and the role of technologies in computationally intensive, interdisciplinary settings. She has published several articles on trust, interdisciplinary interactions around visualizations, and the challenges of current computational methods to traditional thinking about the visual. She has coedited a special issue on computational picturing, and a volume on visualization will appear soon.

Catelijne Coopmans is a Research Fellow at the Asia Research Institute and a Fellow and Director of Studies at Tembusu College, National University of Singapore. Her research draws on ethnography and discourse analysis to examine the promotion and configuration of new imaging and visual applications in business and medicine, with particular attention to the various notions of insight and value that accompany such efforts. She is currently studying research and clinical work related to retinal imaging in Singapore.

Lorraine Daston is Director at the Max Planck Institute for the History of Science, Berlin, and Visiting Professor in the Committee on Social Thought at the University of Chicago. Her work addresses how standards of rationality develop historically through concrete scientific practices such as observation, image making, and archiving. Recent publications include (with Peter Galison) *Objectivity* (Zone Books, 2007) and (coedited with Elizabeth Lunbeck) *Histories of Scientific Observation* (University of Chicago Press, 2011). She is currently working on a book about the history of rules.

Sarah de Rijcke is Assistant Professor in the Centre for Science and Technology Studies, Leiden University. She completed her PhD with honors from the University of Groningen in 2010, where her dissertation focused on different visual ways of knowing the brain, from the seventeenth century to the present, in relation to what constitutes an authoritative image. A dominant theme in her work is the role of representations in the production of knowledge. Her current research focuses on the growing use of assessment procedures and bibliometric indicators in scientific and scholarly research, and its effects on knowledge production.

Joseph Dumit is Director of Science and Technology Studies and Professor of Anthropology at the University of California, Davis. He is the author of *Drugs for Life: How Pharmaceutical Companies Define Our Health* (Duke University Press, 2012) and *Picturing Personhood: Brain Scans and Biomedical Identity* (Princeton University Press, 2003). He coedited *Cyborgs and Citadels: Anthropological Interventions in Emerging Sciences and Technologies; Cyborg Babies: From Techno-Sex to Techno-Tots*; and *Biomedicine as Culture*. He is currently studying how immersive 3D visualization platforms are transforming science, as well as the history of flow charts, cognitive science, and paranoid computers.

Emma K. Frow is a Lecturer in Science and Technology Studies at the University of Edinburgh. Her research focuses on guidelines, standards, and governance in contemporary biotechnologies, with a particular emphasis on synthetic biology. Originally trained as a bioscientist, she has a PhD in biochemistry from the University of Cambridge and worked as a subeditor of research manuscripts for *Nature* magazine from 2004 to 2006. Her article "Drawing a Line: Setting Guidelines for Digital Image Processing in Scientific Journal Articles" was published in *Social Studies of Science* in June 2012.

Yann Giraud is Associate Professor in the economics department at the University of Cergy-Pontoise (France). His PhD dissertation, on the history of economics, studied the changing place of visual representation in the discipline throughout the twentieth century. He has published several papers examining how economists and social scientists have used diagrams, graphs, and pictorial representations at the crossroads of theorizing, teaching, and policymaking. His more recent work explores the role of textbooks in the transformation of American economics in the postwar period.

Aud Sissel Hoel is Associate Professor of Visual Communication in the Department of Art and Media Studies at the Norwegian University of Science and Technology. Her research interests revolve around science images and branch out to include photography, scientific instruments, technologies of thinking, medical imaging, and visualization. Currently she is heading an interdisciplinary research project on neuroimaging, and previously she conducted a project on photography used for identification and control. Hoel has published widely on these topics, including two books on Ernst Cassirer's philosophy of technology, a special issue of *Interdisciplinary Science Reviews* on computational picturing, and a forthcoming volume on visualization.

Martin Kemp is Emeritus Research Professor in the History of Art, Oxford University. He has written and broadcast extensively on imagery in art and science from the Renaissance to the present day. He speaks on issues of visualization and lateral thinking to a wide range of audiences. Leonardo da Vinci has been the subject of several of his books, including *Leonardo* (Oxford University Press, 2004). He has published on imagery in the sciences of anatomy, natural history, and optics, including *The Science of Art* (Yale University Press, 1990) and *Christ to Coke: How Image Becomes Icon* (Oxford University Press, 2012).

Bruno Latour is professor at Sciences Po Paris. He has worked on the question of inscription and visual imagery of science since the 1970s and kept an interest in the semiotics of the visual space of science through international exhibitions, especially "Iconoclash" in 2002, and art form, especially theater. All references and most of his articles can be found on his website, www.bruno-latour.fr.

John Law is Professor of Sociology and Co-Director of the Centre for Research on Socio-Cultural Change at the Open University, UK. He works on noncoherent assemblages and their practices, exploring how these hold together despite, or because of, their noncoherences. He also explores the embedded political and normative performativities of those practices. His work is empirical; he is currently working on agriculture and aquaculture, the social life of research methods, North-South postcolonial relations, and the dysfunctions of British finance and manufacturing.

Michael Lynch studies discourse, visual representation, and practical action in research laboratories, clinical settings, and legal tribunals. His books include *Scientific Practice and Ordinary Action* (Cambridge University Press, 1993) and (with Simon Cole, Ruth McNally, and Kathleen Jordan) *Truth Machine: The Contentious History of DNA Fingerprinting* (University of Chicago Press, 2008). He was Editor of *Social Studies of Science* from 2002 until 2012.

Donald MacKenzie holds a personal chair in sociology at the University of Edinburgh. His most recent books are *An Engine, Not a Camera: How Financial Models Shape Markets* (MIT Press, 2006); *Do Economists Make Markets? On the Performativity of Economics* (Princeton University Press, 2007), coedited with Fabian Muniesa and Lucia Siu; and *Material Markets: How Economic Agents Are Constructed* (Oxford University Press, 2009).

Cyrus C. M. Mody is an Assistant Professor in the History Department at Rice University. He has degrees in engineering sciences and science and technology studies and is the author of *Instrumental Community: Probe Microscopy and the Path to Nanotechnology* (MIT Press, 2011). He is currently working on a second monograph on the microelectronics industry's role in changes in American science and science policy since 1970.

Natasha Myers is an Associate Professor in the Department of Anthropology and a member of the Institute for Science and Technology Studies at York University, Toronto. Her ethnographic research examines forms of life in the contemporary arts and sciences. She is currently completing a book on an interdisciplinary group of scientists who image, model, and simulate subvisible molecular realms through computer-intensive technologies. The book explores research and pedagogy among the protein crystallographers and biological engineers who are reconfiguring twenty-first-century molecular biology.

Rachel Prentice is an Associate Professor in the Department of Science and Technology Studies, Cornell University. She is an anthropologist of medicine, technology, and the body; her interests focus on the assumptions and contradictions in twenty-first-century North American biomedicine. Her book *Bodies in Formation: An Ethnography of Anatomy and Surgery Education* (Duke University Press, 2012) documents how physicians in training come to embody biomedical techniques, perceptions, judgments, and ethics, learning deeply held medical values while learning to practice medicine.

Arie Rip is Professor of Philosophy of Science and Technology, School of Management and Governance of the University of Twente, The Netherlands. He holds degrees in chemistry and philosophy and obtained his PhD at the University of Leiden in 1981, with a thesis on societal responsibility of chemists. He set up a teaching and research program in Chemistry and Society at the University of Leiden, became guest professor in science dynamics at the University of Amsterdam (1984–1987), and then moved to the Chair of Philosophy of Science and Technology at the University of Twente, from which he formally retired in 2006. He works on modes of knowledge production, science institutions and science policy, technology dynamics, and constructive technology assessment. He led the TA program in the Dutch national consortium NanoNed.

Martin Ruivenkamp has a BA in sociology and an MA in social psychology and earned a PhD in 2011 at the University of Twente with a thesis on "Circulating Images of Nanotechnology." He has published on practices of visualization of the nanoscale. Currently, he has a postdoctoral position with the Centre for Society and the Life Sciences at the Radboud University, Nijmegen.

Lucy Suchman is Professor of Anthropology of Science and Technology in the Department of Sociology at Lancaster University, and Co-Director of Lancaster's Centre for Science Studies. Before taking up her present post she spent twenty years as a researcher at Xerox's Palo Alto Research Center. In 2010 she received the ACM Special Interest Group in Computer-Human Interaction Lifetime Research Award. Her writings include *Human-Machine Reconfigurations: Plans and Situated Actions* (Cambridge University Press, 2007).

Janet Vertesi is Assistant Professor in Sociology at Princeton University. She holds a PhD in science and technology studies from Cornell University and an MPhil in history and philosophy of science from the University of Cambridge. Her book *Seeing Like a Rover: Images in Interaction on the Mars Exploration Rover Mission* draws on a multiyear ethnographic study with the NASA team to examine how scientists and engineers work with digital images of Mars. Vertesi has also published on how the iconic London Underground Map affects users' interactions with urban space, and on lunar mapping projects in seventeenth-century astronomy.

Steve Woolgar is Chair of Marketing and Head of Science and Technology Studies at Said Business School, University of Oxford. He has published widely in social studies of science and technology, social problems, and social theory. His current research includes a study of mundane technical solutions to public problems (forthcoming are *Mundane Governance: Ontologies and Accountability*, with Dan Neyland, and *Globalisation in Practice,* with Nigel Thrift and Adam Tickell, both from Oxford University Press); and neuromarketing and the effects of the neurosciences in transforming European social sciences and humanities. In 2008 he was named winner of the J. D. Bernal prize. In 2010 he was elected to the Academy of Social Scientists.

Index

Aesthetics, of scientific images, 32, 39, 163, 179, 181, 183, 235, 241–244, 249, 253, 258–259, 264

Alač, Morana, 18, 65, 66, 78, 83, 149, 259, 262, 302

Almli, C. Robert, 137

Alpers, Svetlana, vii

Amann, Klaus, 18, 20, 39, 40, 63, 81, 254, 262

Amen, Daniel, 305

Ames, Roger T., 340

Analogies in science, 154–169, 196

Analyzability, of visual materials, 18

Anderson, Christopher, 255, 256

Anderson, Warwick, 11

Andre, G., 194

Araya, Augustin A., 218

Arkin, Adam, 166

Aronson, Jay D., 294, 310

Art, 26, 38, 39, 52, 148, 180–181, 183, 187, 191, 194, 235, 241–244
  and science, 147

Artfulness, 18, 39–41, 49, 52

Artifacts, representational, 23–24, 39, 69, 77, 254, 261, 334

Ashmore, Malcolm, 333

Atlases, 131–147, 319. *See also* Databases; Maps and mapping

Audiences, 6, 131, 144, 178, 180, 185, 192, 223–228, 244
  lay, 179–180, 182, 186
  seminar, 109–116
  webinar, 41–42, 47, 49–53

Austin, John L., 62

Authoritativeness, 111, 132–133, 136–140, 145–147, 258, 299, 338, 339. *See also* Credibility; Integrity; Legitimation

Averaging, 77, 135–136, 138–139, 145, 295, 300–304

Bachtold, A., 183, 184

Bainbridge, B., 190, 191

Baird, Davis, 244

Balsamo, A., 99

Barad, Karen, 20, 156, 170, 217, 218, 330, 333

Barany, Michael J., 108, 124, 125

Barnes, Barry, 340

Baro, Arturo, 236

Barthes, Roland, 125

Bastide, Françoise, viii, 192, 196

Baudelaire, Charles, 320

Baumol, William, 276

Beale, Lionel S., 157–158

Beaulieu, Anne, 39, 80, 131, 132, 133, 134, 136, 139, 146, 147, 148, 149, 192, 202, 258, 262, 263, 295, 302, 305

Beckman, Mary, 291, 300, 301, 304

Begley, Sharon, 297, 298–299, 307

Bell, J. F. I., 21

Bennett, James A., 55

Benveniste, Emile, 62

Berg, Marc, 134

Beunza, Daniel, 37, 38, 40

Bijker, Wiebe E., 330

Inside Technology
edited by Wiebe E. Bijker, W. Bernard Carlson, and Trevor Pinch

Janet Abbate, *Inventing the Internet*

Atsushi Akera, *Calculating a Natural World: Scientists, Engineers, and Computers during the Rise of U.S. Cold War Research*

Morana Alač, *Handling Digital Brains: A Laboratory Study of Multimodal Semiotic Interaction in the Age of Computers*

Stathis Arapostathis and Graeme Gooday, *Patently Contestable: Electrical Technologies and Inventor Identities on Trial in Britain*

Charles Bazerman, *The Languages of Edison's Light*

Marc Berg, *Rationalizing Medical Work: Decision-Support Techniques and Medical Practices*

Wiebe E. Bijker, *Of Bicycles, Bakelites, and Bulbs: Toward a Theory of Sociotechnical Change*

Wiebe E. Bijker, Roland Bal, and Ruud Hendricks, *The Paradox of Scientific Authority: The Role of Scientific Advice in Democracies*

Wiebe E. Bijker and John Law, editors, *Shaping Technology/Building Society: Studies in Sociotechnical Change*

Karin Bijsterveld, *Mechanical Sound: Technology, Culture, and Public Problems of Noise in the Twentieth Century*

Stuart S. Blume, *Insight and Industry: On the Dynamics of Technological Change in Medicine*

Pablo J. Boczkowski, *Digitizing the News: Innovation in Online Newspapers*

Geoffrey C. Bowker, *Memory Practices in the Sciences*

Geoffrey C. Bowker, *Science on the Run: Information Management and Industrial Geophysics at Schlumberger, 1920–1940*

Geoffrey C. Bowker and Susan Leigh Star, *Sorting Things Out: Classification and Its Consequences*

Louis L. Bucciarelli, *Designing Engineers*

Christophe Lécuyer, *Making Silicon Valley: Innovation and the Growth of High Tech, 1930–1970*

Pamela E. Mack, *Viewing the Earth: The Social Construction of the Landsat Satellite System*

Donald MacKenzie, *An Engine, Not a Camera: How Financial Models Shape Markets*

Donald MacKenzie, *Inventing Accuracy: A Historical Sociology of Nuclear Missile Guidance*

Donald MacKenzie, *Knowing Machines: Essays on Technical Change*

Donald MacKenzie, *Mechanizing Proof: Computing, Risk, and Trust*

Cyrus C. M. Mody, *Instrumental Community: Probe Microscopy and the Path to Nanotechnology*

Maggie Mort, *Building the Trident Network: A Study of the Enrollment of People, Knowledge, and Machines*

Peter D. Norton, *Fighting Traffic: The Dawn of the Motor Age in the American City*

Helga Nowotny, *Insatiable Curiosity: Innovation in a Fragile Future*

Ruth Oldenziel and Karin Zachmann, editors, *Cold War Kitchen: Americanization, Technology, and European Users*

Nelly Oudshoorn and Trevor Pinch, editors, *How Users Matter: The Co-Construction of Users and Technology*

Shobita Parthasarathy, *Building Genetic Medicine: Breast Cancer, Technology, and the Comparative Politics of Health Care*

Trevor Pinch and Richard Swedberg, editors, *Living in a Material World: Economic Sociology Meets Science and Technology Studies*

Paul Rosen, *Framing Production: Technology, Culture, and Change in the British Bicycle Industry*

Richard Rottenburg, *Far-Fetched Facts: A Parable of Development Aid*

Susanne K. Schmidt and Raymund Werle, *Coordinating Technology: Studies in the International Standardization of Telecommunications*

Wesley Shrum, Joel Genuth, and Ivan Chompalov, *Structures of Scientific Collaboration*

Rebecca Slayton, *Arguments that Count: Physics, Computing, and Missile Defense, 1949–2011*

Chikako Takeshita, *The Global Politics of the IUD: How Science Constructs Contraceptive Users and Women's Bodies*

Charis Thompson, *Making Parents: The Ontological Choreography of Reproductive Technology*

Dominique Vinck, editor, *Everyday Engineering: An Ethnography of Design and Innovation*